MEDICAL INNOVATIONS IN
HISTORICAL PERSPECTIVE

SCIENCE, TECHNOLOGY AND MEDICINE IN MODERN HISTORY

General Editor: John V. Pickstone, Centre for the History of Science, Technology and Medicine, University of Manchester, England

Published

David Edgerton
ENGLAND AND THE AEROPLANE: An Essay on a Militant and Technological Nation

John V. Pickstone (editor)
MEDICAL INNOVATIONS IN HISTORICAL PERSPECTIVE

Series Standing Order

If you would like to receive future titles in this series as they are published, you can make use of our standing order facility. To place a standing order please contact your bookseller or, in case of difficulty, write to us at the address below with your name and address and the name of the series. Please state with which title you wish to begin your standing order. (If you live outside the United Kingdom we may not have the rights for your area, in which case we will forward your order to the publisher concerned.)

Standing Order Service, Macmillan Distribution Ltd,
Houndmills, Basingstoke, Hampshire, RG21 2XS, England

Medical Innovations in Historical Perspective

Edited by

John V. Pickstone
Centre for the History of Science, Technology and Medicine,
University of Manchester

M
MACMILLAN

in association with the
CENTRE FOR THE HISTORY
OF SCIENCE, TECHNOLOGY
AND MEDICINE
UNIVERSITY OF MANCHESTER

© John V. Pickstone 1992

All rights reserved. No reproduction, copy or transmission
of this publication may be made without written permission.

No paragraph of this publication may be reproduced, copied or
transmitted save with written permission or in accordance with
the provisions of the Copyright, Designs and Patents Act 1988,
or under the terms of any licence permitting limited copying
issued by the Copyright Licensing Agency, 90 Tottenham Court
Road, London W1P 9HE.

Any person who does any unauthorised act in relation to
this publication may be liable to criminal prosecution and
civil claims for damages.

First published 1992

Published by
MACMILLAN ACADEMIC AND PROFESSIONAL LTD
Houndmills, Basingstoke, Hampshire RG21 2XS
and London
Companies and representatives
throughout the world

ISBN 0-333-55619-4

A catalogue record for this book is
available from the British Library.

Typeset by Superskill Graphics Pte Ltd, Singapore

Printed in Hong Kong

CONTENTS

List of Illustrations and Tables	vii
Series Editor's Introduction	ix
Preface	xi
Acknowledgements	xii
Notes on the Contributors	xiii
Abbreviations	xv

1. Introduction — John V. Pickstone — 1

2. 'Upon This Principle I Have Based a Practice': The Development and Reception of Antisepsis in Britain, 1867–90 — Lindsay Granshaw — 17

3. The Sanatorium Treatment for Consumption in Britain, 1890–1914 — Michael Worboys — 47

4. From Medical Research to Clinical Practice: Serum Therapy for Diphtheria in the 1890s — Paul Weindling — 72

5. Vaccine Therapy and Laboratory Medicine in Edwardian Britain — Michael Worboys — 84

6. From the Trenches to the Hospitals at Home: Physiologists, Clinicians and Oxygen Therapy — Steve Sturdy — 104

7. X-ray Technology in Obstetrics: Measuring Pelves at the Yale School of Medicine — Anja Hiddinga — 124

8. The Politics of a Spatial Innovation: Fracture Clinics in Inter-war Britain — Roger Cooter — 146

9	Cortisone and the Politics of Drama, 1949–55 *David Cantor*	165
10	Psychiatry in District General Hospitals: History, Contingency and Local Innovation in the Early Years of the National Health Service *John V. Pickstone*	185

Notes and References   200

Index   277

LIST OF ILLUSTRATIONS AND TABLES

Illustrations

3.1	Sanatoria in Great Britain, 1895–1910	57
3.2	Sanatorium beds in Great Britain, 1895–1910	60
7.1	The bony pelvis seen from the side	127
7.2	The bony pelvis seen from above	128
7.3	Nineteenth-century pelvimetry instruments	129
7.4	Making a pelvigram: the semi-recumbent position	135
7.5	Making a pelvigram: making the superior strait horizontal and parallel to the sensitive plate beneath	135

Table

3.1	Sanatoria and consumption hospitals in England and Wales, 1907	59

SERIES EDITOR'S INTRODUCTION

One purpose of historical writing is to illuminate the present. In the late twentieth century, science, technology and medicine are enormously important, yet their development is little studied. Histories of politics and literature abound, and historical biography is established as an effective way of setting individuals in context. But the historical literature on science, technology and medicine is relatively small, and the better studies are rarely accessible to the general reader. Too often one finds mere chronicles of progress, or scientific biographies which do little to illuminate either the science or the society in which it was produced, let alone their subtle interactions.

The reasons for this failure are as obvious as they are regrettable. Education in many countries, not least in Britain, draws deep divisions between the sciences and the humanities. Men and women who have been trained in science have too often been trained away from history, or from any sustained reflection on how societies work. Those educated in historical or social studies have usually learned so little of science that they remain thereafter suspicious, overawed, or both.

Such a diagnosis is by no means novel, nor is it particularly original to suggest that good historical studies of science may be peculiarly important for understanding our present. Indeed this series could be seen as extending research undertaken over the last half century, especially by American historians. But much of that work has treated science, technology and medicine separately; this series aims to draw them together, partly for the interplay of insights, partly because the three activities have become ever more intertwined. This breadth of focus and the stress on the relationships of knowledge and practice are particularly appropriate in a series which will concentrate on modern history and on industrial societies. Furthermore, while much of the existing historical scholarship is on American topics, this series aims to be international, encouraging studies of European material. The intention is to present science, technology and medicine as aspects of modern culture, analysing their economic, social and political aspects, but not neglecting the expert content which tends to distance them from other aspects of history. The books will investigate the uses and consequences of technical knowledge, and how it was shaped within particular economic, social and political structures.

Such analyses should contribute to discussions of present dilemmas and to assessments of policy. 'Science' no longer appears to us as a triumphant agent of Enlightenment, breaking the shackles of tradition, enabling command over nature. But neither is it to be seen as merely oppressive and dangerous. Judgement requires information and careful analysis, just as intelligent policy-making requires a community of discourse between men and women trained in technical specialties and those who are not.

This series is intended to supply analysis and to stimulate debate. Opinions will vary between authors; we claim only that the books are based on searching historical study of topics which are important, not least because they cut across conventional academic boundaries. They should appeal not just to historians, nor just to scientists, engineers and doctors, but to all who share the view that science, technology and medicine are far too important to be left out of history.

JOHN V. PICKSTONE

PREFACE

This collection of essays grew from a series of seminars held at the Centre for the History of Science, Technology and Medicine at Manchester University. They are written by social historians concerned to understand the processes of change in medicine. They emphasise the political and cultural dimensions of innovation and the importance of understanding innovations in their own times and contexts. Together they illuminate major features of the dynamics of medicine from the mid-nineteenth century to the near present.

JOHN V. PICKSTONE

ACKNOWLEDGEMENTS

The editor is pleased to acknowledge the skilful assistance of Joan Mottram and the financial support given by the Wellcome Trust to the Wellcome Unit for the History of Medicine, University of Manchester; also useful comments from Harry Marks. The editors of *Medical History* have kindly allowed the inclusion here of Roger Cooter's essay (Chapter 8), which they published in an earlier form. All the other essays were written for this volume.

The editor and publishers acknowledge with thanks permission from the following to reproduce copyright material.

American Journal of Roentography and Radium Therapy, for Figures 7.1 and 7.2, from an article by Herbert Thoms (1940).

American Journal of Obstetrics and Gynecology, for Figures 7.4 and 7.5, from an article by Herbert Thoms (1926).

Verlag von Gustav Fischer, Jena, for Figure 7.3, from F. Skutsch, *Die Becken-Messung an der lebenden Frau* (1887).

NOTES ON THE CONTRIBUTORS

David Cantor is a research fellow at the Wellcome Unit for the History of Medicine in Manchester. He wrote his PhD thesis on the history of radiobiology in interwar Britain. Subsequently, he was contracted by the Arthritis and Rheumatism Council for Research (ARC) to write their history, which is now in progress. He is now working on a history of images of rheumatoid arthritis. His most recent publications include 'The Contradictions of Specialization: Rheumatism and the Decline of the Spa in Inter-War Britain', *Medical History Supplement* No. 10, 1990, 127–44, and 'The Aches of Industry: Philanthropy and Rheumatism in Inter-War Britain', in Colin Jones and Jonathan Barry (eds), *Medicine and Charity in Western Europe Before the Welfare State* (1991).

Roger Cooter is Senior Research Officer at the Wellcome Unit for the History of Medicine, University of Manchester, and one of the editors of *Social History of Medicine*. His previous publications include *The Cultural Meaning of Popular Science* (1984), *Studies in the History of Alternative Medicine* (1988), *Phrenology in the British Isles* (1989). His *Surgery and Society in Peace and War, 1880–1948* is forthcoming in this series.

Lindsay Granshaw is a Lecturer in the History of Medicine at the Wellcome Institute in London. She is author of *St Mark's Hospital, London: The Social History of a Specialist Hospital* (1985) and co-editor with Roy Porter of *The Hospital in History* (1989). She has also written on the history of surgery and nursing.

Anja Hiddinga is a graduate in biology of the University of Leiden, and was appointed to the staff of the Department of Science Dynamics at the University of Amsterdam in 1981. Her present work is on the development of clinical science, with a special interest in obstetrics. Her publications are in the areas of science studies, history of medicine and women's studies.

John V. Pickstone directs the Wellcome Unit and the Centre for the History of Science, Technology and Medicine at the University of Manchester. He is the author of *Medicine and Industrial Society: A History of Hospital Development in Manchester and its Region, 1752–1946*

(1985). He also writes on the history of bio-medical sciences in France and Britain.

Steve Sturdy is a Wellcome Research Associate at the Centre for the History of Science, Technology and Medicine, University of Manchester, where he is working on the relationship between medical science and medical practice in twentieth-century Britain. He trained in natural science at Cambridge, in the philosophy of science at the University of Western Ontario, and in the history of science at the Science Studies Unit, University of Edinburgh. His recent publications include 'Biology as Social Theory: John Scott Haldane and Physiological Regulation', *British Journal for the History of Science* (1988), and 'Science and the Politics of Medical Practice: A Case Study of the Sheffield Medical School, 1900–1925', *Medical History* (1991).

Paul Weindling has been at the Wellcome Unit for the History of Medicine, University of Oxford, since 1978, where he is now Senior Research Officer. He read modern history at Merton College, Oxford, and undertook postgraduate study in the History of Biology at University College London, leading to an MSc and PhD. His published work includes *Health, Race and German Politics between National Unification and Nazism* (1989), and he has edited *The Social History of Occupational Health* (1986) and co-edited *Information Sources for the History of Science and Medicine* (1981).

Michael Worboys is a Lecturer in History at Sheffield City Polytechnic. He is a graduate of Manchester and Sussex (PhD), where he read biology and the social studies of science. His main research has been on science in relation to British colonial imperialism, and on the development of the biomedical sciences between 1880 and 1914.

ABBREVIATIONS

ACTH	adreno corticotrophic hormone
BMA	British Medical Association
BOA	British Orthopaedic Association
CRA	Clinical Research Association
DAH	disordered action of the heart
DNB	*Dictionary of National Biography*
DSIR	Department of Scientific and Industrial Research
ERC	Empire Rheumatism Campaign
GLRO	Greater London Record Office
GP	general practitioner of medicine
HMC	hospital management committee
kg	kilogramme
LCC	London County Council
LGB	Local Government Board
LMS	London Midland and Scottish Railway
MAP	Medical Advisory Panel
MO	medical officer
MRC	Medical Research Committee/Council
MRHB	Manchester Regional Hospital Board
NAPC	National Campaign for the Prevention of Consumption
NHS	National Health Service
PRO	Public Record Office
RAF	Royal Air Force
RAMC	Royal Army Medical Corps
RHB	Regional Hospital Board
RSM	Royal Society of Medicine
SAMO	Senior Administrative Medical Officer
TAP	Technical Advisory Panel
TB	tuberculosis
TUC	Trades Union Congress
UGC	University Grants Committee
USA	United States of America

1
INTRODUCTION
John V. Pickstone

WHY 'INNOVATION'?

'Innovation' is a fashionable word, but not without reason; we are all rather wary about 'progress'. Our world changes rapidly, not least through new technologies, but we no longer have the high Victorian confidence that change is for the best. We have learned to be wary of utopian technologies – of nuclear power, of high intensity agriculture, even of high technology medicine. And as public programmes recede before the protagonists of market mechanisms, so new technology comes to be identified with new products for markets rather than with grand designs for the better provision of energy, food or welfare. The word 'innovation' suits our time. It evokes particulars and smacks less of 'progress'; we can fairly be suspicious of mere novelty.

'Innovation' is also more helpful than 'discovery' or 'invention', for it distinguishes the complex processes of social and economic change from the perhaps simpler questions of 'creation'. The distinctions are slippery, but 'invention' or 'discovery' refers to the origin of the artefact or concept, 'innovation' to the introduction of that idea or artefact into social and economic systems – for example, into the practice of medicine. Thereafter analysts refer to 'diffusion', if the innovation becomes more widely adopted.

The usefulness of these distinctions is easily seen by reflecting a little on customary modes of writing the history of medicine. We are always told who originated a discovery or technique, we are rarely told how it fared thereafter. The fact, or the supposition, that the novelty became widespread suggests that its usefulness was indubitable. Those who opposed such developments were wrong, probably through conservatism or jealousy, which appear in such accounts as species of original sin. Our aim in this collection is to probe behind such chronicles to produce a more realistic and more helpful kind of history.

The essays presented here have been written with a common agenda in mind, but not to a formula. Some of the authors have been explicitly concerned with models of innovation. Others have studied particular medical innovations as part of their research on larger historical projects – on professionalisation, for instance. We have drawn the essays together because collectively they provide a good sample of how historians of medicine now approach questions of change in medical practice.

I have arranged the essays chronologically, as far as possible, so that they will serve two purposes. By considering their common themes, we may be able to make useful generalisations about the processes of medical change. Secondly, by considering them in sequence, we gain a useful insight into the history of medicine, especially in Britain, over the century 1860–1960. In this Introduction I take the general conclusions of the essays and relate them to various current traditions of scholarship – in economic history, innovation studies, medical policy research and the history of technology. I also try to show how the essays draw on, and in turn illuminate, key features of medicine in the century under discussion – the increasing complexity of medical institutions, the growing influence of medical scientists, and the importance of the two World Wars. But first we may take a glimpse at the essays, beginning with antisepsis in the 1860s and ending with the early years of the National Health Service (NHS).

THE ESSAYS

Wonder drugs and surgery are talismans of the magic of modern medicine. Of these symbols, the first was surgery. In 1900, when therapeutics was still largely a matter of empiricism, surgery was 'rational'. Surgeons understood enough of pathological anatomy to know what to remove or repair; they had the techniques to do so and to leave a patient who was likely to recover. Two techniques have been seen as crucial: anaesthetics and antisepsis (later, asepsis). Anaesthesia was a 'Yankee dodge' from the 1840s; asepsis – the attempt to produce and maintain a sterile environment by ultra-cleanliness and heat sterilisation – was a German practice developed in America around 1900. But antisepsis, using carbolic acid to kill germs in the air and on wounds, was British, the work of Joseph Lister from the 1860s. Furthermore, it was a fruit of science; it had been based on the theory of germs when that theory was new.

In Chapter 2, Lindsay Granshaw takes a long look at Lister and his disciples. She asks the same questions as help structure most of the essays in this volume: What was the 'problem situation'? Who proposed and who opposed the innovation, and why? On what terms was the debate conducted and practice determined? Was the issue resolved? In answering such questions she shows the need for contextual understanding: here the context was the very public debate over insanitary conditions and high mortality rates in large British hospitals around 1860. She shows the importance of technical resource: Lister, in trying to reduce death rates and defend the reputation of hospitals, used a product of Victorian gas technology – carbolic acid, a refinement of creosote. He used it as an antiseptic in the then current sense of the word – an agent which would stop sepsis.

The use of antiseptics on wounds did not necessarily imply a belief in germs, any more than did the use of creosote to prevent the rotting of wood. But Lister believed in germs; he had learned about them from Louis Pasteur, just as his carbolic trials got under way. Among the many nice features of Granshaw's essay is her exploration of the role of this 'science' in the evaluation of a new technique. It is a point to which we shall return, as we shall to her last major conclusion: that this debate was transcended rather than resolved. Antisepsis, around the end of the century, was replaced by asepsis – the practices we still use to 'sterilise' in advance whatever may come into contact with the site of operation. Depending on one's point of view, the new procedure could be seen as a refinement of Lister's technique or as a refinement of the general cleanliness advocated by Lister's opponents. Lister's disciples took the first view, and wrote it into their histories of surgery, though Lister himself had remained sceptical of 'asepsis'.

In Chapter 3, Michael Worboys looks at another technology often linked with 'germs'. He assesses sanatoria as a remedy for tuberculosis, and explicitly uses models of innovation and diffusion drawn from economics and economic history. Sanatoria were a German invention; how were they introduced in Britain? Who developed *private*, fee–receiving sanatoria? What was the market? When was it saturated? For *public* sanatoria, paid for by charity or by government, the economics were rather different; the innovation now depended on the confidence of key medical men and politicians that sanatoria would help solve medical-political problems such as the cost of tuberculosis to a national health insurance scheme. If, in the context of private hospitals, efficacy was ultimately the readiness of patients to pay for treatment, in the public context it was the confidence of officials and politicians that sanatoria were useful. In neither case were 'statistics' more than a minor input to the evaluation; for sanatoria, as for antisepsis, published data were confused and contested. Indeed, as historians have come to see, such debates can rarely be analysed in terms of 'facts', accessible and persuasive for all participants. Instead we must reconstruct the perceptions and intentions of the various actors. We must be aware of changes in meaning, and of the several ways in which a public 'problem' could come to be constituted.

Tuberculosis became a public problem around 1890, when the death rate from this disease had long been declining. (In earlier decades, 'consumption' had been but a way of death.) The discovery of a germ, the evidence from pathological anatomy that lesions did not always progress to death, the known decline in the tuberculosis mortality rate – all gave cause for hope and thus for the construction of a public 'problem'. Germ theory allowed older, 'change of air' practices to be given new meanings and eventually to be sold as part of a public health campaign. The sanatorium came to be seen as the home of discipline, regimen, 'neo-technic'[1] therapies and laboratories; and with this foil, health campaigners struggled to lay open the dark and dirty, crowded and undisciplined recesses of working-class life.

New surgical techniques and sanatoria both had Janus-like qualities. They could be seen as developments of older concerns with cleanliness and fresh air, or they could be seen as the fruits of modern science. Usually, they benefited from both sets of associations. With our next two innovations, the case is rather different. These also concerned the new bacteriology, but they were more specific, more bound up with 'science'; for their advocates, they were novel, rational therapies for particular diseases.

As Paul Weindling shows in his discussion of the 1890s, in Chapter 4, diphtheria anti-toxin was a key technique, both practically and symbolically, for the medical researchers who followed up the work of Louis Pasteur and Robert Koch. Pasteur had shown that the old technique of vaccination against smallpox could be generalised to some other conditions. If weakened microbes were introduced into animals or patients, resistance could be built up, apparently because substances accumulated in the blood which could neutralise the microbes concerned. When researchers reported that some bacteria produced poisonous toxins then it seemed likely that the blood of animals so immunised would contain neutralising 'anti-toxins'. These might be introduced as a therapy into a patient already suffering from the disease. This so-called *passive* immunisation was first developed for diphtheria, a particularly nasty disease of infants; diphtheria anti-toxin thus came to be seen as the first successful and specific therapeutic product of the new medical research. How does such a claim withstand Weindling's detailed investigations?

The use of anti-toxin became widespread within a few years, and it remained so until diphtheria declined. Some doctors were worried about side-effects, and some later doctors have expressed retrospective doubts about the efficacy of anti-toxins, at least under the usual conditions of their use. But most doctors, then and since, seem to have accepted the utility of the anti-toxin. Even so, as Weindling shows, there are grounds to doubt the usual history. The statistics on the effects of anti-toxin on disease-specific death rates are questionable; as in the case of antisepsis, much that was attributed to a specific technique was probably the result of more general factors (such as epidemic waves, changes in virulence, or increased resistance from better nutrition). One could also query the mode of action and the 'meaning' of the anti-toxin. For its advocates it was a therapeutic substance, but for others it was simply a means of reducing the effects of toxins until the body's own resistance could overcome the bacteria. Even here, one sees the tension between those who would stress specific new products and those for whom the new science was elucidating the older theme of resistance, or the healing power of nature. Even when an innovation was generally accepted as useful, and when its use was *relatively* uncomplicated, its meaning and importance were contested as different professional groups 'constructed' the novelty in ways which best suited their own positions.

In Michael Worboys's second essay, on vaccine therapy, in Chapter 5, we encounter a classic instance of rival interpretations. Vaccine therapy, like anti-toxins, was an attempt to use natural agents of the body's defence in order to overcome disease already established in patients. In vaccine therapy, some of the infective material was grown in culture *in vitro* and specific microbes were isolated. These microbes were then killed before being injected back into the patient where, it was predicted, the dead microbes would stimulate the body's immune response. The therapy seemed particularly appropriate when the initial infection was local, as in a boil. In such cases, the body's immune system was but partially activated; injection of killed microbes, however, could fully mobilise the resistance. The method was pioneered by Almroth Wright, one of the best known personalities in British medicine in the decades around the First World War.

For Wright's followers, vaccine therapy was the royal road of modern medicine. Here was a laboratory method of general application, one which was scientific and yet applicable to the needs of individual sufferers. But in the eyes of its detractors, vaccine therapy was 'pseudo-science' of doubtful efficacy. Worboys here examines the spread of the method and its possible importance in the establishment of new laboratories and new jobs in hospitals. He concentrates on elucidating how vaccine therapy became so very controversial, and argues that it was because 'laboratory men' were venturing into the traditional territory of clinicians – the care of individual patients. This potential conflict of interest between clinicians and 'laboratory medics' is a major theme in all our inter-war studies.

With the essay by Steve Sturdy, (Chapter 6), we move away from 'germs' and vaccines towards physiology, instrumentation and quantitative analysis. We see the interaction of laboratory scientists and clinicians as the former began to establish secure, salaried, posts in late Victorian and Edwardian Britain. Physiologists were the vanguard of academic medical science in Britain; from about 1860 they had promoted themselves as liberal educators, training medical students to be analytical observers of nature. By the end of the century most medical schools had physiology laboratories for teaching and for research, but by then medical science, especially through the study of bacteria, was claiming a new utility. As we have seen, from the 1880s 'bacteriology' could be used in diagnosis and increasingly in prophylaxis and therapy. Not to be outdone, physiologists advanced their own claims that many diseases could be understood as *functional* disorders, subject to physiological analysis. These hopes were realised after the First World War, when rickets, diabetes and pernicious anaemia were construed as deficiencies which could be remedied by vitamins, insulin or liver extracts, respectively. Physiologists and the new biochemists appeared as the authors of these impressive 'cures'.

Sturdy here presents a less dramatic, less well-known case, in which an old resource, oxygen, was turned to a new problem – poison gases in the First World War. He shows how the use of oxygen was advanced by physiologists who had been recruited from their laboratories to help in a national emergency. He shows how their perceptions of the problems differed fundamentally from those of their collaborating clinicians. The difference was partly a matter of experience and expectation, partly of vantage point, for the clinicians were often concerned with front-line emergencies, the physiologists with longer-term disability.

As this study illustrates, the European wars were important for medicine. Through them new problems were raised, medical power-structures were changed, new personnel and new roles were introduced and the aims of much medical practice was transformed, at least temporarily. Sturdy's essay shows how a particular innovation can be understood in terms of these general changes. It also shows, as do other related historical studies, that the First World War altered *perceptions* of disease. In conditions of acute man-power shortages, both military and civilian, when doctors were organised by the state, they were called upon to minimise the *disability* among fighting men. The goal of medicine could not be labelling, or simply keeping people alive, for disease and injury now meant a collective loss of man-power; doctors who could claim to reduce disability were thereby privileged. For as long as those peculiar conditions lasted, such doctors had incentive, room and power to innovate.

With Anja Hiddinga's essay, (Chapter 7), we examine another group of inter-war professionals whose self-image and success depended largely on the contemporary appeal of scientific measurement. Hiddinga examines the history of x-ray pelvimetry in America, and shows how Herbert Thoms, a professor of obstetrics in Yale University, attempted to introduce x-ray examination of all pregnant women, to screen for constricted pelves which would lead to obstetrical difficulties. By then the understanding of pelvic abnormalities was at least a century old and x-rays had been used occasionally in this connection from soon after their introduction to medicine (1896). Thoms's programme was a response to the public prominence which the problem of maternal mortality had attained in the USA and Europe during the inter-war years. His programme also reflected the attempts to 'scientise' clinical practice. To most of those who wielded power in medical school administrations and in the related parts of government, 'progress' meant adapting clinical medicine to the concepts, tools and organisation of *science*. That meant laboratory tests and new technology, specialisation of staff and more efficient 'sorting' of patients. In this context, Thoms's innovation seemed highly appropriate.

As must be clear by now, when we discuss innovation in medicine, we are usually discussing changes in the operations of a complex system. We meet the problem in a particularly striking form in Roger Cooter's essay on the establishment of special clinics for the treatment of bone fractures, Chapter 8. Here the *technical* novelty was minimal; plaster-of-Paris came to be seen as characteristic of the new clinics, but it does not seem to have been important for their origins, nor essential to their mode of work. The key factor, Cooter argues, was the professional interests of a would-be surgical sub-profession – a new generation of orthopaedic surgeons, schooled in the First World War and anxious to establish their own expert space within the prestigious general voluntary hospitals. Among their past or potential resources, special military hospitals had disappeared, hospitals for the care of crippled children were too far away from the centres of medicine, and civilian fractures were largely in the hands of general surgeons, usually the overworked junior surgeons in hospital casualty departments. Expertise in fractures was not a promising road to a large private practice; its potential financial backers were public or semi-public bodies – employers, trade unions and municipalities – rather than private patients. Only when such support could be mobilised and organised would orthopaedists have the power to carve out their own domains in the general hospitals where other professional interests were entrenched.

With the two final essays, we move to Britain after the Second World War, to explore how the peculiarities of post-war politics and economics conditioned two medical innovations: the use of cortisone for arthritis and the development of psychiatric units within general hospitals.

Cortisone might have been regarded as a relatively simple case of innovation. The substance was isolated in the USA as part of a research programme on hormones produced by the adrenal gland. It was then found to have beneficial effects on arthritic patients. In simple accounts it might feature as a new 'wonder-drug', introduced into medical practice as quickly as drug companies could manufacture it. Such an account would take 'supply' and 'demand' as relatively unproblematic, possibly by assuming a free market or a sufficiently generous welfare state. David Cantor shows, for Britain, in Chapter 9, that the real situation was more revealing. Supply was a political and economic problem for a country desperately short of dollars and picqued about the 'loss' of penicillin to America. Demand, too, was a political issue, for under the new NHS the government would be judged by the availability of the new drug to patients who desperately wanted it. Using patients' letters and reports of doctor–patient interactions, Cantor explores the meanings then carried by cortisone. Patients had longed for speedy cures; rheumatologists worried about 'quackery'; what then would be the responses to 'spectacular' cures by cortisone? Cantor's novel analysis raises important general questions about medicine and the dramatic; it also shows how such questions should be, and can be, answered for particular times and contexts.

In the final essay, Chapter 10, we move to an innovation which was, in a sense, produced by the birth of the NHS in 1948. John Pickstone here reflects on a peculiarity of the early NHS arrangements: the readiness of the Manchester Regional Hospital Board to develop psychiatric units within general hospitals. He shows that such a move was consonant with the general drift of inter-war policy for psychiatry, which urged more integration with general medical facilities. He shows that the Manchester example became influential in the 1960s, as national policy moved rapidly towards 'community care' and the run-down of large asylums. But he also shows how singular was the Manchester development when it originated about 1950. It arose, he argues, not from leaders in psychiatric reform but from the exigencies of a particular regional board faced with huge, old asylums and acute staff shortages. They found an alternative mode of development in general hospitals which had once been workhouses for paupers. That such hospitals already had large blocks of 'mental' wards was due to the history of Lancashire as a very urban county; that the new policy-makers in Lancashire were open to this solution was in part contingent, but their involvement and their ability to act depended on the introduction and form of the NHS. The case is a good illustration of how long-standing patterns and new forms of organisation may facilitate medical innovations in a way which few could have predicted. It underlines again the complexity of medical systems and the importance of particular, often local, conjunctures.

INNOVATION, ECONOMICS AND MEDICAL HISTORY

Many studies of technical innovation have been written by economists or economic historians. Some suppose a simple world of independent producers and consumers assessing innovation on the basis of known costs and adequate information flow. To the extent that the conditions of uptake are uniform across the potential field, the pattern of uptake will approximate to the *S*-shaped curve characteristic of diffusion processes. The initially slow successes will be followed by a period of rapid adoption until the potential field approaches saturation.[2]

There are some case studies which show such patterns quite well, but in general, as our essays indicate, life is not that simple. The more one explores 'discovery' or 'invention', the more they appear as complex processes, from which the processes of innovation are not easily separated. The more one explores real instances of 'diffusion', the more one sees differentiated fields – for example, with different factor prices – where conditions of uptake will be far from equal. Of course, in as much as demands vary, so the conditions of innovation will vary, and the nature of the actual innovations. In this way, changes in product and process will proceed along with 'diffusion', so that analytical separation becomes difficult and perhaps unhelpful.

It is around this level of complexity that the 'new economic historians' have operated, mobilising economic theory in order to interrogate the past. They have reconstructed the choices of industrialists and investors, to assess the 'rationality' and the consequences of their decisions. On a larger scale they have sought to measure the consequences of innovations for whole economies. What were the economic advantages of steam-power over water-power, for example, and how did the differences vary over place and time? A classic example is the literature on transatlantic differences in technology during the nineteenth century. In the USA, where wood was cheap and labour relatively expensive, coal-based and labour-intensive forms of technology proved much less attractive than in Britain.[3]

Such econometric history works in a concrete and complex world but it is still one in which the inputs to decisions can be recovered, exposed quantitatively, and assessed in terms of economic rationality. Has that kind of approach been used in medical history? Can we, for example, recover and assess the economic rationality of choices in nineteenth century medicine?

The short answer is that some work *should* be possible, but that very little has been carried out. We know scarcely anything about the importance of prices (and other costs) in the choices made by patients about medicines and medical services. There is little systematic information on how doctors' charges compared with the cost of buying medicines directly. Certainly, medicine was a very peculiar market, much influenced by fashion and faith, imponderables, unknowns and uneven distributions; even so, we may argue, medical history would benefit from more systematic attempts to place the medical choices of patients in the context of family expenditures, as well as in the context of medical beliefs or local social relations.[4]

At the level of doctors' choices we have but little more information, for only recently have historians engaged with the therapeutic choices made by nineteenth-century doctors. This literature is as yet almost exclusively on America and, for all its virtues,[5] it says very little about more narrowly economic questions. We are still rather vague about the costs of anaesthetics or antiseptics, about the readiness of doctors to suggest such treatments and the readiness of patients to have such costs passed on to them. One can only hope that historians will take on these questions, aided perhaps by the growth of health economics as a discipline. But in history, as in present-day health economics, a sharper awareness of costs is more likely to underline than to erase the *political* aspects of medical choices.[6]

POLITICS AND MEDICAL INNOVATIONS

A second genre of literature – 'innovation studies' – includes political and sociological as well as economic approaches.[7] Some such researchers have sought to facilitate innovation, or increase the role of 'rational assessment' in the process. Early studies pointed to the roles of 'product champions'; a classic essay on the career of a medical innovation set out seven stages, from 'promising report' to 'discreditation' (random control trials were stage five). Several more recent studies have been concerned with the intricacies of social networks, e.g. among physicians who adopted (or did not adopt) a given innovation.[8]

Such work converges interestingly with that of historians as represented in this volume. And there is room for much more interaction. The possibilities are well illustrated in an essay by Susan Bell on diethylstilboestrol which uses an interactive rather than a sequential model, focuses on the politics of conflicting interests and is fully aware of context and the historical dimension.[9] In such work one sees how far policy analysis has travelled from the simpler, more economic, models of innovation which were once common. Of course, one of the justifications given for this shift of focus is that the object in question – the health services – have grown ever more complex and political. If decisions on new medical technologies are now taken in a series of institutional and governmental committees, then the processes of innovation must be analysed as politics. One could hardly dispute the case.

The historian may, however, dispute the tendency of some policy analysts to assume that medicine was once so simple that economic models of individual choices would be appropriate. It may be easy to guess that medical innovation was once a matter of individual scientific 'producers' and individual clinical 'receivers', linked by the free dissemination of information in medical journals or conferences but, as our essays show, there is little reason to adopt such a model, even for 'archaic medicine'. It is more profitable to accept that decisions, even when they were made by individuals, were usually part of some formal or informal political process, matters of alliances or enmities, of professional co-operation or rivalry. Our historical case studies can then help illustrate how the political frameworks have varied and how they have tended to change over the last century or so of medicine.

For example, in the Granshaw study on antiseptics (Chapter 2), the key *site* of innovating activity was the hospital, usually a voluntary (i.e., charity) hospital; but, as far as we can tell, the key decision-makers were the individual surgeons who decided whether or not to adopt Listerian practices. It may be that in some cases medical boards had some input in such decisions, but generally it would seem that surgical techniques, like choices of therapy, were matters of consultant autonomy. Likewise, with diphtheria anti-toxin or vaccine therapy, responsibility for the *use* of the innovation in particular cases would seem to have rested with individual doctors, though they would have been much influenced by availability of the therapy in any given location, which may have been a matter of corporate decision. Thus, a particular physician in a voluntary hospital could choose to use the kind of vaccine therapy which was commercially available to doctors in private practice. Whether he could use the more sophisticated techniques which depended on laboratory tests would probably depend on whether or not the hospital in question possessed such a laboratory. This was less a question of physical space than of expert staffing; specialist exponents of vaccine therapy would be taken onto the staff of hospitals by corporate decisions, led by medical boards. The initiative might come from such boards (or key members thereof), or it might come from vaccine therapists looking for the prestige (and income) which a hospital attachment could provide.

Even in war-time, decisions on therapy would usually be left to individual doctors, while decisions on supply would be taken higher in the medical hierarchy. Sturdy (Chapter 6) has shown that even where oxygen was supplied for use in military hospitals, doctors were often reluctant to use it. His study also raises key questions about the role of education in conditioning both the uptake of particular innovations and the receptivity to innovations more generally. One of the arguments for 'modern medical education' around 1914, as advocated by Abraham Flexner and associates in both America and Britain, was that doctors educated in medical science would then be less dependent on tradition, more ready and able to seek out and assess for themselves potential improvements in treatment. Sturdy suggests that such training, by exposing students to 'physiological' conceptions of disease, would ensure that inter-war doctors were more likely than their wartime forbears to sympathise with the physiological conception of appropriate therapy.

Similar points could be made from Hiddinga's study of x-ray pelvimetry (Chapter 7), or Cooter's study of orthopaedics (Chapter 8). In both these areas, specialists trained in new methods sought to introduce them into new or re-modelled institutions. In as much as the technique might require relatively complex equipment and/or substantial modification of hospital routines, medical colleagues and hospital governing boards would both need to be persuaded.

Our studies, almost all focused on institutional medicine, do indeed tend to show the increasing importance of corporate decisions, especially where they involved common equipment, service specialties and/or major organisational and administrative changes. But they also show how 'political' were the choices made by doctors, even when acting as individuals. Education, allegiances, and existing professional investments were all major factors in determining whether or when doctors innovated. Whether or not one took up Lister depended, in part, on how one saw surgery and its articulations with the new medical science. Receptivity to diphtheria anti-toxin, vaccine therapy, or oxygen therapy were all similarly conditioned by attitudes towards 'laboratory medicine', sometimes contrasted with clinical art. Attitudes to fracture clinics, or psychiatric units in general hospitals depended on attitudes to specialisation, which were likely to be shaped by existing material interests as well as by general views acquired as a medical student. How a rheumatologist used cortisone was in part determined by its perceived potential for strengthening (or weakening) a marginal medical specialty. Professional interests, whether of doctors as a whole or of specialist groupings, are again revealed as central to these historical case studies.

MEDICAL HISTORY AND THE NEW HISTORY OF TECHNOLOGY

In emphasising the 'social shaping' of innovations and responses, the volume also lends support to a relatively new genre of writing on the history of technology, once a field dominated by technical description and/or technological determinism. The newer writing stresses that technological change has not been 'inevitable': the origins of innovations need to be explained, in part, by social process; the forms taken, the choices between techniques, and the patterns of uptake are likewise revealed as matters of economic and social history. Variety and contingency are stressed, without assuming that technologies are 'of themselves' value-free or value-neutral as to patterns of use.[10]

We have chosen here to concentrate on patterns of choice and uptake, rather than on the factors which helped determine the particular form of innovation. This is a difference of focus and emphasis, rather than one of historical interpretation. It would be possible to reconstruct the variety of antiseptic procedures explored and employed by Lister, and to elucidate the reasons for the choices. Sanatoria, anti-toxins, oxygen and fracture clinics could all be so 'deconstructed', perhaps to good effect. Indeed, the interested reader will be able to proceed some way on this analytical path with the information provided in the essays. To argue, for example, that an innovation was seen by protagonists as a product of laboratory science is to suggest that the form of such innovation was *expressive* of science, perhaps marked by features understood as characteristic of laboratories. Standardisation, quantitative measure of effect, regular regimen, differentiation and organisation of 'therapeutic work' – all could reasonably be seen as aspects of that 'scientific modernism' of which laboratories were paradigmatic from about a century ago. As bearers of such characteristics, vaccine therapy, oxygen treatment or sanatoria would act to reinforce not just the claims of 'scientific medicine' but the more general currents of 'neo-technics', of scientific management and of technical expertise in the diagnosis and solution of social ills. There is certainly room for more historical analysis along such lines, and medical history here could usefully join forces with the better recent work on 'technology and culture'.[11]

But if we wish to understand innovations and culture we need to deconstruct 'problems' as well as 'solutions', 'needs' as well as 'products', and here medical history may be peculiarly useful. As already indicated, there is a substantial literature analysing 'concepts of disease', not least the work on soldiers' heart, shell-shock and functional neuroses as seen in the First World War. However, these disease concepts are still usually treated as independent from, or prior to, instrumental or therapeutic procedures; that may be a mistake, springing from the usual over-emphasis upon knowledge rather than practices. We might better hypothesise that concepts and practices develop interactively, and that we can treat disease-diagnosis-therapy as a complex product. In such a scheme, sophisticated analyses of 'disease' may thus prove useful to analysts of non-medical technologies – for example, when they try to understand how industrial producers, say, diagnosed 'need' in relation to possible 'supplies'. Studies of how patients saw their own conditions may similarly illuminate non-medical consumers. Such studies will surely show that we can fully understand historical formulations of 'needs' or 'remedies' only by attention to their particular contexts.

Does this mean, then, that historical studies such as these can be descriptive or analytical but *not* evaluative? Does the manifest variety of conceptualisation, even at the level of problems, mean that we have no standards by which to judge medical innovations? The question is difficult; that is why it is usually avoided or answered simply from first principles; it deserves more serious consideration.

MEDICAL INNOVATIONS AND THE PROBLEMS OF RETROSPECTIVE ASSESSMENT

As the studies here assembled show, innovations in medicine were but rarely subjected to any rigorous analysis of benefits and costs. Nowadays historians of medicine fall into two main camps on evaluating such matters. Some historians, and most doctors, assume that time is an efficient filter; if procedures or products have persisted, they are probably useful and this credit is read back to the individuals responsible for the innovations. On the other hand, most historians (and perhaps a few doctors), are very wary of any kind of retrospective evaluation. Innovations, for them, are to be understood only in context, in the terms and by the standards of those times, in so far as these can be reconstructed; it is difficult and dangerous to try to find out whether procedures 'really worked'.

One understands well enough why this latter view is common; almost all historians would agree that their *primary* responsibility is to be true to the perceptions and judgements of the historical actors. We must indeed be careful not to 'read back' later knowledge and assume that our actors in the past 'must have known' or 'needed to have known' whatever later generations saw as truth. In teaching the history of medicine, one too often meets the assumption that it was practically impossible to do anything sensible about a given infectious disease, say, before its *cause* (i.e., bacterium) was discovered; and that action thereafter was so easy that all one need explain is the delay. The lesson of history, so often, is precisely the opposite. Many problems were tackled successfully on theories different from ours; the introduction of our theory, however intellectually sweet for those concerned, may then have meant but little to sufferers. It is for such reasons that we must focus on particular contexts and should try to include the judgements of patients as well as of doctors.

But where the historical patient is long beyond recall, can the medical historian speak for him or her, perhaps by making judgements as to the likely effect of medical procedures and the likely consequences for patients? In principle, I see no reason to exclude such attempts. When due allowance is made for changes in perception and belief we may still venture to interpret effects, at least over the relatively small chronological and cultural spans with which we are here concerned. As far as we know, our grand-parents and our great, great grand-parents hoped, like us, to recover from sickness, to suffer less from pain, to die no earlier than need be. In this sense, medicine has general goals, and its efficiency may be judged accordingly.

Yet, I do not claim we have got far in that direction here. As we shall see, Granshaw would be sceptical of any claim that antisepsis was generally more effective than alternative procedures (though both were better than previous practice). Worboys agrees with F. B. Smith that we have few grounds for believing that sanatoria extended the lives of the tuberculous – other ways of using the money might have been more effective. The efficacy of diphtheria anti-toxin was generally accepted; it was certainly less debated than were sanatoria. Vaccine therapy is commonly regarded as a failure but, as Worboys shows, it persisted, at least for chronic conditions, until displaced by antibiotics. Oxygen therapy continued to have a limited role. X-ray pelvimetry never 'caught-on' outside specialist clinics; later fears of radiation hazard would have killed it anyway. Fracture clinics continue to be one way of organising accident cases; but accidents and emergencies continue to raise problems of medical organisation: they do not fit easily into the patterns of specialisation (or the work schedules) of hospital consultants. Cortisone was rapidly accepted as a major resource in the treatment of rheumatism, though the side-effects could be severe. Psychiatric departments continue to be established in general hospitals; most British psychiatry is now practised in general hospitals and by 'community care'. But who will say that patients have benefited hereby?

It would be wrong, in my opinion, to try to banish retrospective judgements, but the present essays fully demonstrate the difficulties of such analysis. They show quite clearly how small has been the role of demonstration and of calculation in the uptake or rejection of these innovations. If we wish to understand how medicine has changed, we have no alternative but to study the real, messy, contested and complex debates by which, over time, some procedures were accepted in preference to others. That is what medical historians do, or at least some of them for some of their time. We do not claim that the work sampled here is 'representative' of present-day historical work, or indeed that the topics covered are a statistical sample of medical innovation. (We, too, collectively and individually, have our own particular interests and concerns.) We do claim that such analyses cannot be ignored by those who wish to understand how medicine has developed or, more generally, how innovations are negotiated in complex social systems. They should appeal to all who are concerned with the institutional and cultural politics of change.

2
'UPON THIS PRINCIPLE I HAVE BASED A PRACTICE': THE DEVELOPMENT AND RECEPTION OF ANTISEPSIS IN BRITAIN, 1867–90[1]

Lindsay Granshaw

Joseph Lister almost single-handedly revolutionised modern surgery through his development of antisepsis in 1865, or so traditional accounts have stated. Like other prophets in their own lands, he was at first rejected, but eventually prevailed: those who opposed or ignored antisepsis were defeated by its obvious truth and efficacy, and Lister was honoured across national boundaries. Meanwhile, antisepsis had been incorporated into surgical practice, with revolutionary effect.[2]

Although the impact of Lister's work should not be underestimated as historical revisionism takes its toll, nevertheless re-assessment of this account is long overdue. To concentrate on Lister alone ignores the context in which his theory was worked out, what Lister's contemporaries were already doing, the way in which antisepsis was promulgated, received and changed, and what its longer-term impact turned out to be.[3]

THE CONTEXT OF THE DEVELOPMENT OF ANTISEPSIS

By the mid-nineteenth century, hospitals in Britain were becoming more central to medicine. From a position in which they were very much on the margins, treating only the very poor, and visited rarely if at all by medical practitioners, hospitals were assuming a much more important role, particularly for the developing medical profession. Doctors increasingly relied on them both for status and for clinical experience. And yet new threats loomed over the hospitals' very existence.[4] It had long been held that sickness could be caused by airborne miasma, which emanated from such decomposing and contaminated sources as manure, sewage or other rotting and foul-smelling material. In the nineteenth century, sanitary reformers applied these ideas with new force not only to the ills of the rapidly growing cities but also to other large congregations of people, such as prisons and hospitals.[5] The very healthiness of hospitals became a widespread lay concern. Such a challenge was not ignored for long by those whose positions depended on the hospitals, and it is in this context that the development of antisepsis should be seen.

Medical men, harnessing the new interest in statistics, initially fuelled the debate. Studies carried out by surgeons between the 1840s and the 1870s initially listed procedures and illnesses, but later analysed causes of death after operation, and subsequently made comparisons between hospitals. It was generally concluded that a 35 per cent post-operative death rate was usual; for amputation at the thigh, a death rate of 65 per cent was common.[6] In the 1840s such levels were regarded as largely tolerable. However, by the 1850s and 1860s, with the public question mark over hospitals, these mortality levels – and more particularly their causes – began to provoke concern. The 'hospital diseases' which afflicted patients only on entry to hospital – hospital gangrene, erysipelas, and pyaemia – were now identified as major killers. According to Thomas Bryant in 1861, 42 per cent of the deaths after amputations at Guy's Hospital had resulted from pyaemia.[7] Miasmatic material would infect one patient, whose emanations might then infect other patients. Poisons could be carried by the blood or the nervous system around the body to produce systemic ills and bodily decomposition. The constitution, sex, age, diet and general state of health of a person were significant in determining susceptibility. Locality, season, and condition of the atmosphere could also be contributory factors.[8]

Even though the statistics did not show hospital disease as the only, or even the main, cause of death (shock scored higher), nevertheless it was hospital disease which came to be seen as intolerable in its incidence. City hospitals, especially large ones, seemed particularly dangerous. Thus Thomas Spencer Wells, surgeon to the Samaritan Hospital in London, reported that the mortality rate in hospitals in the small towns of Oxford, Exeter and Cambridge was around 13 to 16 per cent, while the London figures for amputations stood at around 50 per cent. Wells concluded that hospitals should not be too large or multi-storied (his own was a small specialist institution).[9] Sir James Young Simpson, Professor of Obstetrics at Edinburgh and surgeon to the Edinburgh Royal Infirmary, argued that the scale of the problem was new. Mortality rates after amputation had increased due 'chiefly or entirely to our system of huge and colossal hospital edifices, and to the hygienic evils which that system has hitherto been made to involve'.[10] However, proposing closure or reduction in the size of large urban hospitals became increasingly unpopular by the 1860s among medical men. They now suggested that the apparent difference in mortality between large city hospitals and small country ones was in fact due to the types of patient found in each. The city hospitals received weaker patients, worn out by years of city life. The country hospitals had stronger, fitter patients whose constitutions were more robust. As Timothy Holmes, surgeon to St George's Hospital, argued in 1866:

> No necessity has been shown to exist either for their [the hospitals'] removal into country districts, or for breaking up and disintegrating them into smaller institutions.... No data exists by which we can estimate ... whether this [surgical mortality] is a larger or a smaller death-rate than would prevail in the same class of cases and patients if treated in smaller hospitals, in detached chambers, or in their own houses.[11]

The potential threat to hospitals was exemplified in the 1860s by the forced move of St Thomas's Hospital from London Bridge. Sanitarians like Florence Nightingale argued that this was a perfect opportunity to establish the hospital on a salubrious site in the country, well-ventilated with fresh country air to blow the miasma away. The doctors fought bitterly and in the end successfully against the effectual loss of their hospital.[12] If such drastic measures were to be avoided, with the medical staff losing their status or being separated from their private practice, ways of reducing mortality levels other than by closure had to be sought. Measures included renewed enthusiasm for the long-established practice of whitewashing walls to destroy miasma, removal of privies from wards, separation of medical and surgical patients, building new wards to allow more space around each patient, setting up new ventilation systems, and attempting to prevent contaminated air entering the wards. Wards and beds were disinfected: any substance which banished 'infection' could be termed a 'dis-infectant', and various chemicals including carbolic acid were used for this purpose.

Solutions to the problem of hospital disease were also sought in treatments. In the hope that they would actively assist nature's healing powers, various post-operative and post-accident dressings were used, including nitric acid, arsenic and tincture of iodine.[13] A number of leading surgeons developed their own ways of combatting surgical mortality. Just as cleanliness was greatly emphasised in hospital management, so too cleanliness was stressed in the treatment of wounds. Certain surgeons, such as George

Callender at St Bartholomew's Hospital, developed specific methods of applying cleanliness and published statistics to show that this was making a great difference to their surgical mortality rates. The London surgeons had in many instances developed their own methods before Lister's publications, and had already had considerable influence on others. Thus Lister's theory must be placed in the context of general concern over hospital disease and its likely threat to hospitals. In addition, the reception of antisepsis, and its later development, must be seen in the context of the work already being carried out by Lister's contemporaries.

LISTER'S THEORY OF ANTISEPSIS

Joseph Lister (1827–1912) was born in Essex, the son of Quaker wine merchant, amateur scientist and inventor of the achromatic microscope, Joseph Jackson Lister FRS. Lister studied at University College London, under Sharpey and others, graduating in 1852, and then visited Edinburgh to observe the work of the prominent surgeon, James Syme. Lister stayed on as Syme's dresser, then his resident house-surgeon, and subsequently married his daughter. After their marriage, the Listers toured continental medical centres before returning to Scotland. In 1859 Lister was appointed Professor of Surgery at Glasgow University, and the following year he became surgeon to the Glasgow Royal Infirmary. During the 1850s Lister carried out scientific experiments on inflammation in frogs, for which he was elected a Fellow of the Royal Society in 1860. Three years later he attempted unsuccessfully to return to Edinburgh, as surgeon and professor, and in 1866 he also failed to secure a chair at University College London.[14]

It was while he was in Glasgow that Lister put forward his theory of antisepsis. He later argued that it was concern about the high incidence of post-operative sepsis at the Glasgow Royal Infirmary that sparked his interest, an interest which of course echoed that of other surgeons. The miasmatic theory emphasised the potentially damaging qualities of the atmosphere. Lister was introduced by Thomas Anderson, Professor of Chemistry at Glasgow, to Pasteur's germ theory, which also emphasised dangerous aspects of the atmosphere. In accepting Pasteur's theory, Lister concluded that: 'the essential cause of suppuration in wounds is decomposition, brought about by the influence of the atmosphere upon blood or serum'.[15] Pasteur's researches showed that 'the septic property of the atmosphere depended, not on the oxygen or any gaseous constituent, but on minute organisms suspended in it'.[16]

Pasteur's experiments had emphasised the airborne nature of germs, and Lister long held to this view. Germs could be carried on dust particles and the air could be freed from these agents by filtration, heat or other means. Just as surgeons used various chemical disinfectants to destroy miasma, so Lister employed chemical means to remove the septic property of the atmosphere. What apparently inspired Lister to use carbolic acid as his filter between the air and open wounds was the reported efficacy of the acid in disinfecting sewage in Carlisle. Carbolic acid was already in use in hospitals as a disinfectant, and elsewhere for sanitary purposes. French surgeons had also used the acid in the treatment of wounds, although Lister later claimed not to have known this. Between the summers of 1865 and 1867 Lister applied carbolic acid dressings as filters between air and wound in order to prevent sepsis in twelve patients, apparently with nine successes.[17]

In 1867 Lister presented an account of his theory and method, 'On the Antiseptic Principle in the Practice of Surgery', to the Annual Meeting of the British Medical Association. Referring briefly to Pasteur, he argued that living organisms, carried on dust particles floating in the air, caused putrefaction in wounds: one minute germ could cause bodily sepsis. 'Upon this principle I have based a practice'.[18] In his eyes what was almost taken for granted was acceptance of the germ theory of putrefaction, from which practices consonant with it could be derived. As Lister later recalled, initially he used German creosote as his antiseptic, a substance lent to him by Anderson. However, he then moved on to the use of more soluble and purer forms of carbolic acid, manufactured in Manchester by Frederick Crace Calvert.[19] Lister applied carbolic acid dissolved in linseed or olive oil, or mixed with a chalky paste, directly to the wound to form a barrier to germs. The antiseptic made contact with a wide area of healthy skin surrounding the wound, often causing sloughing, and Lister's use of the chalky paste was developed to try to relieve this. The antiseptic was covered by lint to prevent germs from getting through to the wound, and finally a piece of tin to prevent the carbolic acid from evaporating. The tin was lifted each day so that more antiseptic could be applied. Lister argued that very strict antiseptic precautions were essential in every case in which the skin was broken, lest the air carry a single germ into the wound, causing sepsis. Suppuration induced by the carbolic acid was tolerable; that caused by putrefaction was highly dangerous. He referred to his successful cases treated by these means, concluding that hospital gangrene and pyaemia had been banished from his wards.

Lister hardly sought to explain or justify Pasteur's germ theory in his key talk. However, after encountering scepticism about the principle itself, in a later paper that year he went to greater pains to explain the theoretical basis of antisepsis. 'Decomposition or putrefaction has long been known to be a source of great mischief in surgery, and antiseptic applications have for several years been employed by many surgeons.'[20] A full understanding of this, however, had been lacking.

> In order to prevent decomposition we are guided by the 'germ theory', which supplies us with a knowledge of the nature and habits of the subtle foe we have to contend with; and without a firm belief in the truth of that theory, perplexity and blunders must be a frequent occurrence. The facts upon which it is based appear sufficiently convincing.[21]

Sufficiently convincing to him but not, as it turned out, to others. According to Lister, Pasteur's researches had shown that the atmosphere contained the spores of minute vegetations and infusoria, and that the 'septic energy'[22] of the air was directly proportional to the abundance of minute organisms in it. These organisms were destroyed on boiling, after which they had no effect on putrescible substances such as urine, milk or blood. Lister also mentioned Pasteur's flasks, in which tortuous turns filtered the 'air-dust'. Thus, concluded Lister, it had been demonstrated that the living beings invariably associated with fermentative and putrefactive changes were indeed their causes.

> Admitting, then, the truth of the germ theory, and proceeding in accordance with it, we must, when dealing with any case, destroy in the first instance once and for all any septic organisms which may exist within the part concerned; and after this has been done, our efforts must be directed to the prevention of the entrance of others into it.[23]

To Lister, nothing could be clearer. He had his principle, and upon that he based his practice. He expected others to do likewise.

RECEPTION OF ANTISEPSIS: CARBOLIC ACID, 1867–71

A key feature of the first few years in the reception of antisepsis was the variety of ways in which Lister's statements were interpreted. Key terms such as 'antisepsis', 'carbolic acid', 'antiseptic dressings', 'methods', 'techniques' and 'practices' were in a few cases accepted in the sense in which Lister used them, but more usually were employed differently or ignored. It was his use of carbolic acid as a dressing which became the main focus of discussion in the first years after his initial paper. The *Lancet* indexed most references to antisepsis under 'carbolic acid' until the mid-1870s. Most surgeons identified antisepsis as being simply a new type of wound dressing which involved the use of carbolic acid. Correspondingly, Lister's theory was subjected to considerable modifications. While some surgeons seemed to have accepted Lister's concept of antisepsis based on the germ theory, most adopted only parts of it.[24]

Through his address to the British Medical Association, and its subsequent publication in the *Lancet* and *British Medical Journal*, Lister's views were quickly and prominently displayed to the medical profession.[25] Everywhere it was reported that wounds were being covered with carbolic acid. Wholehearted approval of this practice was not forthcoming, however. Sir James Young Simpson pointed out that two French surgeons, Déclat and Lemaire, had described their use of carbolic acid before Lister.

> The strongest averments cannot alter the simplest dates, and I fear that it will not redound to the credit of English surgery to claim what most certainly does not pertain to it.[26]

In addition, Simpson argued, what he identified as Lister's apparent claim to heal by first intention (without wound suppuration) was hardly original since Simpson himself had long been able to accomplish this by his own method of acupressure, pinning the edges of the wound together. Frederick Ricketts, a Liverpool surgeon, supported Simpson and asked

> why give credit to the re-introduction of an obsolete and inelegant plan, when so simple, effectual, and elegant a method as acupressure obtains such magnificent results?[27]

Simpson and Ricketts touch on three of the most important themes that run through the early reception of antisepsis: Lister's idea (seen here as the use of carbolic acid) was not original; his method was difficult, cumbersome and time-consuming; and most importantly other surgeons had already developed means of combatting sepsis which were not only extremely effective but also much more suited to a busy surgeon's practice.

Another feature of the reception of antisepsis was a significant divide between London and provincial surgeons. The London surgeons, secure in their own positions, did not initially engage in debate over antisepsis which, after all, came from outside the metropolis. It was the surgeons from outside London who now debated Lister's article. In letters to the journals they discussed antisepsis and carbolic acid, they published articles on the subject, and hospital reports show that soon after the British Medical Association meeting some provincial surgeons were using antiseptic techniques.[28] London surgeons may not have taken part in the debate on antisepsis but this did not mean that they had no opinions on the matter. They identified antisepsis, as did the medical journals, as the use of carbolic acid as a specific healing dressing applied in an idiosyncratic way. The *Lancet*, the *British Medical Journal*, and the *Medical Times and Gazette* duly surveyed the opinions of surgeons at the major London hospitals and found great scepticism.[29] The *Medical Times and Gazette* reported that the surgeons at University College Hospital had judged against carbolic acid, those at King's occasionally used the acid but found Lister's method neither very successful nor speedy. At St Bartholomew's James Paget was sceptical, at St George's there was apparently 'nothing special to report on it', at the London Hospital carbolic acid was used only as a disinfectant, at St Mary's carbolic acid had been abandoned 'years ago', and only at St Thomas's was there a slightly more favourable response. A typical judgement was that of Thomas Bryant, assistant surgeon at Guy's Hospital, who argued that antisepsis 'does not seem to do better than many other forms of dressing'. The *Medical Times and Gazette* concluded: 'the verdict of the Profession in London is in favour of carbolic acid as a disinfectant, but ... its value as a healing agent is somewhat doubtful'.[30] Other reports showed similar patterns: London surgeons tended to see the use of carbolic acid as meddlesome and time-consuming, with little reason to adopt it since wounds were doing 'very well on the usual plan'.[31] A sense that London surgeons were not following Lister's principles was creeping into the journals' comments by 1869. The *Lancet* noted that 'Mr Lister's treatment does not find much favour in London. Are the conditions of suppuration different here from those in Glasgow or Dowlais?'[32] In 1869 the *Medical Times and Gazette* judged that in London hospitals:

> The antiseptic method has hardly had a fair trial ... and has been employed with far too little care and attention to small details upon which its success or failure so much depends.[33]

It was in response to claims by proponents of antisepsis that terms were being confused that the *British Medical Journal* sought the assistance of Frederick Crace Calvert in clarifying the difference between 'deodourisers', 'disinfectants', and 'antiseptics'. Deodorisers, according to Calvert,

> simply remove the noxious gases emitted from organic matters whilst in the state of decay or putrefaction, and as they do not possess the property of arresting decomposition or fermentation, it is no proof that the source

of infection has been destroyed; for noxious smells and offensive gases are not in themselves the real source of contagion.[34]

Disinfectants acted by oxidising the organic matter so as to prevent decomposition, but if it was exposed again to the atmosphere it would again begin to decay and putrefy. Antiseptics destroyed all source of decay and decomposition – the very germs of putrefaction.[35] However, Calvert's explanation, grounded as it was in the germ theory, had little effect upon the conception and usage of the various terms.

Antisepsis might be viewed with scepticism, but carbolic acid as a specific aroused continued interest. Contributions in the journals described the use of the acid for a number of different ailments, such as the treatment of cancer or syphilis.[36] Used as a cure-all, some of the acid's dangers, such as its tendency to poison and burn, also began to be reported.[37]

To supporters of Lister, science underpinned his theory and practice; they felt that viewing carbolic acid as a specific illustrated the fact that opponents were unable and unwilling to grasp this scientific basis. Lister himself published accounts of his own tests of Pasteur's flask experiments in order to demonstrate the truth of the germ theory.[38] But opponents claimed that science was on their side, too. Thomas Nunneley, surgeon to the Leeds General Infirmary, attacked antisepsis in 1869 as being unscientific and deviating from progress, asserting that it was 'a professional error, founded on false facts, and supported [only] by plausible assumptions, rather than by accurate observation and true deductions'.[39] He argued that if the antiseptic theory were true, no wound should ever have healed until carbolic acid or a similar substance had been applied. Absolute faith, not scientific rigour, was what antisepsis seemed to demand:

> Surgical science and medical knowledge are reduced to the one plain rule of, *in full faith* – for that is as essential as the acid itself – plentifully imbuing the part with carbolic acid; when, to a true disciple, all wounds, whatsoever and wheresoever they may be, and whatever tissues or cavities be implicated, become as nothing.[40]

Contrasting science with faith, opponents of antisepsis frequently referred to the fact that, to them, unsupported belief in an unproven theory was blindly required by antiseptic practices. Lister's supporters were often termed 'disciples' and critics complained that there was something akin to religion in the pilgrimages made to Lister to watch him at work. Campbell Black, of Glasgow, termed the method 'the latest toy of medical science so-called' and wondered 'exceedingly' who had apprehended 'the intricacies of this surgical arcanum'. He added:

There is nothing more obstructive to scientific inquiry and progress than reasoning from false premises and confounding coincidences with necessary consequences.[41]

Lister received many requests that he should publish statistical accounts of his method, in what was counted as true scientific fashion. He issued one set of statistics only, in the midst of controversy, and refused thereafter to follow up with further figures. He had originally incurred the wrath of his Glasgow colleagues by arguing that it was the high level of sepsis, resulting frequently in ward closures (though not of his own), that had stimulated his development of antisepsis. He then compounded that anger by comparing the results of cases treated along antiseptic lines with cases that pre-dated its introduction, describing the wards in the Glasgow Royal Infirmary as 'some of the most unhealthy in the kingdom'.[42] However, surgeons at the Glasgow Royal Infirmary claimed that improvements had resulted from changes in ventilation, nursing and diet, not antisepsis.[43] Lister replied by saying that the drop in the number of deaths after amputations, from sixteen in thirty-five (46 per cent) to six in forty (15 per cent) had occurred only in his wards. These were the figures now on record, and were to be taken up later by his opponents who maintained that these figures were singularly unimpressive in the light of achievements of other surgeons. Lister refused in future to produce statistics, reinforcing the arguments of those who claimed that his method was unscientific, lacking in persuasiveness and accepted only by true believers.[44] To combat this, Lister's supporters put out vast quantities of statistics, as also did their opponents. Each side then accused the other of issuing non-comparable figures.

To Lister, however, the main problem in 1870 seemed to be to convince surgeons that his method was far more than the mere use of carbolic acid. He stressed that it was not his specific use of carbolic acid (any other germicide would do) but the method by which he employed the acid that was significant. As he wrote:

Want of success in many quarters has not arisen from any unwillingness to try a new mode of practice. On the contrary the publication of my first papers was followed by a very general employment of the material which I happened to select for carrying out the treatment, and which, unfortunately for the principle involved, was then little known to British surgery, so that the striking results which were recorded were too often attributed to some specific virtue in the agent. The antiseptic system does not owe its efficiency to any such cause.... One rule indeed, there is of universal application – namely, this: *whatever be the antiseptic means employed ... use them so as to render impossible the existence of a living septic organism in the part concerned.* But the carrying out of this rule implies a conviction of the truth of the germ theory.... Without this guiding conviction ... many parts of the treatment would be unmeaning; and the surgeon ... would be constantly liable to deviate from the proper course in some apparently trivial, but essential detail, and then, ignorant of his own mistake, would attribute the bad result to the imperfection of the method.[45]

Lister argued that acceptance of the germ theory implied a completely new approach to wound treatment. Instead of applying a dressing merely to keep the wound clean, or more actively to stimulate recovery, his dressings were intended to kill or exclude germs. If surgeons 'lose firm faith in the guiding principle of the treatment, the attainment of a full measure of success becomes with them a matter of impossibility'.[46]

RECEPTION OF ANTISEPSIS: SPRAY AND GAUZE VERSUS CLEANLINESS, 1871–77

In 1869 Lister moved from Glasgow to become surgeon at the Edinburgh Royal Infirmary, succeeding Syme, who died a year later. Two years after his return to Edinburgh, Lister introduced through his surgical address to the British Medical Association elements in his method which were sufficiently distinctive that many surgeons later dated the beginning of antisepsis proper to 1871. These elements were the use of the carbolic acid spray and gauze.

Having experimented with a variety of ways of overcoming difficulties with the carbolic acid, mixing it with oils, forming it into cement with shellac, Lister had now adopted the practice of placing over the wound eight layers of gauze bandage impregnated with carbolic acid. Between the seventh and eighth layers of gauze Lister placed a sheet of mackintosh, to prevent any substance from the wound leaking through and coming into contact with the air. Such contact, Lister believed, would provide a path along which germs could travel. The bandage was applied after washing the wound with carbolic lotion, and the gauze was sufficiently extensive to cover the immediate area around the damaged parts, to prevent germs entering the wound from beneath the sides of the bandage. Lister also advocated spraying the air around the patient with dilute carbolic acid, to kill airborne germs that might enter the uncovered wound during operation.⁴⁷ At first he had diluted the carbolic acid with forty parts of water, but he had found that his hands were constantly in 'a rough and uncomfortable state'.⁴⁸ Diluting it with one hundred parts of water, he argued, still produced a trustworthy antiseptic atmosphere, with hands experiencing no inconvenience and breathing easy. Lister argued that with the spray, incisions could safely be made in the body where no wound existed before. His theory was that since germs were airborne and all-powerful, and during operation might enter the wound, spraying the air would combat this. He noted the irritation and suppuration caused by the carbolic acid, and advocated, as he had earlier, the use of a drainage tube to convey this material away. Antisepsis, as now promulgated, was calculated 'to revolutionize almost every department of surgical practice'.⁴⁹ He could not understand why it had not been overwhelmingly adopted, and put this down to a lack of a scientific mentality in his colleagues.

> I am ready to blush for the character of our profession for scientific accuracy ... and I am tempted to doubt whether some of the commentators can have enjoyed the advantages of sufficient education either in chemical physics or in logic.⁵⁰

He urged his colleagues not to let any other statements found elsewhere shake their belief

> in the truth that putrefaction, under atmospheric influence ... is due to particles of dust ever present in the atmosphere ... and endowed with wonderful chemical energy and power to self-propagation, yet happily readily deprived of energy by various agents.... With this as your guiding principle, you will find yourselves successful with the antiseptic system of treatment, but without it, whatever theory you adopt, you will ever be walking in the dark, and therefore ever liable to stumble.⁵¹

The appearance of the spray and gauze – concrete symbols of a new procedure – had its impact on surgeons' attitudes: increasingly antisepsis came to be viewed as a method rather than merely a dressing. Surgeons began to talk more of 'antisepsis', and less of 'carbolic acid'. The spray and gauze became the hallmarks of Lister's work as carbolic acid alone had been earlier. However, the introduction of the spray and gauze seemed to many to compound the theoretical and practical difficulties already encountered with antisepsis. The spray indicated even more clearly the necessity of belief in the germ theory. While the application of carbolic acid could be seen as little different from applying some other specific, the spray, which had no contact with the wound, could not. It became a symbol of faith in the existence of germs, as airborne causes of disease. Hector Cameron related that

> to the spectator, the spray and the cloud of highly irritating vapour which it emitted to envelope the operator and his assistants, was the most striking feature of the scene. As Lister entered the crowded theatre, his mobile face set and solemn as he bethought himself of the responsibilities he was about to undertake, he was followed in procession by his train of dressers, the first of whom bore aloft the sacred spray. Once the silence was broken by some ribald student whose voice was heard intoning, 'Let us spray!'[52]

Provincial and London surgeons continued to react differently to antisepsis, although now the Londoners began to participate more actively in the debate. It was the provincial surgeons who tended to speak out in favour in both the London and the provincial medical journals. In Liverpool, for instance, Robert Hamilton, surgeon to the Southern Hospital, and Edward R. Bickersteth, from the Liverpool Royal Infirmary, were vocal proponents of antisepsis.[53] In Manchester, William Barlow, honorary medical officer at the Manchester General Hospital, and Edward Lund of the Manchester Royal Infirmary played similar roles, and in other towns and cities leading figures showed a similar interest.[54] In Scotland, most Edinburgh surgeons seem largely to have supported antisepsis by 1869, when Lister was appointed surgeon to the Royal Infirmary.[55] Glasgow was not surprisingly an exception to this pattern. Lister's former house-pupils faced considerable scepticism from other colleagues there.[56]

It was part of Lister's message that extreme care should be taken in dressing wounds to avoid the entrance of even a single germ. But the necessity of paying such attention to detail came to be increasingly regarded as onerous and time-consuming, in an age when surgeons demanded speed: wound dressings should be readily on hand and simple to apply. Simplicity was becoming the watchword of those who criticised antisepsis. John Wood, surgeon to King's College Hospital, judged antisepsis far too complicated, arguing that:

> it is as a practical method of treating open wounds, available under ordinary circumstances in hospitals and private practice, in emergencies, and on the battlefield, that it must be estimated.[57]

In that, he judged it to have failed. Sir John Rose Cormack, an Edinburgh-trained, Paris-based medical practitioner, bore Wood out. During the siege of Paris in 1870–1, he had been unable to use antiseptic techniques because of their complicated nature.[58] London surgeons claimed that such problems with antisepsis were particularly important in the capital: the schedules of medical men in the major London hospitals were especially busy. Normally the dressing of wounds, not regarded as of the same importance as an operation, was left to dressers or house-surgeons. Lister's method seemed to demand that surgeons themselves be responsible for after-treatment. John Cooper Forster, surgeon to Guy's Hospital, argued that:

> to ensure the thorough application of this method necessitates the constant attention of the surgeon who has charge of the case, and practically in the London hospital such entire supervision is a simple impossibility.[59]

The use of the carbolic acid spray added to the difficulties. The carbolic acid irritated the skin of those operating as well as that of the patient. (Later on there were reports of kidney damage from the use of the spray, which animal experiments seemed to confirm.) If the surgeon had any doubt about the efficacy of antisepsis or the validity of the germ theory, the spray became an intolerable burden. This key element of Listerian antisepsis was usually the first to be abandoned.[60]

A further problem was expense. Medical men were reluctant to expose themselves to the criticism of lay hospital management over the cost of antiseptic dressings. Opponents of Lister's method argued that the use of gauze and the emphasis upon the disposal of dressings made the method prohibitively costly. Even in Edinburgh Lister encountered opposition on this front. In 1873, for example, the *Scotsman* editorially denounced the expenditure of over £600 on carbolic acid plaster alone in the Royal Infirmary.[61] The matter was raised again in 1876, when Sir Robert Christison, as manager of the Royal Infirmary, found himself having to justify the large rise in expenditure on antiseptic dressings to the Court of Contributors of the hospital.[62] Dependent as they were on lay men and women for their income, the voluntary hospitals emphasised their financial efficiency. Yet antisepsis was not something in which the governors believed. Whenever epidemics of hospital diseases occurred, the governors urged the disinfection and closure of wards to disperse miasma.

To many medical men, the most telling argument against antisepsis was the apparent efficacy of alternative methods in reducing surgical mortality from hospital disease. Simplicity and cleanliness were the hallmarks of most procedures. As Holmes Coote of St Bartholomew's remarked in 1870, at his hospital and

> most of the older institutions of similar character, the practice of extreme simplicity and cleanliness of dressing has prevailed for several generations.[63]

He himself had been taught by William Lawrence who used 'disinfecting applications',[64] but only if the wound became foul or the discharge offensive. The importance of the environment was stressed rather than necessarily any specific treatment of wounds.

> We may readily admit that there are some hospitals so circumstanced and constructed that the use of disinfectants in any form is an advantage; but the proper inference from such facts would be that the general arrangements of the hospital required improvement and not that any local method of treatment was universally applicable.[65]

Antisepsis was viewed as actually endangering this approach. Coote argued that if surgeons used 'disinfectants' such as carbolic acid, 'faith in their efficacy soon renders the surgeon more or less callous to the principles of general hygiene.[66] For that reason he opposed the 'indiscriminate use' of disinfectants and 'mistrust[ed] greatly the germ-theory and the surgical inferences which have been deduced therefrom'.[67] Holmes Coote suggested that antisepsis was not in fact universally applicable. Possibly it worked in Scotland, but conditions of suppuration, along with the environment, varied from place to place. If a hospital was clean and in a pure environment, then antisepsis would not necessarily be of use. There was a strong sense that there was an additive factor in the putrefaction of wounds: the more dirt there was, the more likely it was that wounds would putrefy.[68]

The term 'cleanliness' was used in opposition to antisepsis: while one surgeon might claim that he practised 'antisepsis', another would state that he practised 'cleanliness'. Cleanliness was not a by-product of certain practices, it was a method in itself, just as much as antisepsis was a method. 'Cleanliness' was implemented by keeping patients, their wounds, and their surroundings dirt-free. This meant that the wound would be washed, possibly with some kind of lotion, before and after operation. It might also mean that patients would be allotted increased space between themselves and adjoining beds. Increasingly, too, cleanliness might imply that the surgeon kept himself and his instruments clean. The most quoted example of an effective alternative to antisepsis in the 1870s became that of George Callender of St Bartholomew's. As Robert Hamilton of Liverpool put it in 1873, those surgeons 'who keep in advance in the field of surgery' would be treating wounds either according to Lister's or Callender's methods.[69] Callender rejected the germ theory in favour of a commitment to 'cleanliness', emphasising that what he practised was an alternative, superior 'method'.[70] He practised scrupulous cleanliness, supplementing this with the use of a camel hair brush to clean the wound.

After operation the wound was carefully dressed, to exclude dirt, and the patient was isolated, to avoid contact with other diseased individuals. In cases of amputation, a rubber tube was implanted in the stump to collect and channel away pus. Tube drainage was closely associated with the Parisian surgeons Malgaigne and Chassaignac, who correlated the incidence of septic infection with the presence of pus in the wound. (Lister also followed this practice.) Callender had picked up elements of other Parisian practices, such as placing cotton wool bandages over the wound, which he argued served both to protect the wound and to help it maintain warmth. He considered the above measures far more important than his limited use of carbolic lotion to wash the wound, and of carbolised oil to bathe it after operation. These elements were also employed by Parisian surgeons, and Callender used them as they did, as healing agents.[71]

Many surgeons, as well as the *Lancet*, quoted Callender's post-operative results as strong evidence that antisepsis was not superior to other methods. Callender published statistics in August 1873 of 200 operations treated according to his method which showed recovery in all but six cases.[72] His mortality rate of 3 per cent thus compared favourably with Lister's published figures of 15 per cent. Callender himself claimed that, although good results had been attained using Lister's method,

> results quite as favourable can be gained by a plan of wound-dressing which is so far more convenient than that treatment, in that it is more simple – one for the carrying out of which the means are always at hand and available for any emergency.[73]

Opponents of antisepsis argued that, if cleanliness and care (rather than the germ theory) were the basis of Callender's success, Listerian antisepsis could not be considered unique or all-embracing.[74]

In such a context, proponents of antisepsis needed to prove their case very firmly, so the critics declared. Neither Lister's brief foray into statistics, nor his later reliance only on case reports, nor the figures issued by his disciples, convinced the sceptics. As the *Lancet* put it in 1875,

> It has been very pertinently remarked that if the special merits of Mr. Lister's plan were really as great as they are alleged to be, they should at the expiration of eight or ten years have declared themselves with overwhelming force and certainty.[75]

Moreover, surgeons frequently argued that their work was largely atheoretical and practical, and that therefore they need pay little attention to antiseptic theory. As the *Lancet* commented,

Happily it is no part of the business of the clinical surgeon to bolster up theories, be they good or bad, or to make facts rigidly conform to them. The germ theory may be perfectly well founded but nine surgeons out of ten do not care much whether it is or not, so long as they cure their cases and reduce their mortality to the lowest possible degree.[76]

Thus, if antisepsis were judged on a practical basis, there seemed to be no compelling reason to adopt it when practices like Callender's were so successful. And, after all, any apparent successes accorded to antisepsis might well be due simply to improved hospital hygiene.[77]

Even where theoretical concerns were admitted, antisepsis was often judged to be lacking. Surgeons were critical of the monocausal explanation of putrefaction implied by the germ theory. Trained to view disease in terms of afflictions which beset different individuals in different ways, medical men retained complex explanations for disease incidence. Yet the germ theory seemed to rule out such subtleties. And when scientific rhetoric was brought to bear on the dispute, Lister's ability to explain even his failures in terms of the germ theory was said to be lacking scientific logic. In what the *Lancet* described as his 'convenient saving clause'[78] Lister ascribed any apparent failure of the antiseptic method to inadequate precautions on the part of the surgeon: if a wound putrefied, then a germ had been admitted through the carelessness of the dresser. The theory thus seemed too broad and irrefutable: any theory which could explain any condition, it was argued, must be false in its premises. What with Lister's failure to provide statistical back-up, two canons of the new scientific stance seemed to have been violated.[79] Sampson Gamgee, from the Queen's Hospital in Birmingham, and a surgeon who had had a long interest in hospital diseases, visited Edinburgh with the avowed intention of forming a scientific judgement on the matter. He criticised Lister for failing to maintain a cool scientific head and adhering too rigidly to Pasteur's version of the germ theory, and concluded that

> what is wanted is a thorough scientific examination of the whole question, discarding the fatal bias of a fixed theoretical preoccupation, and only striving to collect, coordinate, and digest the facts on which alone a comprehensive and sound judgment can be pronounced.[80]

George Thomson, of Oldham, also went to Edinburgh and was far from convinced. He explained that he had used antiseptic methods, but failed to achieve greater success than by other methods. In Edinburgh he saw little difference between Lister's wards and others there, or even those of other hospitals. In at least two cases he found inflammation and suppuration which was then 'explained away' by Lister.[81] According to Thomson, Lister argued that he had rendered putrefaction impossible. Thomson professed himself very disheartened by the experience:

> At first my feeling had been simply that of disappointment, but after hearing Mr. Lister dogmatically give utterance to the same old theories we have all seen in his writings of two years since and more, oracularly regardless of all that scientific research has elucidated since then, I left his wards, as any man who conscientiously desired to find out and uphold the truth must have done in my place, with the last remnant of my belief in Professor Lister dissipated to the winds.[82]

Above all, Thomson commented later, he found deplorable

> the persistent manner in which Professor Lister and his followers make everything fit the theory they postulate and ascribe failure to the imperfect observation of details.[83]

Unlike in Britain, the reception given to antisepsis on the continent was largely enthusiastic. Lister was not averse to quoting this. When Johann Nussbaum introduced antisepsis in Munich in the early 1870s, the incidence of hospital gangrene had apparently declined from 80 per cent to zero, so Lister reported. At Leipzig, Carl Thiersch reported one case of pyaemia in 1874 in his 300 beds. At Halle, Richard Volkmann had practised antiseptic methods since 1872, resulting in the virtual disappearance of hospital diseases. Thiersch and Volkmann stressed the importance of not merely applying antiseptic dressings but of performing the whole operation according to Lister's instructions. At the Charité Hospital in Berlin, and also in Magdeburg and Bonn, there were similar reactions.[84] The news that Lister's treatment was welcomed on the continent was generally disregarded by British surgeons, who held a very low opinion of the cleanliness of hospitals there. As the *Lancet* put it:

> No one acquainted with the filthy and neglected condition of many of the continental hospitals fifteen or twenty years ago, and even more recently than that, will be at a loss to understand why the mortality in these hospitals has so greatly diminished since the introduction of antiseptic surgery. Even the moderate use of clean water and the observance of ordinary habits of personal cleanliness would have sufficed to reduce very considerably the frightful mortality in them.[85]

Lister himself continued, as he had always done, to experiment with his method, introducing avowed improvements and refinements, and in the process shifting the ground on which other surgeons had to work, and for some reinforcing the notion of the difficulty of his method. He experimented with other antiseptics, recommending some in certain circumstances, although generally relying on carbolic acid as his mainstay. He continued to be a prolific speaker and author, assisting in the familiarisation of his work. Lister had by 1875 modified his emphasis on airborne dust, and now stressed the importance of surface dust being kept out of wounds as well. He mentioned the use of carbolic acid to 'purify' the skin before operation, an action akin to that of those who practised cleanliness.[86] In addition, like some of them, he now placed instruments for a length of time in carbolic lotion. He rebuked a follower for picking up a probe which had fallen to the dusty floor and introducing it immediately into a wound.

> I am more and more persuaded, the longer I practise antiseptic surgery, that the chief essential to success is a thorough conviction of the reality of the presence of the septic matter on all objects in the world around us.[87]

It was not just airborne germs which he now feared: germs were everywhere. But his practice was moving closer to that of the proponents of cleanliness, arguing as he did so that he was rendering the environment 'aseptic', without the possibility of sepsis, because all sources of sepsis had been destroyed prior to operation.[88]

IMPACT OF LISTER'S MOVE TO LONDON, 1877-80

The debate over antisepsis was intensified by Lister's controversial invitation to become Professor of Clinical Surgery at King's College, London, in 1877. He was supported by most of the physicians and opposed by most of the surgeons at King's; the *British Medical Journal* came out in his favour while the *Lancet* came out against. Lister's public lack of enthusiasm for moving to London, and his criticism of clinical teaching there, infuriated many surgeons at the major London hospitals, already irritated by Lister's claims over antisepsis. Lister's invitation came from those at King's who were worried about a dearth of students and who thought that Lister's name would help – not a point which the King's surgeons relished. The *British Medical Journal* somewhat optimistically hoped that 'The day of professional jealousies is now ended'.[89] Before receiving a definite offer, Lister told his Edinburgh students that 'There exists nothing in London at the present time which I should consider good enough to call me away from this school'.[90] He described the teaching of clinical surgery in London as a 'mere sham'.[91] In part the *Medical Press and Circular* crowed at such direct language:

> The hospital surgeons and surgical lecturers of Ireland and Scotland have been so loudly bidden and so long expected to bow the knee before the great pre-eminence of London practitioners ... that they will, no doubt, stand helplessly aghast at the audacity of a surgeon who hails from outside the sound of Bow bells, and yet has the courage to administer a most complete and scalding snub to the great ones of the surgical Babel of the south.[92]

On additional reflection, though, the editor added:

> It would certainly seem as if a personal application of the carbolic acid treatment had conclusively stamped out the last germ of modesty [in Lister] and produced a commensurate and contemporaneous hypertrophy of the cranial *locus* of self-esteem.[93]

The *Lancet* grudgingly conceded that Lister had done 'some service to practical surgery' by insisting on 'the importance of cleanliness in the treatment of wounds' although this had been done by 'the glorification of an idea which is neither original nor universally accepted'.[94] The controversy dragged on: John Wood, the internal candidate, was duly appointed. Lister was petitioned but refused to take Wood's vacated position as Professor of Systematic Surgery. A second chair of Clinical Surgery was created, embodying the kind of teaching of which Lister approved, with two wards and a right for him to bring his own house-surgeons from Edinburgh. Lister accepted. Although one house-surgeon between them had apparently sufficed the three King's senior surgeons previously, Lister brought from Edinburgh four assistants, a house-surgeon and three dressers.[95] The *British Medical Journal* noted that 'It is well known that antiseptic surgery has established for itself a firmer hold in all the other great cities of Europe than in London'[96] and it was in a missionising spirit that Lister now came to London.

Lister's arrival in London marks a turning point in attitudes to antisepsis in Britain. Lister was present at numerous meetings in the capital, he spoke on many occasions, he was teaching within a London school, he had brought disciples with him, and other followers, such as William MacCormac, were gaining appointments at London hospitals. Meanwhile Callender, whose publication record was not great and who had tended to rely on others to spread the word of his work, died in 1878. Continental experience, too, began to carry more weight, assisted and reflected in the holding of International Medical Congresses, the sixth of which was in Amsterdam in 1879 and the seventh in London in 1881. The germ theory, too, was gaining wider acceptance, not just within medicine.

Confirmatory evidence of the power of antisepsis seemed to come from a variety of sources. Surgeons wrote laudatory letters to the medical journals detailing astonishing successes apparently achieved with Lister's method. Thus L. J. Hobson, of the Newcastle Infirmary, submitted to the *Lancet* a table of mortality from major operations which showed that a 59.2 per cent death rate of 1873 had by 1878 declined to 4 per cent.[97] The lay managers of the Glasgow Royal Infirmary compared the results from antiseptic and non-antiseptic practitioners there, concluding that the former experienced roughly half the surgical mortality of the latter.[98] Reports of antisepsis were also appearing in the textbooks. Druitt's *The Surgeon's Vade Mecum: A Manual of Modern Surgery* (1870), its 10th edition, only briefly noted the use of carbolic acid in amputations of the thigh, with no discussion of antiseptic theory, or mention of Lister's name;[99] however, the 11th edition (1878) discussed 'Professor Lister's beneficent scheme' in considerable detail.[100]

Discussions of antisepsis were also entering London teaching. In 1878 *Guy's Hospital Reports* published a prize essay on antiseptic surgery by one of its dressers, Henry Davy. The essay depended largely on the lectures of one of the surgeons, H. S. Howse. In the printed version of the essay, however, Howse added numerous footnotes, toning down his pupil's enthusiasm for antisepsis, and in particular stressing that the theory could be ignored as of dubious validity, although the practice was useful.[101]

Surgeons who remained sceptical of antisepsis were now shifting the emphasis of that opposition. Lister's work, they said, if not necessarily accurate, had nevertheless proved a valuable stimulus to surgeons. Sir James Paget argued that between the mid-1850s, when the problem of wound sepsis was gaining increasing attention, and the mid-1870s there had been massive improvements in surgical care that owed much to antisepsis. Besides the potency of the antiseptic treatment 'for the repression of nearly the whole of the influences of air and infective disease', the development of antisepsis had provoked 'very keen rivalry into which our surgeons have thrown themselves, so that, if any cases under them do well it should be cases after operation'.[102] Thus at St Bartholomew's in the period 1847–57 mortality had been around 15 per cent. From 1857 to 1867 it averaged 10 per cent, and between 1867 and 1877 it had dropped to 2 per cent. The downward trend preceded Lister's introduction of antisepsis, and the final figures were contemporaneous with the development of Callender's methods of cleanliness.

Even with a greater appreciation of his work, surgeons expressed renewed doubts in Lister's version of the germ theory, though some might in principle accept the existence of germs. Open wounds did not always putrefy, while in other cases a patient would die of pyaemia or septicaemia without any external wound at all. John Wood, Lister's colleague at King's, argued that in ninety-nine cases out of a hundred germs seemed to have no effect. And then, 'how is it that a patient may die of pyaemia or septicaemia, self-poisoned, without any external wound at all?'[103] Jonathan Hutchinson, of St Bartholomew's Hospital, concluded that Lister had an exaggerated notion of the injurious effects of 'common air'. He argued that:

> It was air laden with specific germs – those of erysipelas, hospital gangrene and pyaemia – which in former times was so hurtful. From the risk of these, the isolation-ward, and the frequent use of disinfectants for sponges, bands and dressings has to a large extent freed us.[104]

Hutchinson's views represent a very important strand of thinking in this period. He accepted the germ theory, and that specific germs caused specific disease (as Koch seemed to have proved with anthrax in 1876), but he expressed his belief in the germ theory in a way that had strong similarities with an older additive view of dirt – the dirtier the wards, the more likely it would be that wounds would putrefy – and its corollary of emphasis on cleanliness. Those who had previously professed reliance on cleanliness rather than on antisepsis began to fit the germ theory into their alternative methods; they did not find that belief in the germ theory necessitated acceptance of Listerian antisepsis. They allowed that germs caused sepsis, but they argued that these were carried by diseased patients, or on dirty instruments or on other materials. Strict cleanliness would thus result in a germ-free environment. Unlike Lister, such surgeons did not believe that germs were always present, nor did they emphasise that the presence of one germ could result in sepsis. They therefore did not necessarily feel obliged to apply antiseptic substances to wounds at all times, and certainly not to use the spray. Thus Messenger Bradley, of the Manchester Royal Infirmary, emphasised that just as a hospital was made healthy through cleanliness, so too was a wound.

> I am daily more and more convinced that such matters as the mysterious spray, and the patent character of the dressings, act as a real hindrance

to the comprehension and general adoption of the true principles of antiseptic surgery.[105]

Even Lister, then, did not understand true antisepsis.

Druitt's *Vade Mecum* (1878) emphasised the differences between the 'disciples of Lister', who did not care whether the ward was 'aesthetically clean' so long as the patient was 'surgically clean', and proponents of cleanliness, for whom an 'aesthetically clean' ward was necessary if sepsis were to be avoided.[106] For those who favoured cleanliness, antisepsis merely 'neutralises the evil influences of dirt and crowding'. What rendered antisepsis unnecessary was 'absolute cleanliness, free ventilation and isolation from infection'.[107] There were a number of leading surgeons, it was stated, who

> act upon the principle, not of destroying septic germs, but of taking care that there shall be no germs to destroy, by avoiding all conditions of contamination to a patient under operation, whether arising from the place or from the persons assisting.[108]

Bradley himself claimed that it was 'wiser and easier ... to be *ante*septic than *anti*septic'.[109] One of the 'antiseptic' precautions which he took was to clean his instruments thoroughly with soap before using them. Too many other surgeons were now gaining comparably low mortality rates after operation for Lister to seem unique. As Bradley put it: 'so much for the infallibility claim and statistical question' of Listerian antisepsis.[110] The essential ingredient of Lister's success must in fact be cleanliness.

Lister not surprisingly refused to accept that there was persuasive evidence against his method. At the International Congress on Medical Science held in 1879 in Amsterdam, he asserted once more that those who claimed similar successes to his 'under ordinary modes of dressing' could hardly know 'what the successes of the antiseptic surgeon really are'.[111]

The certainty and frustration of proponents of antisepsis came through in the words of newly appointed St Thomas's surgeon William MacCormac in 1879.

> The antiseptic method initiated by Professor Lister fifteen years ago occupies at present so prominent a position in surgical practice as to compel the attention of all practical surgeons; and the time has surely arrived when its pretensions, and the theory on which these pretensions are founded, may claim a definite expression of opinion. Foreign surgeons, especially German ones, will say that this is already done; but in England there is by no means the same unanimity on the subject.[112]

Certainly Lister's colleagues at King's were not about to grant his superiority. John Wood declared that his own 'antiseptic bark' had suffered shipwreck 'on the shoals of expense'.[113] He reported that Sir William Fergusson and others believed that 'the pure waters of Damascus were as good as and better than all the carbolised waters of Israel'.[114] In Wood's own view 'the full tide of hospital hygiene' had already reached the London hospitals before the wave of antisepsis arrived, 'and topping the wave was the foamy crest or rather spray of Listerism'.[115] Wood himself was willing to accept parts of antisepsis while retaining doubts about current formulations of the germ theory.

However, proponents of Listerian antisepsis were modifying their own views as well. As Mr Corley put it in 1880, there were four links in the chain which ended in septicaemia: germs, an open wound, 'a peculiar condition of the wound ... and a peculiar condition of the patient'.[116] In other words, sepsis did not always result simply from the admission of a germ; the soil on to which it fell had to be fertile for sepsis to develop, a point increasingly taken up in the 1880s.

1880s: COMPROMISE

The 1880s saw a considerable shift in the position of antisepsis. Convergence between opponents and proponents of antisepsis can be seen, with both modifying their positions. A compromise was being developed according to which parts of the theory behind antisepsis were amalgamated with parts from the theory of cleanliness, producing something which was closer to cleanliness than to Lister's antisepsis and leading the way in the development of aseptic surgery. The discussion over wound healing had never remained static. Just as the proponents of cleanliness were now increasingly fitting the germ theory into their thinking, so the proponents of antisepsis were shifting ground, too. Lister, who had been so adamant in the face of those who rejected antisepsis because they rejected the germ theory, seemed to be more willing to amend his notion of the theory at the instigation particularly of continental supporters. After all, they had so enthusiastically taken up antisepsis and yet they amended its practice as their conception of the germ theory changed from that of Pasteur's airborne germs to the specific and not necessarily all-powerful germs of Koch. Moreover, the means they used to kill germs were not always Lister's: steam sterilisation seemed to many to be preferable to chemicals. The older notions of the additive nature of the problem, and also the soil on to which it fell, seemed again to fit new conceptions of the germ theory.

Lister himself began to change position quite markedly. In a debate in 1880, he declared that:

> every good surgeon was, whether consciously or unconsciously, an antiseptic surgeon ... Scrupulous cleanliness, frequent changes of water dressings or poultices ... were antiseptic means – that was to say, means tending to guard against the evil effects of septicaemia.[117]

Drainage, free ventilation, the judicious selection of cases for operation could also be termed antiseptic. John Erichsen declared himself 'somewhat staggered' to hear that

> under the word antiseptic were to be now included every possible hygienic and other precautions taken in the hospital, even to the exclusion of unfit cases for operation. Used in that sense, the word simply lost all significance.[118]

Unabashed, a year later Lister declared that while he remained convinced of the importance of micro-organisms he warned against 'a tendency to exaggeration in this direction'.[119] Inflammation was certainly caused by micro-organisms, but he also detailed other mechanical and chemical causes.

Opponents of antisepsis had made the point that if what Lister said was true, then no wound should ever have healed in pre-antiseptic days. By the 1880s those who made this point focused on the major abdominal operation of ovariotomy. Lawson Tait from Birmingham and others asked how ovariotomy could have been achieved successfully prior to antisepsis.

> If the views of the germ theorists were correct, no surgical operation could be successful without rigid antiseptic precautions; a cut should be followed by septic poisoning, and surgery must long since have become an extinct art.[120]

In the 1860s and 1870s Lister had tended simply to discount reports of success without antiseptic methods. In 1881 he tackled the points made by the ovariotomists. Seeking to neutralise their influence, he now declared that their very success had been a major reason for what he saw as the stormy passage of antisepsis in Britain. Spencer Wells and Thomas Keith 'had achieved results which astonished the world before strict antiseptic treatment was thought of'.[121] In line with his statement of 1880, he explained that Keith had 'always paid the most scrupulous attention to cleanliness', including the important 'antiseptic precaution' of purification of sponges by boiling.[122] But he also queried how 'the success of ovariotomy performed without the use of antiseptic means [could] be reconciled with the truth of the antiseptic principle'.[123] He now thought that germs were not necessarily all-powerful and that the body itself had powers to resist them. He argued that abdominal operations differed from the treatment of other wounds because of 'certain peculiarities' of the abdominal cavity: serum from the wound was absorbed as fast as it was produced and therefore there was no opportunity for putrefaction. Moreover the peritoneum had 'high vital power'[124] which helped to resist septic agencies. He had conducted a number of animal experiments from which he concluded that normal healthy undiluted blood serum was resistant to bacteria, in contrast to what he had earlier thought. Blood clots in particular exercised such activity. He also concluded that putrefaction resulted from the presence of septic matter in concentrated form, rather than that a single germ could cause bodily sepsis. If that was the case sepsis was not almost inevitable once the body had been opened, and Lister questioned the necessity of the spray.

> In other words, Is there sufficient chance of the air of an operating theatre or private room containing septic matter which can prove effective in blood serum to make it needful to regard the question of contamination from the atmosphere at all?[125]

In fact, echoing a number of continental surgeons, if further investigation were to confirm the conclusion

> to which our recent facts seem to point, and it should indeed be proved that all idea of atmospheric contamination of our wounds during operations may be thrown to the winds, then no one will say with more joy than myself 'Fort mit dem Spray'.[126]

Lister was thus modifying his view of the causation of sepsis, particularly in relation to the role of the atmosphere. However he also sought to explain others' successes in his own terms. If success had been achieved, then the surgeon, perhaps unbeknownst to himself, was operating antiseptically. Referring in 1890 to successes in abdominal surgery by Bantock and Lawson Tait, 'without, it is said, the use of antiseptic means', he claimed that their means were in fact antiseptic. Both were scrupulously careful in the purification of their sponges, both observed the strictest cleanliness

> which is surely an antiseptic precaution – for it owes its virtue to the fact that it presents the septic organisms in the smallest possible numbers

and thus reduces their power for evil to the utmost that can be done by any measures that are not germicidal.[127]

Although now including in 'antisepsis' the measures that those pursuing cleanliness followed, Lister still maintained that it was wiser to ensure the entire absence of microbes from hands and instruments by the use of a germicide than by 'the most perfect cleanliness in the ordinary sense of the term'.[128]

The revisions in approach both by those who supported antisepsis and those who had formerly argued for cleanliness can seen be very clearly in the debate over military surgery. In 1870 it had been said that antiseptic practice was too complicated for military surgery. By 1880 it was suggested that this was precisely where antisepsis had greatest application. However, to proponents of antisepsis in military surgery, Listerism did not mean all that Lister had so often emphasised that it meant, and yet they did not regard themselves as deviating from Lister's main message. At a meeting in 1884 to discuss antisepsis in military use, attended by both Lister and the Director-General of the Army Medical Department, Lister argued that the application of antiseptic principles to military surgery was 'in a more hopeful position now than at any former period'.[129] He emphasised that the spray was the least important part of the package: he advised those in military surgery to do without it. He was clearly willing to modify his method according to circumstance. Sir James Hanbury, Deputy Surgeon-General, saw antisepsis as particularly important in Britain's current conflicts. In European wars there were an 'abundance of railways, towns and villages ... volunteer ambulance corps, philanthropic societies of all kinds, and ... all the resources of civilisation, to fall back upon'. Britain was now engaged in much less 'civilised' wars, 'in tropical or semi-tropical countries, more or less remote, against uncivilised or semi-civilised natives'.[130] Antisepsis had a crucial role to play. To him, and to most of his colleagues, antisepsis did not mean a method so much as a dressing. And the main question was how these dressings could be carried and most easily made available when needed. It was agreed that the soldiers themselves should carry them: 'every wounded soldier his own antiseptic dresser'.[131] Hanbury recommended that one pocket should be used exclusively for the dressing.[132] Surgeon-Major C. H. Godwin argued that it should be sewn under the flap of the coat.

> Soldiers have no room in their pockets for half they want to carry; and they are not likely to keep a packet of dressings in their only pocket, to the exclusion of tobacco and other things they value.[133]

This discussion, and Lister's acceptance of it, represents an important shift from the position that only full antiseptic precautions would secure a patient against sepsis: now such major adjustments could be made to antisepsis that it could be practised even by the wounded soldier himself.

Corroboration of antisepsis was increasingly accepted from other sources, too. By the 1880s, greater credence was given to continental surgeons, and the surgeons themselves frequently reported their work on visits to Britain. A decade earlier little interest would have been shown in what a continental surgeon had to say. Few now dismissed their claims by referring to horrendous continental conditions. Germany was fast establishing its pre-eminent position in medicine, and in particular in medical science, and was now patently commanding greater respect. Even though German surgeons' practices differed from Lister's in many instances, their praise had its effect. In 1881 it was reported without detraction that Johann Nussbaum had such a high regard for antisepsis that he maintained that examining a wound with a non-disinfected finger should be considered negligent in law.[134] In 1884 it was reported that the operator and assistants in Germany and Austria in the twenty-four hours prior to the operation would make no post-mortem or anatomical dissections, or attend any infectious cases. On the day of the operation, 'care is taken that no scurf shall fall from the head, and the beard is carefully attended to'.[135] At operation, the coat was removed, and also collar and waistcoat in some clinics, and a clean apron or a short-sleeved linen jacket was put on, arms being bare to the elbows. Hands and arms were washed, first in hot water, and then in a corrosive sublimate solution or else a carbolic solution, and 'the nail-brush is diligently employed for the entire hands and arms, not for the nails alone'.[136] A basin of antiseptic solution was frequently used during the operation, and if anyone touched an 'unpurified' object, he must again wash his hands 'and not merely dip them'.[137] The room itself was thoroughly washed with carbolic soap just before the operation, and the spray was often used for some time in advance. Instruments were scrubbed with carbolic solution, and left to soak. Bottles, syphon-tubes, syringes, and trays were washed with antiseptic. The patient was cleansed before being brought in, and if a limb was to be operated on, it was drenched in an antiseptic solution for some hours previously. On the table, the patient might be wrapped in a blanket covered by waterproof sheeting, which was sponged with antiseptic solution. During the operation nothing was allowed to touch the patient 'unless both it and the hands of the assistant are absolutely beyond suspicion'.[138]

According to the *British Medical Journal* in 1890, it was evident that 'antiseptic or rather aseptic surgery is held in the very highest repute' in

Berlin.[139] Extreme care and lavish expenditure had been incurred to ensure that all wounds were aseptic. As perusal of operating theatres showed, surgical instruments were kept in cupboards constructed solely of glass and iron, and therefore completely washable, scalpels were composed of one piece of metal, so that dust could not collect between the blade and the handle, the operating table was constructed of one slab of green glass about an inch thick, placed upon an iron frame. In offices adjoining the theatre there were ovens for the sterilisation by dry or moist heat of everything, including bandages, cotton wool and other dressings, that was to be applied to wounds or for use during the operation. Special clothes were worn by the surgeon, even goloshes.[140] The British medical profession now took note in a way that it had not before.

Lister continued to adapt both his media and his message. Finally abandoning the spray in 1887 and publicly ruing that he had ever recommended it, he continued to try out various non-volatile antiseptics including corrosive sublimate and latterly double cyanide of zinc and mercury. The *British Medical Journal* commented that it was unfortunate that the term 'antiseptic' had been applied in the first place by Lister to a particular way of treating a wound, when 'asepsis' might have been more appropriate. To many, as a consequence, antisepsis meant carbolic acid spray and gauze even though these were now being abandoned by Lister himself.[141] Certainly his experiments were in keeping with his colleagues' interests: the journals bristled with reports of the different powers, advantages and problems of various substances used as antiseptics, as well as the media through which these might be used. Lister's persistence in his research now won him the *Lancet*'s praise. The history of Lister's work, it declared,

> is in itself a revelation of the scientific acumen, perseverance, and minute attention to details which are required for the perfection of such a work, and it furnishes a valuable lesson to any surgeons who may be inclined to forget the extreme importance of real and thorough scientific knowledge in their branch of the profession. No one without great chemical knowledge could have even attempted with any chance of success such labours as those by which Sir Joseph Lister has revolutionised surgery.[142]

Quite a change from the *Lancet*'s criticisms of a decade earlier.

Until about 1880, Lister's perceived inflexibility and dogmatism, his unwillingness to admit the methods of others, had seemed to consolidate opposition to antisepsis. The 1880s, however, saw a marked change in his approach: supported by increasing external approbation, he now overtly took on board others' conclusions. By 1891 he was confident that all now 'recognised the truth' on which antiseptic surgery was based; it no longer concerned him that 'the practice which has resulted from a recognition of this truth varies greatly in the hands of different surgeons'.[143] In a major shift from his earlier standpoint, he now accepted that it was impossible to exclude all microbes from wounds, but that this did not matter as only certain specific bacteria were dangerous, and they could be killed by carbolic acid. Chemical germicides remained vitally important: Lister would go only so far along the aseptic road, even if atmospheric dust was no longer to be dreaded.[144]

CONCLUSION

By the 1880s, before the hagiographers had got to work on Lister, he was being lionised at conferences and in the very medical journals that had previously been so critical. Yet the practices that surgeons now followed were often very different from those which he had proposed, often more closely derived from the 'cleanliness' techniques of his opponents. Lister had always ensured that he had a tight group of devoted followers around him, and those followers were now gaining positions of influence in the surgical community. Moreover it was Lister on whom the German surgeons had heaped their praise, even when rapidly deviating from his methods. And it was German opinion and practice which was commanding increasing respect in Britain, not least for its association with science, with which Lister too was so strongly identified. As the germ theory came to be accepted in almost total fashion as the key to all diseases, it was Lister (and Pasteur) who received the reflected glory, not Callender or Pouchet.

By the 1890s Lister was given public credit as the innovator he saw himself to be, and that has echoed down to the present within a medical profession which has stressed that science underlies all its work. But the evolution of surgical practices was far from being the result of Lister's efforts alone. Lister established his reputation, but his techniques were far from universally accepted. Those who stressed cleanliness and those who argued for antisepsis adapted their theories and practices to those of their opponents, and new generations of surgeons moved beyond the old warring camps to produce a new synthesis. 'Cleanliness' and the germ theory were merged into the theory and practice of asepsis.

3
THE SANATORIUM TREATMENT FOR CONSUMPTION IN BRITAIN, 1890–1914

Michael Worboys

INTRODUCTION

Sanatorium treatment for pulmonary tuberculosis[1] was developed in Germany from the 1860s and introduced into Britain in the 1890s. After 1898 it attracted considerable public and medical attention for it claimed to cure the major killer disease of the age. Sanatorium provision grew rapidly during the years 1900–10, though the treatment was surrounded with controversy as enthusiasts and critics debated its efficacy. By 1910 its rate of growth had eased and confidence was ebbing. Yet in 1911, the treatment was endorsed by the state, when provision for 'Sanatorium Benefit' was surprisingly included in the National Insurance Act and this ensured the renewed growth in provision after 1918. Sanatoria and sanatorium treatment have been at the centre of the two recent studies of tuberculosis by F. B. Smith and by Linda Bryder.[2] Smith's study is impressionistic and does not provide an account of the actual development of the treatment; however, I would accept his verdict that it did not reduce the incidence of the disease and may even have increased it through cross-infection and the diversion of resources from more valuable measures. Bryder's concern is with the anti-tuberculosis movement and service which grew up in Britain after 1898. She stresses the role of sanatoria in the early twentieth century in promoting the idea that the scourge of tuberculosis was beatable and in spearheading philanthropic and local authority efforts. However, her account is largely written from within the anti-tuberculosis movement and does not discuss the evolution of the treatment before 1914.

This essay uses the Schumpeterian distinctions between the invention, innovation and diffusion of new products and processes to analyse the development of the sanatorium treatment in Britain.[3] Its invention or first creation is discussed in relation to changes in the understanding and treatment of tuberculosis during the period 1870-92. This part of the discussion is international in scope, because treatment was developed in Germany and transferred from there. The innovation or first public use of the treatment in Britain is followed through the activities of the individuals who established private sanatoria, and through the wider anti-tuberculosis movement which led to the building of public sanatoria. The discussion of diffusion or spread looks at the growth of these institutions, at the evolution of the treatment, at its competitors, at the controversies over its value and at its incorporation into the 1911 National Insurance Act.

INVENTION

Consumption was the largest single cause of death and chronic illness in Britain during the nineteenth century. When figures became available in 1838, they showed an annual mortality rate of 380 deaths per 100 000 population. Although the rate had fallen to 140 per 100 000 by 1900, consumption remained the second largest cause of death, producing an annual toll at the turn of the century of 50 000 deaths and an estimated quarter of a million sufferers.[4] For most it was a lingering illness, where one's condition worsened over months or years and ended almost certainly in death. Given the huge potential 'demand' for care and treatment and the impact of the disease on the public's health, one might expect that consumption would have been an area of major medical activity. In fact, it was ignored at almost every level of medicine. Voluntary hospitals shunned consumptives as expensive incurables and though there were by 1900 some fourteen specialist consumption hospitals, between them they provided 1000 beds. Consumptives were, however, a major category of inmates in workhouses and Poor Law Infirmaries. These were mostly people who had been pauperised by the disease and were in its advanced stages, awaiting death.[5] In general practice, all that could be offered to patients were tonics, symptomatic and palliative treatments, and the recommendation of a 'change of air'. If regular medicine could not or would not meet the demand, 'irregulars' might have been expected to fill the gap. This they did to some extent, but contemporaries observed that the range of patent remedies for consumption was less than for other, less common or serious diseases.[6] Although a major cause of death during the nineteenth century, consumption was not a 'public health disease'. It is now widely acknowledged that the public health movement, reforms and administration were driven mainly by epidemic and 'filth' diseases, not endemic diseases.[7] However, some sanitarians did claim that environmental measures, like soil drainage and the improvement of ventilation, especially in factories, were the causes of the decline in consumption in the second half of the century.[8]

If the limited provision of care and medical interest in consumption was not due to lack of 'demand', then it might be explained by problems of 'supply' – medicine's inability to produce and deliver treatments or preventive measures. Indeed, the common view amongst medical practitioners was that consumption was incurable, and even untreatable. Before 1880 there was a strong consensus that the disease was the product of an hereditary diathesis or constitution, whereby the presence of tubercles was a consequence of disease rather than its cause.[9] It was even thought that consumption was not so much a disease, but 'merely an expression of lowered or defective vitality' or 'a mode of dying'.[10] It must be remembered that at this time notions of heredity were fluid, so that it was possible to acquire as well as to inherit a 'tubercular diathesis', 'defective vitality' or a 'weak constitution'.[11] The most fashionable treatment, a change of air and climate at the resorts of the Mediterranean and the Alps, aimed to promote vitality or constitutional progress by reforming the body, mind and behaviour.[12]

After 1880 several factors combined to begin to change the supply side and convince medical opinion that consumption was both preventable and curable. First, the downward trend in mortality was clearly evident and was seen by many as due to medically-inspired sanitary improvements. The second cause of optimism came from pathological anatomy, where autopsies and necroscopies were finding that the incidence of tubercles was much higher than the incidence of consumption.[13] The majority of people with tubercular lesions thus did not develop consumption or, as it was optimistically put, they had enjoyed a 'spontaneous cure'. However, Koch's announcement that a bacillus was the essential cause of consumption did not cause that great a stir.[14] A shift in pathological understanding had been under way since the 1870s, with the acceptance that tubercular diseases were due to inflammation not degeneration. The question then was, what produced the inflammation, with bacteria one of many candidates proposed. In 1881, Joseph Coats concluded a discussion on tuberculosis by observing that 'pathology undoubtedly points to a virus ... and clinical facts to the state of the system'.[15] What was novel about Koch's work was his absolute conviction in the contagiousness of tuberculosis. But this claim was largely ignored in Britain until the late 1890s, because it conflicted so much with clinical experience. In 1885 a Collective Investigation by the British Medical Association (BMA) found that only a quarter of doctors believed consumption to be truly contagious, the remainder questioned 'whether the bacillus alone produced and propagated the disease'.[16] One common argument was that 'physico-chemical changes must precede botanical aggression', another was that the bacillus was a necessary, but not sufficient cause of consumption.[17] It would be wrong to suggest any radical break in understanding due to Koch's work.

The 1880s saw the proliferation of treatments and new advice on how to avoid consumption. Until then, therapy had involved a range of symptomatic measures to control coughing, night sweats and tiredness, together with a judicious combination of sedative and tonic medicaments and measures to counter wasting.[18] C. J. B. Williams of the Brompton Hospital claimed that the introduction there of cod-liver oil in the 1880s had brought about a therapeutic revolution by building up patients' strength.[19] The most novel treatments were those which sought directly to destroy bacteria in the patient, mostly by the transfer of techniques from antiseptic surgery.[20] Means of delivering antiseptics to tubercular lesions included inhalation (by the use of masks soaked in carbolic or creosote), the direct introduction of antiseptics into the lungs by injection, and the intravenous injection of antiseptics in the hope that they would be carried to the lungs by the circulation. Vaccines to prevent or cure consumption were produced by many workers, especially in Germany, France and Italy. The best known of these was 'tuberculin', announced in a blaze of publicity by Koch at the International Medical Congress in Berlin in 1890.[21] It is often suggested that the high hopes for this substance were soon dashed, and that it came to be used only as a diagnostic aid. This is very misleading; tuberculin continued to be used by a large minority of doctors throughout the 1890s and came back into favour on several occasions, usually when a new formulation was produced or when immuno-therapy was in vogue.

Alongside attacks on the 'seed' of consumption there emerged regimes to reduce the fertility of the human 'soil' in which it flourished. Most notable was the new impetus given to 'climatic treatment' and its transformation into the 'open-air' or 'hygienic' treatment. A 'change of air' remained the preferred treatment throughout the 1880s and 1890s.[22] However, it was gradually re-defined at two levels. First, the 'air' had to be pure and germ-free and secondly, the therapeutic regime was seen as building up resistance to infection.[23] The interest in the purity or aseptic quality of the air led to the publication of tables giving the comparative levels of 'aerial sewage' in different localities. It had long been known that the levels of bacteria were highest in city centres, lower in the suburbs and at their lowest in rural areas, at high altitude, by the coast and at sea. Whatever the location, levels were higher indoors than they were out. An important part of the 'open-air' treatment became removal of the patient 'to an area free of the active virus', and the prevention of the 're-breathing' of air became important in reducing the number of bacilli inhaled and in preventing re-infection.[24]

It is important to remember that many germ theories of disease and the recognition of the presence of germs in the environment posed the problem of why people do not suffer constantly from infections. In the case of tuberculosis, the simultaneous recognition of the importance of human immunity had perhaps a greater practical impact than the discovery of the bacillus. Once established in the body the bacillus proved very difficult to attack from outside, hence the interest in prevention and in assisting the body's own defence mechanisms. This approach was also facilitated by the fact that the new ideas on immunity were readily seen in terms of the older notions of 'diathesis' or constitution.[25] Regimes which aimed to promote 'constitutional progress' by the management of diet, exercise, medicines and mind, were now presented in terms of 'hardening off' or building up resistance to infection. Incidentally, sanatoria served as laboratories for the confirmation of germ theories: none of their already weakened patients ever caught 'colds' or suffered from 'chills' due to exposure to the elements – or so it was reported.

The ætiology of consumption, as of other bacterial diseases, was discussed in a series of overlapping models and metaphors. Along with the older notions of predisposing and exciting causes came new military metaphors of invasion, defences and resistance, notions of bacteria as poisons, with antidotes, and the botanical analogy of 'seed' and 'soil'. This latter metaphor was particularly useful in explaining the limited infectivity of consumption.[26] It was soon recognised that the tubercle bacillus was not very robust or virulent and was in fact an unlikely cause of a major endemic scourge.[27] It was readily killed by drying or sunlight and once infection had taken place, the bacillus multiplied only slowly and took a long time to produce lesions, giving the body's defences the chance to produce 'spontaneous cures'. Sufferers, it seemed, had either to be initially infected with a large number of bacilli, or to be continually re-infected, alternatively they had to be 'open' to infection because of lowered vitality. Epidemiological evidence, clinical experience and now germ theory all combined to suggest the same predisposing factors: heredity, poverty and poor nutrition, housing conditions, occupation, alcohol consumption and moral and social behaviour. The attention given in the open-air treatment to carefully supervised diet, exercise and discipline shows that it was designed to render the human 'soil' infertile, hence arresting the further development of the disease.

There was initial confusion about the nature of the open-air treatment and sanatoria. It seemed to many that, as consumption had been shown to be an infectious disease, sanatoria were a new kind of isolation hospital. Indeed, in the last quarter of the nineteenth century the term 'sanatorium' was commonly used for isolation hospitals, especially those for smallpox.[28] However, its inventors stressed that sanatoria for consumption were not institutions of isolation and recuperation; they were curative. The 'cure' required a lengthy residence in an institution where the patient, under strict supervision, spent as much time as possible in the open-air, took an ample and rich diet, plus rest or exercise as recommended, with his or her condition constantly monitored by thermometry, auscultation and the bacterial examination of sputum. Sanatoria sought early or incipient cases where the patient's infectivity was low. Advanced cases were avoided if possible; these were inevitably terminal and highly infectious, and for these isolation was seen to be appropriate, though few facilities were provided outside of the Poor Law. A surprising feature of the discussion of the 'open-air' treatment was the lack of interest, let alone firm ideas, about the properties of the air or its mechanism of action. There was mention of the disinfecting powers of ozone and sunlight, of changes in the depth and regularity of breathing, of the linings of the lungs being toughened and of improved circulation, but none of these was seen to have any specific effect. What was said about the open-air and outdoor life was that it provided a near-aseptic environment, and hence individual isolation, for each patient. Therapeutic action was usually attributed to dietary management, rest and exercise, hygienic training or medical treatment. The sanatorium was not, as many did and still do suppose, an isolation hospital where pure air was the therapeutic agent; rather pure air provided the isolation and the hospital regime the therapy.

The change from the climatic to the open-air treatment developed in Germany in the 1880s. Increasingly sanatoria were built in 'imperfect climates' where, by ensuring that patients spent as much time as possible in the open-air and were subjected to 'carefully arranged hygienic management', very favourable results were reported.[29] The open-air treatment in Germany is usually traced back to Brehmer in 1859; however, it was Dettweiler's Curanstalt at Falkenstein and Walther's Sanatorium at Nordrach which attracted British doctors.[30] These were private sanatoria catering for fee-paying patients in direct competition with Alpine resorts. It was Dettweiler who also founded the first sanatorium for the poor in Germany in 1892. This stripped-down version of a private sanatorium was one catalyst of the wider anti-consumption movement in Germany, which eventually provided both insurance and philanthropic sanatoria for the working class. Another major catalyst of the German movement was the problem consumption caused for the workers' insurance scheme. As the largest cause of disability and death amongst adults, consumptive claimants threatened financial ruin to the scheme, hence the support given by the German state to sanatoria for the working class.

INNOVATION

The standard British account of the sanatorium treatment is that it was invented by Bodington, a little known English doctor, in 1844 and innovated, following the usual time-lag, later in the century. The lineage suggested here is bogus. German inventors of the treatment owed nothing to Bodington and the British innovators in the 1890s were following German precedent, not some previously unrecognised British breakthrough. What was imported into Britain in the 1890s was a rapidly changing treatment, which had been and was being continually re-invented. Also, it was no longer a single innovation; it was linked with wider anti-consumptive measures which involved the state, employers, trade unions, insurance companies, charities, the Poor Law, public health authorities, hospitals, rival specialists, pharmaceutical companies and the German working class.

As in Germany, the open-air treatment was introduced into Britain in two stages, in first private and then public sanatoria. It was expensive and inconvenient for Britons to use institutions in Germany, so when climate and topography diminished in importance, there was an opening for a cheaper, more accessible local alternative. A few British doctors, having gained first-hand experience in Germany as visitors or patients, began to introduce the open-air treatment into Britain. Whereas German sanatoria were objects of national pride, those in Britain began as small private hospitals which attracted little attention.[31] The first institution offering the open-air treatment proper was opened by Dr Jane Walker at Downham Market in Norfolk in 1892 – a small converted farmhouse, which by 1897 had treated only seventeen patients. The growth in the number of private institutions was slow, but by 1900 there were twenty-three, providing just over 400 beds.[32] Most were mixed institutions, with the ambience of residential hotels, and they often doubled as convalescent or rest homes.[33] Many were in resort areas, like Bournemouth and the Isle of Wight, while others, seemingly seeking continuities with the older German sanatoria, were on high ground or in pine forests.[34] Often an appropriate name compensated for topographical deficiencies, as in 'The Firs' on the Norfolk coast, the 'Swiss Villa' at Swanage and 'Nordrach-on-Mendip' in Somerset. All except one of these institutions were owner-managed by medical practitioners. Of the eighteen proprietors on whom it has been possible to find detailed information, seven had visited Walther at Nordrach and three of these, Drs Gwynn and Thurnam at Nordrach-on-Mendip and Dr Mander Smyth at Ringwood, were former patients who had been cured there.[35] Five of the group were established consultants at chest hospitals, who also oversaw their own sanatoria. Another five sanatoria were established by general practitioners in resort towns, a move which would have attracted extra patients in one of the most competitive areas of medical practice. The general qualification seems to have been some experience of the treatment of consumption and perhaps a desire for personal advancement by engaging in advanced work.

All of the private sanatoria were small businesses needing capital outlay, usually in a country house or hotel. Most charged between three and five guineas per week, which meant that they were affordable by the middle classes who were seen as their main clientele. Most seem to have been successful, if success is defined by survival and growth. There was little of the rapid turnover of new businesses often seen in an innovative area. Over three-quarters of the institutions established by 1900 lasted for ten years or more, and over half survived to the outbreak of the First World War.[36] However, these institutions were rarely full; they advertised widely to attract patients and some offered additional treatments, like the Weir Mitchell rest-cure.[37] The number of private sanatorium beds did not increase after 1904, suggesting that effective demand had been met, if not exceeded. Another means of providing the open-air treatment privately was to supervise patients at home, in a converted room or in the garden. The number of people who received domiciliary treatment in tents, sheds, shelters or on balconies is impossible to estimate, though the medical literature regularly discussed cases from the late 1890s onwards.[38]

For a hospital doctor the simplest and cheapest means of introducing the open-air treatment was to adapt part of an existing institution. Established specialist voluntary hospitals for consumption, like the Royal National, the National Sanatorium at Bournemouth, the Royal Victoria Hospital for Consumption in Edinburgh, Eversfield Hospital at St Leonards and the Manchester Hospital for Consumption at Bowdon all put some of their patients on to the new treatment; for example, Bowdon added verandahs to the south side of the building in 1898.[39] In fact, the arrival of the open-air treatment changed the meaning of the word 'sanatorium' from a place of rest or recuperation to that of a specialist institution for the treatment of consumption.[40] It was not only at specialist hospitals where the 'open-air' treatment was tried. For example, Frederick Burton-Fanning used the convalescent home of the Norfolk and Norwich Hospital at

Cromer from 1895, Hector McKenzie used the balconies at St Thomas's, London, in 1898 and in 1899 Noel Bardswell treated patients in the grounds of the Royal Infirmary in Sheffield, then one of Britain's worst cities for air pollution![41] Sanatorium wards were added to many Poor Law infirmaries and the treatment was tried in at least one asylum.[42] Many of these trials led to the establishment of independent sanatoria and some of the practitioners involved became leading figures in the anti-consumption movement which emerged in the late 1890s.[43]

A national anti-consumption campaign in Britain was first mooted at the Annual Meeting of the BMA in 1896, following similar initiatives on the continent.[44] A formal organisation, the National Campaign for the Prevention of Consumption (NAPC) was formed in 1898.[45] Its three main objectives were: (1) to educate the public as to the means of preventing the spread of consumption from those already suffering from the disease; (2) to extinguish tuberculosis in cattle; (3) to promote the erection of sanatoria for the open-air treatment of tubercular disease.[46] On the final point, it particularly sought to provide sanatoria for the working-class and the poor, who were a danger to themselves, their families and society. In this context the open-air treatment was an integral part of an attempt to make consumption a public health disease. The movement's general rhetoric often sounded contagionist, but this was always qualified. In fact, it advanced a kind of contingent contagionism, whereby consumption was seen to be infectious only where unhygienic conditions and behaviour prevailed. Armstrong's suggestion that consumption came to be seen as 'a disease of social contact' does not go far enough, for what was dangerous was not social contact *per se*, but certain kinds of contact, with certain classes, in certain environments.[47] On notification and isolation the NAPC was ambivalent. It supported notification but did not campaign strongly; it favoured the isolation of advanced cases, but not in sanatoria. Other medical practitioners, especially Medical Officers of Health, Poor Law and Asylum Medical Officers, tended to adopt a more straightforward contagionist stance.[48] It seems also that the complexities of contingent contagionism were lost on the general public, who increasingly saw the disease as very contagious, shunned the tuberculous and viewed sanatoria as dangerous places.

The NAPC programme stimulated wide interest in sanatoria and many articles appeared in journals and in the popular press.[49] In its early years many senior medical figures toured the country to speak on behalf of the Association and most, like John Burdon Sanderson, stressed the importance of providing for the consumptive working man and the poor.[50] In order to reach this class, many new methods of propaganda were developed and tried, including touring exhibitions and films.[51] The national organisation concentrated on health education and political lobbying, leaving to local branches the promotion and establishment of sanatoria. In the event, these local voluntary efforts built very few sanatoria; most 'public sanatoria' were built by local authorities. Indeed, the two are difficult to separate as voluntary sanatoria usually enjoyed some form of local authority support and many local authorities relied on the availability of voluntary beds.[52]

The first two public sanatoria were opened in 1900. The Westmoreland Sanatorium at Meathop was established by Dr Paget-Tomlinson of Kendal; each of its seventeen beds was subscribed by local Poor Law authorities. A similar arrangement was operated in Durham by Dr John Gray. An editorial in the *British Medical Journal*, welcoming these developments, seemed to see them as isolation hospitals, for it spoke of the threat posed by the unhealthiness of the poor, who were 'so dangerous to others'.[53] Some Poor Law Guardians established purpose-built sanatoria, for example, Liverpool (Heswall) in 1902 and Bradford in 1903, while others added sanatorium wards to their Infirmaries. Given that consumptives were a large group in many Poor Law institutions, it was not surprising to find Guardians active. Given the extra cost of boarding a patient in a sanatorium, it seems likely that those Guardians who took up the treatment were looking for cures; these would carry high short-term costs, with the prospect of long-term saving to the rates and long-term benefits to the patient.

Charitable public sanatoria were organised in one of three ways. First, an existing voluntary hospital, usually a specialist consumption hospital, could launch an appeal for an annexe or separate sanatorium. Annexes of the Liverpool and Manchester consumption hospitals were built at Crossley (Delamere Forest) in 1901 and 1906 respectively, with money from local plutocrats. Most famously, the Brompton Hospital, London, established a sanatorium at Frimley in Hampshire.[54] The second route to a voluntary sanatorium was for a local branch of the NAPC to collect or raise funds and build their own institution. For example, the Bristol and Bath branch did this and were able to open a sixty-six bed sanatorium at Winsley in 1904. Finally, eminent medical men of reputation and standing could launch local appeals for a sanatorium which they would oversee, if not supervise directly. In Norwich, Dr Fanning-Burton, who already ran a private sanatorium, led an appeal which resulted in the building of the Kelling Sanatorium. Similarly, in Nottingham, W. B. Ransom organised a scheme which resulted in the opening, in 1902, of the Ransom Sanatorium. Between 1900 and 1903, sixteen voluntary sanatoria opened, though they provided only 460 beds; after 1903 the rate of growth increased steadily (see Figure 3.1).

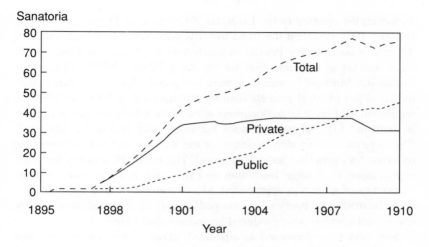

Figure 3.1 Sanatoria in Great Britain, 1895–1910

The motives behind the foundation of public sanatoria were mixed. Some were common to general hospital philanthropy as it had evolved by the end of the nineteenth century. There was a reliance on single large donations of money or large houses, or both. In Lancashire, local industrialists secured the establishment of the Crossley Sanatorium, the Wilkinson Sanatorium in Bolton and the Aitken Sanatorium in Bury.[55] Bryder suggests that sanatoria attracted philanthropic support because they were sold as modern curative institutions that would combat an economically important disease.[56] Their position in remote locations meant that it was not conspicuous charity, so it does seem that the appeal was to medical progress and the public good. Most local appeals were initiated by medical practitioners and made three common claims: (1) the need to invest in the new possibility of curing and eradicating consumption; (2) the wider role of the sanatorium in educating the poorer classes in self discipline and attention to hygienic laws; and (3) not being left behind by the activities of other towns and cities. Medical progress, reform of the working class and civic pride were a potent combination around 1900 and throughout the Edwardian period.

The major problem facing those founding and running public sanatoria for the poorer classes was finance. Special buildings were often deemed necessary and the treatment was lengthy and expensive. Most private sanatoria had converted existing buildings, as had the first public sanatoria.

However, the opening of the Liverpool Sanatorium at Delamere Forest in October 1901 signalled the trend towards purpose-built accommodation. There was considerable interest in sanatorium design and location, but an ideal was set in a competition for the King Edward VII Sanatorium at Midhurst.[57] Most public sanatoria were less grand, style was dictated more by the depth of local pockets than by architectural ambition or medical requirements. Single large donations allowed the Liverpool Sanatorium to be built quickly and gave Dundee the impressive Sidlaw Sanatorium.[58] The more modest wooden structures at the Ransom and Kelling sanatoria reflected less generous local responses.[59] The trade-off between the high capital costs of a large institution and its lower running costs through economies of scale was appreciated. Most sanatoria were as large as they could be made, and many were enlarged cheaply by the addition of chalets and shelters, which were produced by industrialised methods.

Sanatoria were promoted as advanced, curative institutions, but they were also powerful symbols. For the NAPC's wider educational programme, sanatoria and their regimes demonstrated hygienic living. They stood as the antithesis of working-class life as feared by other classes. They were spacious rather than overcrowded, clean rather than dirty. The dietary was wholesome, recreation was 'dry' and 'decent'; and everywhere was open to observation; there was no space for the 'unseen life' of the working classes, no 'residuum', no domestic squalor, no meetings in smoke-filled rooms, perhaps no alternative culture.[60] While there are some suggestions that sanatoria were seen as providing holidays, most evidence shows that the working class avoided what they saw as harsh and strict regimes, involving personal economic and social losses and exposure to contagion. The popular fears of contagion were also evident in local resistance to the siting of sanatoria.

By emphasising the infectious nature of tuberculosis and by associating it with ignorant, feckless behaviour, 'The campaign against consumption ... ended in a war against the consumptive'.[61] The publicity of the NAPC helped create the twentieth century stereotype of consumptives as dangerous and deficient; it made them social outcasts.[62] Given the incidence of consumption, the public fears about the disease and the weight of anti-consumption propaganda, it is likely that the NAPC's campaigns were amongst the most powerful assaults on working-class culture mounted in the Edwardian period.[63] One enthusiast thought that, together with the wider open-air movement, sanatoria would in the future 'be a means of opening up endless possibilities in bringing about such practical reform as to revolutionise social, municipal and national life'.[64] Arguably, this attempt at 'cultural control' was of equal, if not greater, significance than the better-known Edwardian campaigns about motherhood and physical degeneration; indeed, it may be artificial to separate them.

DIFFUSION

Most historians who have discussed the growth in the provision of sanatoria have used official figures, which were given only for 1907 and 1911.[65] My data are based on numerous contemporary sources and provide a continuous series from 1895 (see Appendix). For the two years for which the Local Government Board provided figures, mine are quite different.[66] For 1907, it is clear that the Bulstrode Report adopted a very loose definition of what counted as a 'sanatorium'. For example, it counted all the consumption hospitals, many of which were unable to offer the open-air treatment, and it counted children's sanatoria which were a very different kind of institution. Also, in the category of 'kindred' institutions, were included several homes for the dying, the Royal Sea Bathing Hospital and the Hahnemann Home! Perhaps the most interesting thing about the list is what was officially thought to be 'kindred'. By comparison, I have adopted an exclusive definition, counting only specialist, independent institutions, providing the open-air or sanatorium treatment for adults.[67] Using my definition the Bulstrode Report would show 1748 beds, not the 3989 reported; my more thorough survey revealed a higher total of 2088 beds (see Table 3.1).

Similarly, the official figure of 5500 beds in 1911 should be treated with caution.[68] That count was an exercise to ascertain what beds could be made available to insurance patients receiving the planned Sanatorium Benefit. Many beds in isolation hospitals were thus counted, along with proposed temporary accommodation; there was also some double counting of the beds which local authorities maintained in voluntary sanatoria.

When we look at the diffusion curves showing the growth in provision, several important features stand out (Figures 3.1 and 3.2). First, they show

Table 3.1 Sanatoria and consumption hospitals in England and Wales, 1907

Type of institution	Bulstrode	Worboys
Private	583	670
Public	1165	1418
Total	1748	2088

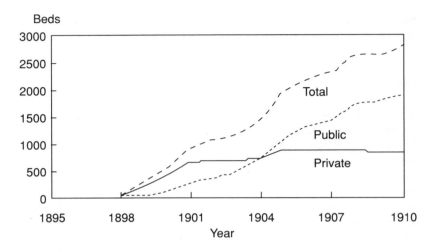

Figure 3.2   Sanatorium beds in Great Britain, 1895–1910

the 'S-curve' typical of many technical and social innovations. Secondly, as well as showing an overall 'S-curve', there are within this two other 'S-curves', for the two main kinds of sanatoria – private and public – reaching their plateaus at different times. Lastly, the fact that there is an 'S-curve' means that there was a definite slowing down, if not levelling off, in the growth of provision before 1911. This contradicts the assumption made in other studies of a continuous increase in provision.[69] The demonstration of a standard 'S-curve' suggests that diffusion of this innovation might be explained in the ways common in economic and social studies. The slow initial growth of private sanatoria can be explained by lack of information, German practice being slow to become known in Britain. However, lack of information was no constraint on the growth of public sanatoria: there were private models, the NAPC disseminated information and the public and professional press discussed the treatment extensively. There was little initial resistance to public sanatoria, which did not have to displace any established treatment and which were initially viewed very optimistically. The main factor holding back growth was undoubtedly the logistic and economic problem of founding a new medical institution.

The eventual rapid growth of sanatoria would seem to suggest that there was a ready demand for the treatment in Britain. Private sanatoria 'sold' their cure to paying customers, though often they were purchasing something recommended by a third party – their doctor. For this market, effective demand was met by 1904. In the public sector, by contrast, notions of supply, demand, the market and consumers are more problematic. The suppliers of public sanatoria for the working class were either local voluntary committees, local authorities or medical practitioners, the very same groups that also created the demand. There is little evidence of any demand directly by potential (working-class) patients. The open-air treatment had thus to be actively publicised and promoted both within medical and philanthropic circles, and to the public at large. In this particular case, supply had problems creating its own demand. The open-air treatment also spread in the sense of being adopted as a therapy for other diseases. It was used in cases of pneumonia and in isolation hospitals, where it was seen to provide effective individual isolation and control of fever.[70] It was even tried in asylums as a treatment for insanity.[71]

At the beginning of 1905, when the rate of increase in beds was at its greatest, an editorial in the *British Medical Journal* observed that sanatoria had 'so far met with no serious opposition'. This statement proved premature. It was made with reference to an article, by the Nottingham physician, W. B. Ransom, which demonstrated the success of the treatment, but which sparked off a controversy which persisted until at least 1911.[72] It is clear from Ransom's remarks that despite the absence of published criticisms, there had been grassroots opposition from general practitioners and consultants in general hospitals. Once the matter was aired there was no shortage of opinion; the attacks were often bitter and the enmity felt by those working in and supporting sanatoria was an important force in the subsequent development of the innovation. The main point discussed in 1905 was, quite simply, whether the treatment worked and whether it was worthwhile. Opponents claimed that any improvements brought about by the treatment were short-lived, relapses were the norm. The statistics on cure-rates, so carefully compiled by the sanatorium superintendents, were dismissed by critics as valueless or used to show how poor long-term survival rates really were.[73] There were also accusations that superintendents could not be objective; some criticisms verged on accusations of profiteering and professional malpractice.[74] Provision of sanatoria for the working class was opposed on economic grounds, particularly the futility of helping people who had weak constitutions, unhygienic habits, and would perforce return to conditions which had produced the disease in the first place. Some claimed that, if after-care could be provided, then there might be some point in trying to arrest the progress of the disease, but if so, might not the money spent on sanatoria be better used on general housing and sanitary improvements?[75]

The emergence of 'root and branch condemnations' of sanatoria drew an immediate and concerted response from supporters.[76] Ransom, like many of his colleagues, was puzzled as to why so many doctors were willing to abandon consumptives. Most supporters argued that in the absence of an effective alternative, sanatoria were the best hope; some went on to suggest that it was unethical not to try the best means available. In general, supporters argued that the treatment had been misunderstood, that it was not being given the opportunity to succeed, and that results were encouraging.[77] To begin with it was stressed that the treatment was based on more than fresh air; it consisted of a scientifically regulated combination of air, diet, rest, exercise and other treatments.[78] It was also claimed that too many advanced cases were being admitted to sanatoria and that the average length of stay, of 12–15 weeks, was simply not long enough, even with early cases.[79] Despite these unfavourable conditions, it was claimed that the results were impressive, with 'arrest' rates of up to 50 per cent reported. How much more, they asked, might be achieved under optimum conditions? Needless to say the argument over the statistics was inconclusive, with both sides claiming victory. The quantity of statistics, and the effort put into their collection, was indisputable; clearly there was something to prove.

The controversy initiated by Ransom continued in the columns of the *British Medical Journal* throughout 1905 and was halted only by an editorial decision. A new arena was soon established when the *Lancet* started its own debate in January 1906, with a review of the treatment and collection of representative opinions.[80] Again a long correspondence followed. In 1907, controversy was rekindled by an intervention from Karl Pearson, whose criticism of the statistical assessment of the results was linked to the eugenic contention that consumption was an hereditary condition and hence incurable.[81] In 1908 it was hoped that the long-awaited publication of the Local Government Board's report on sanatoria by H. T. Bulstrode would settle matters. This had begun in 1902 as an inquiry into the desirability of local authorities building sanatoria for the working class, but as controversy extended it came to be seen as an official enquiry into all aspects of the tuberculosis question.[82] The report was generally supportive of sanatoria, yet its detailed verdict was equivocal: results looked promising, but it was still too early to judge. Most neutrals and supporters saw the report as favourable, but critics were able to seize on the fact that there was no clear endorsement.[83]

It was observed on all sides that the aims of the treatment shifted during the 1900s; in the words of one observer, the 'open-air treatment of tuberculosis became the treatment of tuberculosis in the open-air'.[84] The new emphasis was reflected in the gradual shift from the term 'open-air treatment' to 'sanatorium treatment'. This further reinforces the claim that it was the institutional regime and not the mystical properties of ozone or whatever that was the basis of the treatment. Institutions were increasingly depicted as 'educational', providing patients with 'the rules of living' necessary for them to pursue their own longer-term cure.[85] Greater attention was given to discipline and to work. Work was originally introduced to reduce costs and to ease patients' transition back to normal life.[86] Later 'work therapy' was rationalised by a few sanatorium superintendents in terms of its psychological value to patients in countering sloth and its immunological value in stimulating resistance by auto-inoculation.[87] Bryder has suggested that work was 'Possibly the most important feature of the treatment in early twentieth century British sanatoria', her chapter on the topic is entitled, 'The Pickaxe Cure for Consumptives'.[88] This undoubtedly overstates the case. First, it is difficult to generalise about sanatoria and it should not be assumed that all sanatoria followed the practices of élite institutions like Frimley or Midhurst which Bryder discusses.[89] Secondly, 'work therapy' was part of a more general feature of the treatment – namely discipline and surveillance. All authorities stressed that the treatment was successful only with constant medical supervision and with strict adherence to the rules by patients.

As early as 1898 Mander Smyth had spoken of sanatorium methods as 'strictly analogous to the splint' and in 1900 Jane Walker described it as 'a minute and elaborate system made up of endless details, all of which are important'.[90] The treatment came to be described as a timetable in which every waking minute was occupied. The schedule went beyond medical routine and personal hygiene, to requirements on social activities and relationships and codes of behaviour. Part of this was undoubtedly a direct concern with the orderly operation of the institution. The problems of running a sanatorium full of patients, who were mostly young adults, often not incapacitated and with time on their hands, are not difficult to imagine.[91] Most sanatoria segregated male and female patients, though it was difficult to prevent mixing, not least at night when not only the doors and windows were open, but the walls too![92] Some patients were dismissed for disciplinary reasons and most found the regime harsh; sanatorium superintendents were often said to be autocrats or martinets.[93] The strict hygienic regime could be met with active as well as passive resistance; in 1908 the Bradford sanatorium at Eastby was reported as being 'in a state of insurrection'.[94] Discipline was also a means of countering the swings of mood thought to be typical of the tuberculous mind. It countered the 'happy go lucky attitude' characteristic of some stages of the disease and the melancholy typical of others.[95]

In 1910 the controversies surrounding sanatoria were given several new twists. First, tuberculin treatment was once again in vogue and the proposals of the Australian doctor, W. Camac Wilkinson, to establish tuberculin dispensaries began to attract attention and financial support.[96] Wilkinson attacked the sanatorium treatment for not working, for offering treatment to only a tiny proportion of working-class sufferers and for ignoring aftercare. Tuberculin dispensaries, it was claimed, could treat patients more cheaply, allow them to continue to work and would not disrupt family life. Supporters of sanatoria countered by arguing that there was no conflict between sanatoria and tuberculins; in fact, tuberculins were widely prescribed in sanatoria which were undoubtedly the best environment for their administration.[97] Wilkinson, however, was not prepared to compromise; he pressed his claims, in time persuading some local authorities to establish dispensaries.[98] Secondly, the statistical evidence began to mount against the treatment. Pearson re-entered the fray, this time using purported statistical inadequacies in an influential pro-sanatorium book by Latham and Garland, once again to advance his hereditarian ideas. Soon after this, an actuarial study found no evidence that sanatoria had altered the long-term survival prospects of consumptives.[99] If this was not enough, a number of 'horror stories' were published in the popular press, followed by letters and articles by those who had 'suffered'.[100] People now talked openly of the treatment's failure.

The increasing variability of the treatment, together with uncertainties over its value and an unfavourable popular image, sapped medical and philanthropic confidence. Finance was a continuing problem – for the sanatoria after the initial enthusiasm had died down, and for patients who had to endure long-term loss of wages and subsequent difficulties in finding work. The further establishment of sanatoria was inhibited, so that by 1908 the growth of both institutions and beds had begun to level off. The time-lags in the founding of new institutions would seem to suggest that the doubts and difficulties had begun to have their effects much earlier.[101] The slow-down in the growth rate may have meant that effective demand had been met; while some public sanatoria reported waiting lists, most were never full and many advertised for patients. There was, however, no lack of potential patients. Assuming an average stay of three months and an average occupancy rate of 75 per cent, it can be estimated that the number of patients treated in sanatoria in Britain rose from 1500 in 1900 to 12 000 in 1913. This meant that by the latter date only 4 per cent of total sufferers, or 20 per cent of the 60 000 'early' cases which the sanatoria sought, were receiving the treatment.[102]

Given the uncertainties, lack of public confidence and virtual cessation in the growth of provision, it was surprising to say the least to find that the government proposed a 'Sanatorium Benefit' in the National Health Insurance legislation in 1911. Moreover, money was also promised for the construction of many new sanatoria for insurance patients. This reprieve had little to do with the merits or otherwise of sanatoria, it was an incidental effect of a new initiative in social policy. The main reason Sanatorium Benefit was included in the legislation was that it had been an integral part of the German insurance system, on which the British scheme was partly modelled.[103] Lloyd George, whose father had died of consumption, had been impressed by the work of the sanatoria when he had visited Germany in 1908. Further justification could be found in the last official word on the matter, by Bulstrode. There may also have been political pressure for action, as Wilkinson's campaign for tuberculin dispensaries for the working class was matched by Lister and Garland's attempts to mobilise the support of state and labour organisations.[104] The sanatorium clauses of the Bill were approved by Parliament, only after many reservations had been expressed both in the House and outside.[105] The main question was, why was the government supporting this treatment alone and ignoring others? Also, why this disease and not others, like cancer? Alternative claims were advanced: supporting tuberculin dispensaries, medical research, prevention by improving housing and after-care. Lloyd George compromised with the opposition by incorporating research and a comprehensive tuberculosis service into the scheme; it was suggested that in the end only the eugenists were not satisfied.[106] It was a compromise of no little consequence. From it emerged the Medical Research Committee (later Council), and a tuberculosis service which anticipated many of the features of the later welfare state.

The details of Sanatorium Benefit and the related schemes were worked out by an expert committee headed by Lord Astor. This was 'packed' with supporters of sanatoria and they recommended that local authorities organise their own institutions for residential care and develop dispensaries and after-care agencies.[107] After 1912, medical opinion, as at the International Congress on Tuberculosis, turned decisively against tuberculins and once again endorsed sanatoria; this was unsurprising as by this time most tuberculosis experts worked in sanatoria. However, the legislation brought a new constituency to the support of sanatoria – the Medical Officers of Health – and it was these who then led the planning of local schemes.[108] Local authorities were required immediately to provide sanatorium treatment, a situation they met by converting accommodation in Poor Law infirmaries and isolation hospitals, and providing tents and home shelters. By June 1914, in England alone, 9200 beds had been approved in 259 residential institutions, plus 180 beds available in military hospitals.[109] This was an over threefold increase in beds in three years. Only some two-thirds of these beds were used, but the immediate jump in provision shows the impact of the National Health Insurance Act and the vigour with which its provisions were pursued by Medical Officers of Health and local authorities. In the longer term there was an even larger expansion between the wars, such that by 1938 there were 420 specialist residential institutions and some 38 000 beds in Great Britain.[110] This amounted to an over fourfold increase between 1914 and 1938, for a disease whose incidence approximately halved in the same period. The treatment remained controversial, but with the endorsement and support of the state its growth was assured.

CONCLUSION

The sanatorium treatment was introduced into Britain through the adoption of the Nordrach method from Germany in two stages. First, in an adaptive way, in private sanatoria akin to small nursing homes and hospitals, then in public sanatoria for the working class, as part of a national anti-consumption campaign. The success of the innovation lay in the coupling of scientific push and social pull – of new bacteriological and immunological research and of fears about the dangers from and costs of consumption amongst the working class. During its innovation and diffusion the rationale and practice of the treatment was 're-invented' many times. The role of the 'open-air' was progressively played down and other features emphasised – indeed, there developed great variation between the regimes of different institutions. Such shifts can be seen only if we pursue the study of medical innovations beyond first use to their diffusion; this latter process should be studied qualitatively as well as quantitatively. Equally, attention must be given to the medical and socio-political environment into which innovations diffuse; that is, to existing ideas and practices, responses and resistances and to the differing contexts of uptake. Sanatoria were not assessed in isolation, but in relation to other 'cures', like tuberculin, and in relation to whether cures were possible at all. The opposition it met undoubtedly contributed to the shift from 'cure' to 'education' and in time eroded confidence and support for the treatment. Until rescued by the National Health Insurance Act, the future of sanatoria looked bleak and it was possible that an alternative treatment or strategy, such as tuberculin dispensaries or collapse-therapy, might have been substituted and formed the basis for a very different anti-tuberculosis programme from that which developed between the wars.

Sanatorium treatment was often described as a 'specific treatment' for consumption. It has been shown, however, that the innovation was highly social, largely a matter of regime. It required a new cadre of specialists and new institutions; a 'new way of life' was created and taught to patients. We see here, in extreme form, a feature common to all medical innovation – that technical change and social change are inextricable. In as much as innovations are born of, and remain part of, wider sociotechnical systems, the study of single innovations will involve restrictions which may be misleading. The turbulent development and discontinuous growth of the sanatorium treatment in Britain before 1914 also demonstrates that innovation and diffusion are historical processes, whose outcomes are contingent.

APPENDIX: NUMBER OF BEDS IN INDIVIDUAL SANATORIA, 1898–1911

I Private Sanatoria

Name of institution	Country	\multicolumn{14}{c}{Number of beds reported annually}
		1898	1899	1900	1901	1902	1903	1904	1905	1906	1907	1908	1909	1910
Alderney Manor	E			25	25	25	25	25	28	28	28	28	28	28
Ashover	E													10
Belle Vue	E		10	10	10	15	15	15	20	20	20	20		
Brinklea	E		10	10	12	12								
Brookside	E	8	8	9										
Caverhill	S							14	14	14	14	14	14	14
Cotswold	E	10	30	30	37	37	37	37	37	39	39	39	39	39
Crooksbury	E			12	12	12	22	22	24	24	24	24	24	24
Dartmoor	E						20	20	20	20	20	20	20	20
Dean Head	E													
Dunstone Park	E					8	10	10	10	10	10	10		
East Anglian	E				35	35	35	35	35	35	35	35	35	35
Ellerslie	S								8	8	8	8	8	8
Firs	E			6	6	6	20	20	20	20	20			
Fritton	E				7	7	7	7	7	7	7	7	7	7
Grampian	S				20	20	20	20	20	20	20	20	20	20
Hailey	E			12	12	12	24	24	25	25	25	30	30	30
Holme Chase	E				8	8	8	8	8	8	8			
Inglewood	E		10	10	20									
Kingwood	E					12	12	12	12	12	12	12	12	12
Knocksmaltach	S			6	6	6	6	6	6	6	6	6		
Linford	E		15	15	15	15	19	24	24	24	24	24	24	24
Maldon	E		12	12	12	12	12	12	12	12	12	12		
Maltings Farm	E							16	16	16	32	32	32	32
Mendip Hills	E			20	20	20	20	20	20	20	24	24	24	24
Midland	E				20	20	20	20	20	20	20	20	20	20
Moorcote	E				12	12	12	15	15	20	20	20	20	15
Mundesley	E		15	15	15	15	15	15	15	20	20	20	20	20
Nordrach in Wales	W								23	23	23	23	23	23

Name of institution	Country	1898	1899	1900	1901	1902	1903	1904	1905	1906	1907	1908	1909	1910
Nordrach-on-Dee	S				36	36	50	52	54	54	54	54	54	54
Nordrach-upon-Mendip	E		40	40	40	40	40	40	40	40	40	40	40	40
Ochill Hills	S					16	16	16	60	60	60	60	60	60
Ockley	E						12	12	12	12	12	12	12	12
Overton Hall	E	8	16	16	16	16	16	16	16	16	16	16		
Painswick	E			12	12	12	12	12	12	12	12	12	12	12
Pinewood, London	E				60	60	60	60	60	60	60	60	60	60
Poole, South Lodge	E					14	14	14						
Rudgwick	E	8	8	8	12	12	12	12	14	14	14	14	14	14
Stourfield Park/Home	E	4	45	45	45	45	45	45	45	45	45	45	45	45
Sunny Nook	E				6									
Swiss Villa	E				4	4	4	4						
Udal Torre	E											13	14	14
Vale of Clywd	W		20	20	20	20	20	20	20	20	20	20	20	20
Whitmead Hill	E		20	20	20	20	20	20	20	20	20	20	20	20
Woodburn	S	20				20	20	20	24	24	24	24	24	24
Woodhurst	E					16	16	16	16	16	16	16	16	16
Annual total of private beds		58	259	401	618	655	731	771	834	848	864	859	786	796
Annual total of private sanatoria		6	14	23	33	35	36	38	37	37	37	36	31	32

Note:
E England
S Scotland
W Wales

II Public Sanatoria

Name of institution	Country	Type	1898	1899	1900	1901	1902	1903	1904	1905	1906	1907	1908	1909	1910
Abergele	W	LA													50
Argyll County	S	Vol												30	30
Ayrshire	S	Vol									12	40	40	40	40
Birmingham, Saltley Grange	E	LA											40	40	40
Birmingham, Yardley Road	E	LA													50
Bournemouth, National San.	S	Vol				62	62	62	62	71	71	71	85	85	85
Bellefield	S	Vol								30	30	30	30	30	30
Benenden	E	Vol										32	64	64	85
Blencathra	E	Vol								20	20	20	28	28	28
Bridge of Weir	S	Vol	38	38	38	38	38	80	80	92	92	120	120	120	120
Chilton Hill House	E	Vol													18
Coppins Green	S	Vol									10	10	10	10	10
Crossley	E	Vol									90	90	90	90	90
Daneswood	E	Vol					22	22	22	22	22	24	24	24	24
Devon & Cornwall	E	Vol							15	16	16	16	35	35	35
Durham, Stanhope	E	Vol			12					44	44	44	45	45	45
Durham, Wolsingham	E	Vol							20					30	30
Eastby, Bradford	E	LA						32	32	32	32	32	33	33	33
Eldwick	E	Vol											24	24	24
Engel Home	E	Vol						15	15	16	16	16	16	16	16
Eversfield	E	Vol						55	55	55	55	55	55	55	55
Fairlight	E	Vol										16	22	22	22
Frimley, Brompton Hos.	E	Vol								100	100	100	108	108	108
Heswall	E	LA					24	24	24	24	24	24	24	24	24
Hillside	S	Vol				20	20	20	20	20	20	24	24	24	24
Hull & East Riding	E	Vol					30	30	30	30	30	30	30	30	30
Kelling	E	Vol						19	42	50	56	56	56	56	56
King Edward VII	E	Vol									100	100	100	100	100
Lanfine	S	Vol						25	32	32	32	32	32	32	32
Leeds	E	Vol						40	30	30	30	30	30	30	30
Liverpool	E	Vol				40	40	40	40	40	40	40	40	40	40
Maitland Cottage	E	Vol		6	6	8	8	8	8	16	16	20	24	24	24

Name of institution	Country	Type	Number of beds reported annually											
			1898	1899	1900	1901	1902	1903	1904	1905	1906	1907	1908	1909	1910
Manchester, Bowdon	E	Vol													
Northampton	E	Vol			50	50	50	50	50	50	50	50	50	50	50
Northumberland	E	Vol													40
Northwood, Mt Vernon	E	Vol							100	100	100	100	50	50	50
Nottingham, Ransom	E	Vol					24	32	32	32	32	32	100	100	100
Seaforth	S	Vol											32	32	32
Sheffield, Crimicar	E	LA											18	18	18
Sheffield, Moor End	E	LA										20	20	20	20
Sidlaw	S	Vol							40	40	40	40	40	40	20
Southrepps	E	Vol								7	7	7	7		
Strinesdale	E	Vol											18	18	18
West Wales	W	Vol											30	30	30
Westmoreland	E	Vol			17	17	17	17	17	25	25	44	44	44	44
Wilkinson	E	Vol										25	25	25	25
Winsley	E	Vol							66	66	66	66	68	68	68
Worcestershire	E	Vol					16	16	16	16	16	16	26	30	30
Annual total of public beds			38	44	123	255	371	567	833	1076	1294	1472	1777	1834	1973
Annual total of public sanatoria			1	2	5	8	13	18	22	27	31	35	42	43	46
Combined private and public totals			1898	1899	1900	1901	1902	1903	1904	1905	1906	1907	1908	1909	1910
Annual total of all sanatorium beds			96	303	524	873	1026	1298	1604	1910	2142	2336	2636	2620	2769
Annual total of all sanatoria			7	16	28	41	48	54	60	64	68	72	78	74	78

Note:

E England
S Scotland
W Wales
LA Local authority
Vol Voluntary

4
FROM MEDICAL RESEARCH TO CLINICAL PRACTICE: SERUM THERAPY FOR DIPHTHERIA IN THE 1890s
Paul Weindling

Serum therapy for diphtheria, introduced between 1890 and 1895, is often ranked as one of the first triumphs of medical research: laboratory animal experiments had provided a treatment which could be used in clinical practice. This is held to have boosted the status of such ancillary medical disciplines as public health and paediatrics by placing therapy on a properly scientific basis. Scientifically, diphtheria research provided the opportunity for a synthesis between German bacteriology (notably of the Koch school) and the French research at the Pasteur Institute on vaccination and immunisation. In terms of therapy, diphtheria exemplified a novel approach which was expected to apply to a wide range of diseases ranging from syphilis to cancer. Socially, it came at a crucial time in the effort of laboratory-trained researchers to take a lead in medical education, clinical practice and professional organisations. Because the programme of specific therapies for specific diseases required large-scale research institutions, manufacturing facilities and financial resources, serum production and distribution provided the opportunity for a re-casting of relations between the medical profession, state authorities and the nascent pharmaceutical industry. The sensational news of a cure mobilised the profession, state authorities and the public in a campaign to ensure early diagnosis and treatment for diphtheria.

In this essay, I intend to scrutinise some of the claims made for serum therapy, and to consider whether groups of hitherto marginal laboratory researchers exaggerated the efficacy of anti-toxin sera for polemical purposes. It is, therefore, important to recognise that in tackling diphtheria they were dealing with a highly emotive disease which commanded public attention. During the 1890s high death rates from infant and childhood diseases contrasted with the decline already evident in *adult* mortality from infectious diseases. For European children aged between one and ten, diphtheria was one of the top killers, accounting for roughly one-fifth of deaths. Despite diagnostic difficulties in distinguishing diphtheria from such throat infections as croup, thrush and quinsy, and also from scarlet fever and measles, diphtheria had come to be a focus of medical attention.[1] But even after the advent of bacteriological diagnosis, the diversity of clinical signs remained puzzling, and the bacilli exhibited varying patterns of growth. The disease was endemic in major towns and cities where damp and overcrowding (partly attributed to compulsory schooling) were held to be causes. The incidence of diphtheria fluctuated in epidemic waves.

Though the 1880s had seen bacteria identified for such major diseases as tuberculosis and cholera, many epidemiologists continued to see epidemics as resulting from environmental conditions, or from innate biological characteristics of the disease. Advocates of bacteriology and serum therapy argued that the introduction of such therapy caused a rapid fall in diphtheria mortality, but other commentators were sceptical. It was argued that serum therapy had been introduced at a high point of naturally occurring cycles, so that graphs depicting a spectacular fall in diphtheria mortality during the 1890s were misleading. Here Arthur Newsholme's study of diphtheria, published in 1900, is particularly helpful. Diphtheria mortality rates in London had been rising until 1893, but in Paris a long-term decline had been evident since 1877. Newsholme commented: 'I am, however, strongly of the opinion that although a proportion of the decreased mortality from diphtheria in Paris is almost certainly due to the general employment of the antitoxin treatment, another influence is also at work viz. the ebb and tide of epidemic prevalence'. He reminds us that the use of the anti-toxin treatment was limited until the autumn of 1894.[2]

Newsholme's comments suggest that we, too, should be cautious in accepting proof of a rapid 'Pasteurian revolution' in the therapeutic applications of microbiology.[3] Mortality rates did decline, but this was an uneven process, and there continued to be fluctuating waves of the disease. Moreover, Newsholme's adherence to a localist interpretation of epidemic disease indicates how progressive medical experts in 1900 could still be sceptical of bacteriology and germ theory. His local approach was justified by the widely differing incidence of the disease. In Berlin diphtheria mortality had reached a high point of 211 per 100 000 inhabitants in the mid-1880s with a trough of 60.1 in the early 1890s, whereas in London rates stood at 42.5 in 1894 and 55.6 in 1895. Evaluating the efficacy of serum therapy is problematic in that the disease cannot be assumed to have had either a constant incidence or virulence. The epidemiological evidence thus counters glib assertions of a 'bacteriological revolution' having a rapid impact in reducing the incidence of infectious diseases. Unfortunately graphs beginning in 1894 have been routinely reproduced by historians of medicine and public health experts such as McKeown.[4] These preliminary observations should serve as a cautionary background against which to judge the use of serum therapy.

FROM ANIMAL TO CLINICAL RESEARCH

The 1880s saw the rapid development of bacteriology. Doctors came to accept Koch's principles that diseases were caused by a specific organism, which could be isolated and cultured, and that a disease could be experimentally reproduced by the inoculation of pure cultures. The diphtheria bacillus, first recognised by Klebs in 1883, was cultured by Koch's assistant Loeffler in 1884. But whereas Klebs had insisted on the variety of manifestations of the disease and had compared outbreaks in the different local contexts of Basel, Prague and Zürich,[5] the Koch school insisted on a uniform type of the disease, irrespective of geographical context.[6] Loeffler's application of Koch's rigorous principles for the isolation and replication of a causal organism led him to emphasise the singularity of diphtheria as a disease entity.

By the 1890s bacteriologists had achieved some success with disinfectants, but specific therapies still eluded them. Koch's introduction of tuberculin therapy for tuberculosis in 1890 aroused enormous professional and public acclaim, but the results proved inconclusive, and the prestige of bacteriology began to suffer. University academics were often suspicious of a subject for which new research institutes had been created outside the university system. On occasion, the élitist medical faculties could be intensely hostile to bacteriological researchers such as Koch, Behring and Ehrlich. Bacteriology was dismissed as lacking principles of general medical validity; doubt was thrown on the analogy between animal and human organisms, and on whether bacteria could account for an array of signs and symptoms.[7]

So great were public hopes that new cures could be derived from laboratories that major new scientific institutions were founded for research rather than teaching. Public acclaim for Pasteur's new anti-rabies serum, first used in 1886, meant that it was possible for the Pasteur Institute to be established on the basis of public subscriptions; the Institute opened in 1888 as both a public dispensary and research institute.[8] Emulating this success, in 1891 the Prussian state established an Institute for Infectious

Diseases for Koch. This was innovative in having extensive clinical facilities alongside the research institute. Both institutions angered the vested interests in medical faculties, and there were tensions between clinicians and the researchers whose claims for laboratory-based medicine were judged inconclusive.

Diphtheria research gave a further boost to the fortunes of the Pasteur Institute, where an alliance formed between marginal groups of bacteriologists and child health specialists. New diphtheria wards were added to children's hospitals, and paediatrics gained academic and professional status. The connections between laboratories and children's clinics were especially close in Paris, where the Pasteur Institute was situated close to the Hospital for Sick Children. Laboratory researchers (with connections to both institutions) were alerted to the crying need for research on childhood infections. In 1887, for example, Alexandre Yersin gained clinical experience of diphtheria when a very junior 'externe' at the Children's Hospital. He was then able to persuade his scientific supervisor, Emile Roux, that diphtheria was of interest in two ways: as an infection which also involved an intoxicant which damaged the nervous system.[9]

Despite deep scientific, personal and patriotic conflicts between Koch and Pasteur, the development of diphtheria serum therapy was the result of co-operation between younger researchers in Paris and Berlin. Loeffler was able to separate the poison from the bacillus, and he characterised it as an enzyme. (The term 'toxin' was not yet used.) His strategy for eradicating diphtheria was compulsory isolation of cases, disinfection of housing, and treatment by 'internal' disinfectants such as gargles.[10] In 1887 Roux and Yersin, in Paris, discovered that the broth used in culturing the bacillus had poisonous properties. They concluded that to counter the damage to the organism from the bacillus it was necessary to halt its progress as soon as possible.[11] They confirmed the bacteriological discovery of Klebs and Loeffler, and that the disease was to be diagnosed bacteriologically from a specimen taken from the 'diphtheritic membrane' in the patient's throat, rather than by observation of clinical signs and symptoms. As Behring later remarked when receiving the Nobel Prize in 1901: 'Without this preliminary work by Loeffler and Roux there would be no serum treatment for diphtheria'.[12]

Behring and a Japanese co-worker, Shibasaburo Kitasato, made observations similar to those of Loeffler and Roux regarding the culturing medium for tetanus bacilli. Though Behring's early experiments had been a response to Loeffler's call for internal disinfectants, he soon embarked on a new course of research. In 1890, while still a military medical officer, he injected into healthy animals the poison or 'toxin' produced by the culture of diphtheria bacilli. The animals, he claimed, thereby produced blood serum which was able to kill diphtheria bacilli. Behring and Kitasato described this process as the formation of antibodies to a toxin – the formation of an 'anti-toxin'. That bacteria or their products could be neutralised by sera from immunised animals seemed to open a general pathway to the control of infectious diseases. A wave of research followed, which effectively established the conceptual foundation of a new science of 'immunology'.[13] The intellectual and social presuppositions leading to this discovery merit further historical analysis.

From 1891 until 1893 there were clinical trials in selected children's hospitals where clinical specialists had begun to forge an alliance with the new type of scientific medicine. One such hospital, the Kaiser und Kaiserin Friedrich Children's Hospital in Berlin had been founded by a citizens' committee in 1890, for the treatment and isolation of children with infectious diseases. Koch's Institute for Infectious Diseases in Berlin was allocated a ward at the Charité state hospital, and facilities in the new municipal hospitals. But the first reports on the use of serum therapy were not encouraging. When Otto Heubner achieved a rate of 62 per cent cure at Leipzig, this was regarded as being inconclusive, and as possibly due to spontaneous recovery. At Berlin only eleven children were initially treated with the new serum. In the summer of 1893 Behring and his co-workers admitted that the results were disappointing.[14]

Despite these teething troubles the efforts at clinical trials were intensified, and more encouraging results were obtained by Roux in Paris, researching on the effects of the toxin on large animals. These were preliminary to clinical trials carried out between February and July 1894 at the Pavillon Trousseau of the nearby Hospital for Sick Children.[15] After clinical trials comparing traditional therapies with the serum therapy, Roux claimed to have reduced mortality from 50 per cent to 25 per cent.[16] He used increased doses of serum on larger numbers of patients and pioneered large-scale production of the serum by substituting horses for guinea pigs. This method was taken up by a number of manufacturers such as Hoechst in Frankfurt and Schering in Berlin.

As long as there was close co-operation between researchers and manufacturers of sera, the quality of sera was assured. But as competition increased between manufacturers it became necessary for there to be some type of regulatory authority, as well as standardisation in the measurement of the strength of sera. Here the initiative was taken by the Prussian state and by researchers such as Ehrlich and Koch. Ehrlich's skills were applied to quantifying the dosage; he replaced the imprecise definition of a unit of serum as the lethal dose for a guinea pig by a new method of accurately determining the value of sera in pure form so as to maintain constant standards; he calculated the value of 'immunity units' per gramme of dried sera, abandoning the use of living cultures as too unstable. A turning point came when Kossel, an assistant of Koch, treated 223 children: of the 78 children whom he treated on the first or second day, 97 per cent recovered.[17] The principle was thus established of early diagnosis and treatment with substantial doses. Such results had significant therapeutic and institutional repercussions.

PROFESSIONAL IMPLICATIONS AND PUBLIC SUPPORT

Until the advent of serum therapy, management of such infectious diseases as diphtheria was problematic for children's hospitals and death-rates were staggeringly high. Serum therapy marked a turning-point as children's hospitals established new departments and facilities in which participating physicians were to make outstanding careers, supported by philanthropists, and by municipal and state governments. For example, Heubner, the Leipzig paediatrician, was appointed by the Prussian state to the first full chair of paediatrics in Berlin in 1894, in the face of opposition from the medical faculty.[18] Behring, who had incurred jealous opposition from Koch, was appointed to a university chair in 1894, despite bitter hostility from the medical faculty in Halle. The state saw scientific innovations as a way of stemming case-mortality rates from diphtheria which were as high as 50 per cent in children's hospitals.

Behring assumed a dual role as a leading medical researcher determined to make his professional reputation with serum therapy, and as an ideologist demanding professional and public support for laboratory-based therapies. In 1893, he proclaimed serum therapy as a radically new approach to medicine, one that replaced social remedies against epidemic diseases by instead concentrating on the healing powers of the blood. He compared his view of the spread of contagious – or 'catching' – diseases (ansteckende Krankheiten) to a forest fire, and argued that the term 'infectious disease' (Infektionskrankheit) was derived from alien Latin roots. Behring considered that Germanic folk beliefs about the spread of diseases coincided with his views on the new therapy, and he attacked the 'liberal' approach to disease, as exemplified by Rudolf Virchow, the pioneer of cellular pathology and a politician who had argued for liberal social reforms as the necessary precondition for improvements in health. Behring, by contrast, sought to strengthen laboratory-based medicine. The medical researcher would have a new position of authority within the medical institutions of the state, and would ensure that the medical profession and the pharmaceutical industry would collaborate closely. He expected that technical innovations in medicine could eradicate epidemic diseases, which liberals and the newly legalised socialist party attributed to poverty and political repression. Even before the efficacy of such innovations as serum therapy was proven, the expectation was clear that such advances would boost professional authority and defuse social conflicts.[19]

Public appeals were instituted, partly to finance the new therapy, partly to improve the image of medical science, for in the 1890s fears were widespread that hospital patients were used as human guinea pigs. In France and in Germany, newspapers cashed in on the public enthusiasm as they took a lead in co-ordinating fund-raising efforts. From September 1894, *Le Figaro* organised a national subscription to finance production of the serum.[20] Roux had expected that it would take years of struggle to introduce serum therapy, and he expended much effort to build up good relations with the Department of the Seine and the municipality of Paris. He was surprised by the wave of generosity that broke over France; money poured in from social groups ranging from workers' organisations to banks.[21] Indeed, such was the popular support that production of serum lagged behind public demand. The Pasteur Institute benefited from the humanitarian ideology of scientific research for the public good, and medical research seemed to provide an attractive cause for traditionally antagonistic social groups. This was particularly important at a time when supporters of the Third Republic were seeking a new broad-based stability and consensus. Medical research thus promoted social cohesion as well as opening up career opportunities to economically hard-pressed scientists.

By the end of 1894 the Pasteur Institute had established a national network for the production and distribution of serum, which was to be free of charge. It established a special *Service de Sérothérapie* under Louis Martin, who conducted further research on the culturing medium, oversaw the stabling of horses, and tested the sera.[22] A distribution network for the serum was provided by the bureau of public assistance established in July 1893. The lavish public subscription meant that substantial profits went to the Pasteur Institute, as did additional grants from state sources as well as from foreign governments such as Hungary and Belgium, which commissioned the Institute to produce serum. [23] Whereas in France the Pasteur Institute spawned a centrally directed but independent network of research institutes, in Germany there was a high degree of competition between research institutes. During the 1890s Behring and Ehrlich were both to acquire autonomous institutes as a result of their pioneering research on sera.

In Germany, the distribution of sera depended on public subscription, and on academic and commercial competition. As in France there were public campaigns to raise subscriptions for the new therapy. In 1894 a philanthropic committee for the distribution of the therapeutic serum for the poor was established,[24] and Ehrlich and Wassermann, another talented young researcher at Koch's Institute, held lectures to raise subscriptions and to train the doctors who provided free treatment under Poor Law schemes. Poor Law doctors were asked to refer sick children to the children's hospital and the three municipal hospitals for treatment. There were also in Berlin five 'depots' – accident clinics where the serum could be obtained free of charge.

Anti-diphtheria measures proved to be a crucial link between the short-lived public panic over, and academic acclaim for research on, rabies during the 1880s, and the long-term development of a national system of dispensaries against tuberculosis.[25] As in France, newspapers organised the raising of funds at the same time as sensationalising the news of the serum. One of these, the *Berliner Lokal-Anzeiger*, administered a fund and published details of each contribution, the first being from August Scherl, the entrepreneurial publisher of the newspaper. Another subscription list on behalf of the children's hospital was administered by Rudolf Mosse, the liberal publisher of the *Berliner Tage-Blatt*. By the autumn of 1894 researchers felt confident that they had an efficacious remedy. State authorities were cautious because the early 1890s had seen a rush of spurious claims for the discovery of specific remedies for diphtheria, but official attitudes changed with the success of the clinical trials, and the Reich Health Office throughout the years 1894 and 1895 was supportive of research on serum therapy.[26] In November 1895 the serum was finally pronounced by the Reich Health Office to be harmless and suited for the widest use.[27] The rapid French and German successes contrasted with development in Britain, where it took over a decade for free distribution of anti-toxin to be arranged.[28]

Research and development of diphtheria serum provided opportunities for the take-off of several pharmaceutical companies, notably Hoechst, Merck, Lingner and Schering in Germany, and the Wellcome Laboratories in England.[29] Academic jealousies were reinforced by commercial rivalries. Shortly after Behring's announcement of the discovery in April 1893, a veterinary physiologist Hans Aronson announced a similar cure; Behring worked with the resources of the Hoechst company, and Aronson with the Schering company: Aronson was also backed by Virchow – by now the arch-enemy of Koch and Behring.[30] The result was acrimonious academic and commercial competition. By June 1894 Aronson had acquired thirty horses for production of serum, the numbers reaching sixty by September of that year.[31] The varying strengths and qualities of the various sera produced by different firms were debated in the medical press.

As a tense situation of academic and commercial rivalry arose in Germany, state intervention was proposed in October 1894.[32] The state authorities of the Reich Health Office introduced the novel requirement for standardisation and the routine testing of sera. Throughout the autumn of 1894 the higher education official, Althoff, presided over discussions with Koch, Behring and Ehrlich, and used the opportunity to found an institute for Ehrlich. An institute for serum testing was quickly opened in the Berlin suburb of Steglitz, and this was given additional state and private finance when it moved to Frankfurt in 1899, as an Institute for Experimental Therapy. This novel institute acted both as a supervisory body to maintain the standards of sera, and as a research centre where such notable innovations as Salvarsan treatment for syphilis were pioneered. The German state and municipal authorities had apparently realised that the momentum in clinically-oriented research had to be maintained, and that this could best be achieved in a non-university context such as Frankfurt. For their part, commercial companies loudly advertised their certificates of approval from Ehrlich's Institute, indicating that the serum was of adequate strength. Thus, whereas in France a single central (but private) institute benefited, in Germany the discovery stimulated the development both of the pharmaceutical industry and of state-sponsored research institutes.[33]

CRITICS OF 'SERUM THERAPY'

Some of the doctors and researchers trained in bacteriology remained severe critics of serum therapy. Researchers focused on such side-effects as vomiting, dyspepsia, hypertension and paralysis, and on the fatalities resulting from the use of the serum. There was, for example, widespread medical controversy and public outrage when the son of a leading Berlin medical professor, Paul Langerhans, died after a preventive injection of serum. Such disasters provided an additional incentive for routine testing of batches of sera, and in the Langerhans' case, Ehrlich was able to show that fatalities had not resulted among other patients injected with the same batch of sera. The ensuing controversies acted as a spur to research by Charles Richet and Portier, who in 1902 introduced the concept of *anaphylaxis* to explain a condition of hypersensitive reaction by the body to sera. Clinicians distinguished between the immunising qualities of the serum, which they regarded as impressive, and the healing powers, which were open to doubt. Virchow made this point in October 1894, but it was best put by the bacteriologist Hans Buchner, the discoverer of the bactericidal powers of blood leucocytes. In 1887, he questioned whether the antitoxin could really destroy the poison of the Klebs–Loeffler bacillus; he considered that the serum could prevent only further poisonings by cells. Thus the serum conferred immunity, but was not a therapy as Behring had claimed. Buchner placed greater emphasis on spontaneous cure and the body's independent capacity to form anti-toxin; he conducted research on bactericidal qualities of body fluids – for example, on the constituents of blood.[34]

The response of the body was also the concern of Adolf Gottstein, a lapsed bacteriologist, who in 1894 published a major scientific and statistical review of serum therapy, which launched him on a brilliant career in public health, emphasising welfare institutions for infants and children. He noted that the diphtheria bacillus could be present in the noses and conjunctiva of healthy children, so the virulence of the bacillus and the response of the infected body must vary from case to case. In his opinion, the contagionist focus on the distribution of bacilli outside the body overlooked the essential area of concern. It seemed more important to consider why the normal functioning of the mechanisms of immunity broke down than to focus solely on the infective agents. Gottstein condemned research on animals as inappropriate, because the animal and human organisms were so different. He argued that it was unnecessary to isolate the infected and to keep children away from school, for they might be harbouring only benign bacilli and have already built up resistance to infections.[35]

These criticisms of serum therapy had major implications for public health measures. They suggested a greater attention to the mechanisms of resistance and to promoting overall fitness. Such positive measures as dietary supplements, improved housing and education in personal hygiene were promoted to strengthen 'the human constitution'; improved welfare would promote 'natural immunity'. Public health and welfare measures thus had as much a rationale in bacteriology and immunology as did the new serum therapy, and public health officials faced a choice of 'natural' or 'artificial' methods of controlling and preventing diseases. The implications of the resulting measures were substantial, both for the child health movement, and for public campaigns against other infectious diseases such as tuberculosis. Indeed, the arguments against serum therapy were subsequently deployed against Koch's tuberculin as a remedy for tuberculosis.

Serum therapy for diphtheria underwent a sequence of stages of development resulting in the toxoid serum that was appropriate for the mass immunisation campaigns of the 1920s and 1930s. Behring was then commemorated as the saviour of the nation's children, and Nazi health officials hailed serum therapy as a major breakthrough in the application of laboratory sciences for child health.[36] But, as we have seen, the introduction of serum therapy had not been a simple 'discovery' by a solitary single microbe-hunter. During the 1890s doctors had found themselves linked in a complex web of social interests including pharmaceutical companies, specialised children's hospitals, and municipal, state and voluntary philanthropy. There were distinct differences between Behring, who co-operated with Hoechst with an eye to commercial exploitation of the serum, and the Pasteur Institute, which itself undertook production of the sera, so gaining money, public prestige and further opportunities for medical research. Diphtheria anti-toxin came to benefit both commercial interests and public health laboratories; with amazing rapidity, researchers became involved with commercial production. At virtually the same time, valid objections were raised by critics of serum therapy, who thus renewed the emphasis on welfare measures as part of public health.

Measures to combat diphtheria proved to be a crucial turning point in the institutional and professional development of clinical research and modern immunology. Bacteriological understanding of infectious diseases boosted the status of medical research; yet scrutiny of the scientific research and clinical trials on serum therapy suggests that a strong dose of ideology accompanied the development and distribution of the sera. Medical researchers such as Behring probably overstated the achievements of the sera, but heroic claims helped mobilise necessary financial resources, and public, professional and state support. The ideology of laboratory research shaped the contemporary meanings of 'cure' and 'curative serum'. Appeals couched in the apparently value-neutral terms of laboratory research served the interests of hitherto marginal groups of young researchers seeking a stable institutional basis for their careers. When denounced by anti-vivisectionists or the anti-vaccination lobby, laboratory researchers could claim to be in the vanguard of the struggle to benefit mankind of eradicating disease. Humanitarianism gave medical research a powerful popular appeal among a wide range of social groups.

The expectations that the methods of serum therapy would be extended from diphtheria to the eradication of all disease proved excessive, though the knowledge about 'immunology' had a permanent value. Even for diphtheria, the transition from the early animal experiments to a safe and standard remedy took far longer and was more complex than the early researchers had expected; and the result – the 'curative' serum – might have been more accurately described in less appealing terms – such as 'preventive' or 'neutralising'. There was, however, a secure and lasting contribution to both therapy and preventive immunisation. The development of diphtheria serum therapy was indeed a major medical innovation, but the process of innovation was far more complex than hitherto appreciated, not least because of competition between researchers and between conflicting ideologies of medicine and public health.

5
VACCINE THERAPY AND LABORATORY MEDICINE IN EDWARDIAN BRITAIN*

Michael Worboys

INTRODUCTION

The use of vaccines to prevent some infectious diseases was well established at the turn of this century and had been recently extended by the work of Louis Pasteur and others.[1] With general diseases, like smallpox, vaccines appeared to have no value when given after an infection was established; it seemed that they acted only by preparing the body in advance. However, in the 1900s a British doctor, Almroth Wright, proposed that vaccines could be used to cure certain bacterial diseases, especially localised infections. Serious and chronic local infections, such as tuberculous joints and glands, boils and other skin infections, were very common in the nineteenth and early twentieth centuries; they caused enormous suffering and many deaths and did not respond well to treatment. Wright's claim was based on the belief that local infections produced an immune response only in the area affected, the rest of the body's defences being unaware of the reaction. If this was the case, it ought to be possible to use vaccines to prepare and augment the resisting power of the rest of the immune system, thereby conferring on the body as a whole the power to overcome the local infection. Wright summarised his ideas as follows: 'to exploit in the interest of the infected tissues, the unexercised immunizing capacities of the uninfected tissues'.[2] Or, more graphically, that therapeutic vaccines would mobilise the unalerted reserves of 35 billion white-blood cells (leucocytes), which by dint of their sheer numbers and enhanced power, would defeat the most intractable infection.

Although vaccine therapy was one of a number of transient treatments for infections produced in the first quarter of this century, most of which with hindsight are seen as 'failed' innovations, a major claim for its historical significance has been made. This is, that it transformed the position and prospects of the clinical laboratory in Britain before 1914.

W. D. Foster, who wrote many valuable books on the history of the biomedical sciences, claimed that with the advent of vaccine therapy:

> No hospital could afford to be without a bacteriology department and a bacteriologist; careers in the subject became possible ... and the status of the laboratory worker vis à vis their clinical colleagues was gradually raised almost to parity.[3]

S. C. Dyke noted that before the First World War, '"Clinical Pathology" became equated in the minds of many with the preparation of vaccines – as indeed it then very largely was'.[4] Such assertions were first made by Leonard Colebrook in assessments of Wright's work in 1947 – indeed, Colebrook went further suggesting that 'beyond all question ... Wright's conception of immunization turned the direction of medical effort all over the world for 20 years'.[5] Such views of the historical significance of the treatment were also found amongst contemporaries. For example, Hilary Putnam, an American who visited Wright's out-patient clinic at St Mary's Hospital, London, in 1907, wrote on his return:

> One can hardly avoid forecasting that in 1917 clinics will resemble this one more than today in definiteness of diagnosis and of remedy; that central laboratories then will identify microorganisms for the general practitioner ...; and ... vaccines will be furnished by some central laboratory.[6]

J. M. Beattie, Professor of Pathology at Sheffield, observed in 1910 that, 'the work of Wright and his school, no matter what may be our views as to the methods employed, or the deductions drawn from them, ... must be regarded as epoch-making'.[7]

Wright's work seems to have been highly regarded by leading continental medical scientists, including Metchnikoff and Ehrlich; Wright was the only British worker acknowledged by the latter in his famous address to the International Medical Congress held at London in 1913. Significantly, Wright was chosen as the first head of bacteriology at the National Institute of Medical Research, which was set up by the Medical Research Committee (later Council) (MRC) in 1913; indeed Wright thought he might head the Institute.[8] Those who worked with him in the 1900s thought that, in time, he would become recognised as the 'British Pasteur'.

The claims made for vaccine therapy relate directly to a major change in twentieth century medicine – the rise in the authority and importance of the laboratory.[9] This can be seen in every aspect of medicine, from the emergence of research laboratories as the sources of new knowledge, through to the laboratory-produced treatments. How this change occurred has been little studied and even the periodisation of the transformation is unclear. The implication of studies, like those of Reiser and Harvey, is that this change occurred steadily and incrementally through this century, driven by a steady stream of technical and cognitive innovations.[10] The alternative possibility – that the change was discontinuous, wrought by a few key innovations or other changes – has not been seriously considered. Yet certain innovations – insulin therapy in the 1920s most readily springs to mind – could have been decisive in showing the value of laboratory research, in creating work and in necessitating the provision of better clinical laboratories. Similarly, the First World War legislation on venereal disease and the tests it required had an abrupt impact on the work of clinical laboratories. Could the now discredited practice of vaccine therapy have also decisively advanced and shaped laboratory medicine in Britain in the Edwardian era?

Medical historians seem not to have recognised, let alone taken seriously, the statements of Foster, Dyke and Colebrook. In the main, those historians who have discussed vaccine therapy have been critical. Parish described it as 'a lamentable episode'.[11] Foster himself described it as 'bogus' and 'valueless to the point of fraudulence'.[12] Dowling described it as a field 'opened wide for those who could profit by deceiving, and the charlatans took full advantage of the opportunity'.[13] More recently, Keating, while trying to be agnostic, could not resist impugning Wright's motives, describing his attitude to clinical trials as 'supercilious'.[14] Keating's discussion also follows previous historical accounts of immunology in the period 1880–1910 in emphasising theory rather than practice.[15] The secondary literature would suggest that the paramount concern of immunologists was the relative merits of phagocytic, humoral or other theories. This is clearly misleading; early immunology was dominated by practice, not theory. What mattered to most contemporary medical practitioners and scientists was not so much how the body resisted infection, but how they could produce anti-toxic sera, vaccines or constitutional measures to raise the body's immune responses. Wright himself always stressed the practical rather than the theoretical significance of his work. Indeed, he went further and claimed that vaccine therapy would bring about the complete reorganisation of clinical medicine around applied bacteriology, or more generally the laboratory. In the words of Wright's own much-quoted prediction, 'The physician of the future will be an immunizator'.[16] Lowy has noted that immunology had such a high profile that it was used in literature, before 1925, to represent prestigious scientific research. In fact, both the examples she discusses concern not immunology in general, nor even sera, but vaccine therapy.[17]

My emphasis here is on the relationship between vaccine therapy and the system of clinical laboratories which was developing in the years around 1900. In this respect I share the interests of Foster, Dyke and Colebrook; however, I do not share their conclusions, except Foster's about the status of clinical pathologists. I will show that vaccine therapy was not critically important to the growth of clinical laboratories; St Mary's apart, the bulk of the workload of these laboratories was diagnostic testing, hence laboratories would have grown at much the same rate and in the same way without vaccine therapy. However, vaccine therapy made the role of laboratory medicine a central issue.[18] Diagnostic laboratory procedures were not that controversial, vaccine therapy was. This was because Wright's innovation clearly put the bench above the bedside and took the laboratory into the key domain of the clinician – therapy. The methods Wright advocated so forcefully offered – as well as a diagnosis – a therapeutic agent, a regime and an evaluation, all controlled by laboratory-based doctors. Vaccine therapy, if not vital to the growth of laboratories, was then pivotal to hospital laboratory pathologists, to whom it brought enhanced status and income. Here was a radical new model for medicine, which promised a re-ordering of practices and structures. I will argue that in this potential transformation lies the major historical importance of vaccine therapy. This essay begins by considering Wright's career and identifying his particular style of laboratory medicine. The introduction of vaccine therapy is then considered in two stages, the 'Tuberculo-opsonic' phase from 1902 to 1906 and the phase after 1906 when vaccine therapy was extended to a wide range of diseases. The development of vaccine therapy in the years before the First World War, both as a specific therapeutic innovation and as a symbol of 'laboratory medicine', are then assessed in turn.

ALMROTH WRIGHT AND THE ORIGINS OF VACCINE THERAPY[19]

Almroth Wright's early career was typical of the relatively small number of British medical laboratory scientists of his generation. It involved study abroad and several years of hand-to-mouth existence in short-term appointments. Many young medical men soon abandoned research for the security of clinical practice; Wright was one of the very few who followed a life-long research career, which continued into the

1940s. He obtained his MB from Trinity College Dublin in 1883 and went straight to Germany for two years, where he studied under Cohnheim, Ludwig and Weigert. On his return he continued research part-time at London University's Brown Institution, while studying for the Bar and working as a clerk at the Admiralty! In 1886, when C. S. Roy, the head of the Brown Institution went to the Chair of Pathology at Cambridge, Wright went with him as his assistant. Unable to obtain a permanent university position in Britain, in 1888 he went to the University of Sydney as Demonstrator in Physiology. He stayed for two years, before returning to London to work at the newly established laboratories of the Royal College of Physicians and the Royal College of Surgeons. In 1892, on the recommendation of the head of the laboratories, G. Sims Woodhead, Wright was appointed to the Chair of Pathology at the Army Medical School, Netley.[20] This bold decision soon produced results, though perhaps not the ones the Army Medical Service was expecting. Wright quickly turned his department at Netley into one of Britain's leading centres for medical research, with work on serology, bacteriology, immunology and tropical medicine.

At Netley, Wright established a 'style' of work which he followed until at least 1918. It had four main features. First, he sought to confront major disease problems, which in the Army at that time meant infectious diseases. In 1894 he supported Haffkine's work on a prophylactic vaccine for cholera and he developed this line of work to produce his own anti-typhoid vaccine. Later, at St Mary's, he worked on the major scourges of tuberculosis and pneumonia. He served on the Plague Commission in 1898 and did important, but again controversial, work in the First World War on wound infection. The second characteristic of his work was a commitment to active immunity and to biological rather than chemical approaches to the treatment of disease. He tended to stress the limitations rather than the advantages of the passive immunity gained from anti-toxic sera and he had little time for chemotherapy.[21] Thirdly, most of his research was done on human subjects rather than laboratory animals. Why this was so is not clear, but his biological approach may have led him to stress the specificity and uniqueness of immunological, and indeed all vital, phenomena. Though a licensed vivisector Wright, like others in his laboratory, increasingly avoided animal experimentation.[22] Lastly, and relatedly, Wright was an acknowledged master of laboratory technique. He invented numerous new procedures in clinical pathology, especially for studying the blood and immune responses. He also pioneered micro-techniques which allowed the rapid monitoring of a patient's condition and *in vivo* experiments.

This style of work first bore fruit with a prophylactic vaccine for typhoid. Its introduction at the beginning of the Boer War and the controversy which followed has been discussed many times.[23] The important points to note are that typhoid fever was the major problem of military medicine at that time; that he developed a killed-vaccine which produced active immunity; and that elaborate tests had established the best way to kill the bacilli, the optimum dosage and the effects of the vaccine. For a short period after inoculation the subject's resistance to typhoid was shown to be depressed rather than raised, but eventually resistance was enhanced; a pattern later referred to as the 'negative' and 'positive' phases. The failure of the Army readily to adopt his vaccine, together with other difficulties, led Wright to resign from Netley in 1902 and accept the post of head of pathology at St Mary's Hospital, London. Again, Wright turned an unexceptional department into a centre for medical research.

Before leaving Netley he had already published on the treatment of boils and other skin infections with a vaccine of killed staphylococci.[24] An explanation of the action of vaccines, preventive and therapeutic, in terms of the role of opsonins was published in the same year – 1902.[25] Wright supposed that vaccines worked by either directly or indirectly stimulating the activity of white blood cells. He proposed that this stimulation came from substances, produced by serum, which attached themselves to microbes and made the organisms more attractive for the white-blood cells to ingest. He termed these substances opsonins, from the Greek word for food relishes. Though this model was a significant theoretical innovation, linking humoral and phagocytic ideas, Wright never made much of this. In developing vaccines, Wright also invented techniques for measuring changes in the 'resisting or opsonic power' of the blood, which allowed the monitoring of patients' immune status and their responses to both preventive and therapeutic vaccines. These methods involved obtaining an index of serum levels of opsonins; the effects of the vaccines could thus be monitored by laboratory measurements of the 'opsonic index'. It was supposed at the outset that opsonins were specific – that is, different opsonic 'relishes' were required to stimulate the ingestion of different microbes. The practical implication was that vaccines should ideally be made from the actual bacilli causing the infection. Later these were known as *autogenous* vaccines, to differentiate them from *stock* vaccines, bulk-produced from bacteria collected from hospital patients. The opsonic index test involved comparing the anti-bacterial activity of normal blood with that from a patient.[26] The same bacterial source was mixed with the two samples, incubated and then counts were made of the number of bacteria ingested by white-blood cells in the two samples. Normal blood was given an opsonic index of 1.0, whereas those of patients suffering from infectious diseases invariably showed levels below 0.8. The procedure was technically difficult and time-consuming, but it was regarded as essential to monitor opsonic changes, not least because of the existence of the 'negative phase'.

THE EVOLUTION AND DIFFUSION OF VACCINE THERAPY

Vaccine therapy was initially developed on staphylococcal infections, but it was only in 1903, when it was extended to the treatment of local tuberculosis, that it attracted attention and was taken up by others. The treatment involved using tuberculin, a well-known 'cure' for tuberculosis which had been introduced by Robert Koch in a blaze of publicity at the 1890 International Medical Congress.[27] The timing, circumstances and background to the announcement insured that it received world-wide attention. Medical practitioners, not to mention patients, flocked to Berlin to learn of the new breakthrough. However, hopes were soon dashed; clinical results showed little advantage from the treatment, while in some cases there were serious side-effects and a worsening of the disease. It is usually suggested that these adverse reactions and subsequent criticisms led to the rapid abandonment of tuberculin therapy. In fact, tuberculin was not abandoned as a therapy; it continued to be used, alongside its now better known use as a diagnostic skin test for tuberculosis, until the 1920s.[28] During the 1890s and early 1900s many new therapeutic tuberculins were produced, including Koch's own New Tuberculin, and even some tubercular anti-toxic sera. For a long time the formulation of tuberculin was shrouded in mystery, but it eventually emerged that it was a killed vaccine. Given Wright's interests it was not surprising to find that he turned to the treatment of localised tuberculosis and the use of tuberculins. Tuberculosis was considered the major disease problem of the age and while most attention was focused on the pulmonary form, localised forms were prevalent and serious. Also, Wright now had an explanation of the earlier failure of tuberculin. First, it had not been recognised that consumptives were immunologically vulnerable because of their depressed opsonic indices.[29] For the tuberculous, then, all but the smallest doses of tuberculin would be dangerous; the adverse reactions were now explained as being due to repeat doses having been given in the 'negative phase'. Armed with this knowledge, Wright planned new methods of tuberculin therapy; beginning by establishing the opsonic index and building this up with injections of minute doses of tuberculin.

This work was reported in 1904 and 1905 and was the subject of a special meeting of the Royal Medical and Chirurgical Society in November 1905.[30] It attracted considerable medical and public interest, for as Bulloch remarked at the meeting, 'at the present time ... the eyes of the civilized world scan the scientific horizon with anxiety for the long-expected "cure for consumption"'.[31] This new endorsement of tuberculin therapy was taken up by practitioners treating pulmonary as well as nonpulmonary forms of the disease. New tuberculin trials were reported from the Mount Vernon Hospital, the Royal National at Ventnor and the Brompton Hospital, London; there was also a re-assessment of the work of those, like R. W. Philip in Edinburgh, who had remained faithful to tuberculin.[32] Many physicians adopted the treatment and voluntary hospitals began to admit tuberculous patients, often for the first time.[33] For example, in January 1906, Lorrain Smith, Professor of Pathology, suggested to the Medical Board at Manchester Royal Infirmary that Wright's new treatments ought to be made available in Manchester.[34] In 1906 attention was also drawn to the treatment by George Bernard Shaw's *The Doctor's Dilemma*, in which the principal character, clearly based on Wright, developed a new cure for tuberculosis.[35]

Alongside the work on tuberculins, though initially less well publicised, vaccines were being developed for chronic local infections. This work differed from that on tuberculins in that it did not use a standard or commercially prepared product; instead autogenous vaccines were made with bacilli isolated from each patient. As well as preparing the vaccine, the laboratory had to provide for its standardisation so that carefully controlled inoculations could be given. These procedures, together with the measurement of opsonic indices, meant that the true practice of vaccine therapy involved extensive and intensive laboratory work.[36] The novelty of the techniques, together with the requirements of time and equipment, meant that the innovation was slow to diffuse from the relatively small group of workers associated with St Mary's. Most doctors who used the treatment at this time learnt it by working at St Mary's or with one of Wright's disciples.[37]

Vaccine therapy was soon featured in all forms of medical communication. The *Medical Annual* carried a review article in 1907 and the May 1908 edition of *The Practitioner* was devoted entirely to vaccine therapy.[38] In 1907, Ronald Allen, then at Guy's, published the first textbook on the treatment.[39] This was reprinted in 1908, 1910 and 1912. Other books followed, mostly by followers of Wright at other London teaching hospitals: J. W. H. Eyre (Guy's), R. Hewlett (St Bartholomew's), W. d'Este Emery (King's), amongst others. The treatment was also widely discussed at medical meetings. Most of the major London and provincial medical societies received papers on the subject.[40] The 1909 Annual Meeting of the British Medical Association had a new Section for 'Haematology and Vaccine Therapy' and in 1910 the treatment was discussed at the Royal Society of Medicine (RSM) for six sessions. In 1912, Allen founded the *Journal of Vaccine Therapy*.[41]

As reports of favourable results flowed out of St Mary's, money flowed in. In 1907 a Department of Therapeutic Inoculation was established, independent of the Pathology Department and eventually the hospital itself, with £17 000 raised by subscription.[42] It was overseen by a committee chaired by Arthur Balfour and contained other senior politicians and businessmen. Teaching fees of £1000 per annum were anticipated. The Department began to attract medical men and patients from throughout Britain and from abroad; lucrative private practices also grew. In its first year the department treated 550 patients, had a salaried part-time staff of fifteen and soon demanded larger facilities. In April 1909 it took over thirty-one beds in the Clarence Wing of the hospital; these 'research wards' were probably the first formal clinical research facility in England. The most important source of revenue was a contract with the Parke Davis Company, who from February 1908 were to be sole agents for stock vaccines produced at St Mary's.[43] With these vaccines, doctors without immediate access to laboratory facilities could adopt the new treatment for common infections. Initially staphylococcal, streptococcal, gonococcal, tubercle and typhoid vaccines were produced; a year later these were supplemented by two vaccines for acne and one for cancer.[44] Some indication of the workload and resources available is given by the fact that each year the Inoculation Department offered around sixteen assistantships to students; each post was worth £100 per annum – a steady stream of disciples thus ensued.[45] Other pharmaceutical companies were not far behind Parke Davis. By 1910, vaccines were also available from companies, like Allen and Hanbury's (who marketed vaccine produced at the Lister Institute of Preventive Medicine), Burroughs and Wellcome, W. Martindale (Wimpole Vaccines) and the Roborat Company.[46] They were also made by the Clinical Research Association (CRA), a London-based commercial agency, which provided mail-order pathological and laboratory services to general practitioners and some local authorities.[47]

In succeeding years most of the major London hospitals established similar departments or facilities for vaccine therapy. At Guy's, work had begun with tuberculins in 1904 and a Vaccine Department was formally constituted in February 1908.[48] St Thomas's opened an Out-Patient Clinic for vaccine therapy in 1907 and later separated its Vaccine Laboratory from other clinical pathology work.⁴⁹ The Westminster opened a Department of Bacterio-Therapeutics in 1909. St Bartholomew's, the London and the Middlesex did not create special departments, though their pathologists did do vaccine work.⁵⁰ Vaccine therapists were also employed at specialist hospitals, like St John's Hospital for Diseases of the Skin.⁵¹ In the provinces and in Scotland the treatment was not institutionalised to the same degree. Only in Manchester and Liverpool were there formal departments. At the Manchester Royal Infirmary work began with tuberculins in 1906 and expanded subsequently.⁵² The Vaccine Department at the Royal Southern Hospital in Liverpool was opened in 1909.⁵³

An important factor in the successful diffusion of any innovation is the education of users to change their practice, or adapt the innovation to existing practices. Alongside formal and informal medical channels of communication, companies like Burroughs and Wellcome, tried to educate users by producing booklets which were both promotional and instructive. Allen and Hanbury's eased the technicalities by colour coding the different dilutions of vaccines. Their success is difficult to judge; the companies certainly thought they were worthwhile, for their booklets were revised and re-circulated annually.⁵⁴ As it diffused, the practice of vaccine therapy was altered significantly. First, few practitioners bothered to monitor with opsonic measurements; most claimed that body temperature was as good an indicator, or that the index was unreliable and a waste of time and money.⁵⁵ The technical difficulties of the test must have been a bar to its use; its accuracy was also criticised.⁵⁶ Secondly, and relatedly, stock vaccines began to compete with autogenous vaccines; the advantages here were again time and money. Stock vaccines could be bought directly from suppliers, often by mail-order, and came in standardised doses ready for immediate use. One report in 1913 cited five guineas as a price for an autogenous vaccine, as against 1s.6d. to 5s in the Wellcome Catalogue for an equivalent stock vaccine. Thirdly, the therapy was extended from a remedy for specific infections to a more general treatment for any or unknown infections. This was most notable in the production of a third type of vaccine – mixed or polyvalent vaccines, sometimes called 'phylacogens', for non-specific infections. These contained strains of many common bacteria and offered what one contemporary termed a 'blunderbuss' approach to treatment. Vaccine therapy was now passing from the initiates at St Mary's to the untutored physician or general practitioner, who seemed unable or unwilling to indulge in the labyrinthine world of 'true' vaccine therapy.

There is little doubt that from 1906 onwards there was a considerable medical vogue for vaccine therapy. A survey of metropolitan hospital laboratory work in 1912-13 showed that at Guy's and St Bartholomew's it accounted for about 10 per cent of all laboratory specimens. At Guy's in 1910, every consultant used vaccines – a total of 269 in-patients, or an average of one in twenty, were so treated.[57] In 1913, H. W. Batty Shaw of University College Hospital, observed that recent years had seen 'the spectacle of the appearance at almost every hospital of a new sort of member of staff – the vaccinist or inoculator'.[58] It has not been possible to confirm this claim from hospital Annual Reports or lists of staffs, as there is no evidence of the appointments of specialist 'vaccinists'. What Batty Shaw was referring to was probably the appointment, as medical officers, of young enthusiasts fresh from the Inoculation Department.

For hospitals other than St Mary's and Guy's, it is difficult to sustain the suggestion of Foster that vaccine therapy was the mainstay of clinical laboratory work. The only numerical evidence he cited came from the growth of work in one hospital, St Thomas's.[59] However, my own analysis of the work of this laboratory over the period 1896–1914 shows that the increase in work began before the advent of vaccine therapy and was based on a wide variety of work.[60] By 1912, the department, which was based in the Medical School, had a vaccine therapy section, but also sections for chemical pathology and research.[61] Indeed, that the growth in demand for laboratory services was broad-based can be shown for St Mary's itself, where in 1908 a Clinical Investigation Department was established to cope with work in morbid anatomy and pathological chemistry being neglected by Wright and his colleagues. The 1912-13 survey for the Local Government Board showed that at the London and Westminster Hospitals only 2 per cent of laboratory specimens were for vaccine work. There is then little quantitative evidence to support claims that vaccine therapy was decisive in the general growth of hospital clinical laboratories before 1914.

Vaccine therapy was, however, a development everyone mentioned when discussing laboratories. Thus, the *Annual Report* of the St Thomas's laboratory in 1909 noted:

> The bacteriological investigation of acute and chronic infective diseases and the treatment of these infections with suitable vaccines made in the laboratories occupies a considerable amount of time. A regular out-patient department was established in 1907, the patients coming to the laboratory for treatment. The total number of patients who have received treatment is about four times as great as in 1907.[62]

The statement was less than honest, for a summary of the work done by a Grocer's Research fellow, based in the Department and published in the same *Report*, says that stock vaccines were used and that opsonic measurements were not taken.[63] The 1912-13 survey noted, 'In the last few years ... work has very greatly increased in the light of the increasing knowledge of bacteriology and its applications in the treatment of infective disease'.[64] Two features of vaccine therapy seem to have created an exaggerated sense of its importance at the time and in the accounts of Foster *et al*. The first was its 'striking novelty' and the 'evangelicalistic fervour' with which it was promoted.[65] The second was that the technical and clinical nature of the work ensured that it was performed or supervised by the senior pathologist and not delegated to junior staff. Indeed, many pathologists built up private practices and reasonable income from the therapy, using the laboratory facilities at their hospitals.[66] The retrospective views of Dyke, Colebrook and Foster may be clinical pathologists remembering what they did, which was not necessarily the same as what their laboratories were doing.

How widely the therapy was used in private practice is very difficult to judge. Some vaccinists, like R. W. Allen, gave up hospital posts for full-time private practice.[67] In Manchester, the pathologist Edward Loveday combined hospital work with a lucrative private vaccine practice.[68] Evidence on the use of vaccines amongst general practitioners is also difficult to obtain. Impressionistic evidence, from advertisements in journals and annual publications, suggests that vaccine therapy was not something general practitioners were expected to use. Income from tuberculins and stock vaccines produced at the Lister Institute and sold mainly to general practitioners shows a steady rise, from under £200 in 1905 to nearly £1600 in 1911. Yet by comparison, in the same year the Lister's income for investigations and diagnostic tests was £4808, and for anti-toxins £9581.[69] Despite these figures and typical of the hype which surrounded the treatment, in 1910 the *Chemist and Druggist* reported that amongst the old school of pharmacists there was 'apprehension that ... drug treatment would be superseded by vaccine treatment and their occupation would be gone'.[70]

Such worries proved unfounded. The evidence of income at the Lister Institute shows that tuberculin and vaccine sales peaked in 1911. A similar pattern is discernable in most metropolitan hospitals. Attendances at the out-patient clinic at St Thomas's declined after 1909, while at Guy's the number of in-patients treated fell after 1910. The number and frequency of publications on the treatment also fell after 1910; indeed, the *Journal of Vaccine Therapy* folded in 1913, after only six quarterly issues.[71] All this is further evidence against vaccine therapy having been the motor of the growth of clinical laboratories before the First World War. It would be wrong, however, to conclude that vaccine therapy was a 'failure' by 1914. It continued to be practised during and after the First World War. Macfarlane and Hare have recently shown that Alexander Fleming's main job at St Mary's in the 1920s was as vaccine production manager.[72] In 1931, Parke Davis were still marketing thirty-three different types of vaccine prepared at St Mary's. In the same year it published the 17th edition of its annual volume on vaccine therapy, which was claimed to have gone through 150 000 copies.[73] Details of the development of the treatment in the 1920s and 1930s have been discussed by Keating, though indicators of its prevalence remain elusive. After 1918, its leading proponent, after Wright, was Fleming, who in the late 1930s was arguing for combined antibiotic and vaccine treatment.[74] The Second World War was to make Fleming famous for his association with penicillin, but his last pre-war publication was entitled 'Recent advances in vaccine therapy'.[75]

VACCINE THERAPY AND LABORATORY MEDICINE

The evidence presented above suggests that vaccine therapy, if judged by the extent of its practice and duration of popularity, was a significant, though minor, medical innovation. However, the treatment was promoted as more than a single innovation, it was a new kind of medicine. In 1905, Wright had attracted considerable attention with an article in the *Liverpool Daily Post*, entitled 'The World's Greatest Problem', which attacked his fellow doctors for their neglect of research and for seeming to accept their ineffectiveness in combatting infections.[76] Throughout the article, the implicit model of the value and nature of medical research was vaccine therapy. In 1907, Wright made this explicit in another public statement, this time on nothing less than 'The Future of Medicine'. Here he noted that 'opsonins were ushering in a new era of therapeutics' and went on to make his claim about the physician of the future being an 'immunizator'.[77] In 1907, R. W. Allen made similar claims.

> The medicine of the future is the medicine of vaccines and of sera. The empiricism of the past will give way to methods based upon scientific knowledge and the public will no longer look upon medicine with a sceptical eye and dose themselves with ineffective nostrums.[78]

This missionary zeal was also common amongst other St Mary's-trained vaccinists. Throughout the years 1900–10 Wright consistently attacked all branches of the medical profession, especially physicians and surgeons, for their ignorance and ineffectiveness.[79] He described himself 'as the ill bird whom a sense of duty impels to foul its own nest'; a renowned 'controversialist', Wright was seemingly bent on confrontation.[80] He was one of the few laboratory scientists of sufficient stature to take on clinicians. He was a world-renowned scientist and he enjoyed considerable independence, due to the financial and institutional support he had marshalled for his work. Wright, his followers and his supporters saw vaccine therapy as nothing less than the vanguard for the reconstitution of medicine, leading it to be more scientific and research-based, with the laboratory its central agency and symbol.

Studies of the late Victorian and Edwardian marriage of medicine and science now recognise that this union was not without its problems.[81] It is now argued that the reaction of many clinicians to science, the laboratory and technology was ambivalent, even hostile.[82] It is certainly possible to find evidence of differences of opinion over the relative merits of clinical signs and symptoms, as against laboratory tests. However, it seems that this issue was more in the nature of an underlying tension than a conflict. This was because the two 'sides' – bedside and bench – were not that distinct, nor were they evenly matched. Most medical bacteriologists were clinically trained and held only part-time appointments. A significant number of physicians, surgeons, general practitioners and state medical officers dabbled in bacteriological technique and speculated on aetiologies. An indication of the extent to which laboratories were used is that in the early 1900s over 4000 doctors, approximately one in five, subscribed to use the diagnostic laboratories of the CRA.[83] In most settings, including the hospital and in public health, bacteriological diagnosis was normally described as an adjunct to clinical methods.[84] Besides, clinicians were far too powerful to be challenged by the lower-status and usually more junior staff in the laboratory. In some universities, bacteriological laboratories were run by senior and often distinguished scientists; however, these facilities were often isolated institutionally and had few links with, let alone influence on, clinical teaching or practice.[85] With regard to laboratory diagnosis, it was clinicians who requested tests in the first place and who interpreted and selectively absorbed the findings. This view has been supported by the work of Howell and Jacyna, who show that before about 1910 laboratory results were rarely seen as authoritative.[86] Clinicians were also – in many cases, rightly – concerned about how samples were taken, the quality and competence of laboratory personnel and, hence, the reliability of the reports they received. There were also practical difficulties in the way of using results – they took time to be produced and when they were, the options open to clinicians remained for the most part unchanged, at least until vaccine therapy arrived.

Several features distinguished vaccine therapy from other bacteriological and laboratory activities. The position and character of Wright was undoubtedly important. Wright was a international scientific celebrity; it was said that his laboratory and clinic was 'a place of scientific pilgrimage'.[87] Through contacts with Ehrlich, St Mary's enjoyed a near-monopoly of the earliest supplies of Salvarsan in Britain, even though Wright himself had no time for this latest German chemical remedy. Arguably, it was Wright's promotion of tuberculin treatment, which came to be seen as a special case of vaccine therapy, that led to the inclusion of medical research in the 1911 National Insurance legislation.[88] In the House of Commons, it was Arthur Balfour who argued most strongly for state support of medical research into tuberculosis and against the whole anti-tuberculosis effort going to the sanatorium treatment. In fact, it was Wright's other powerful ally, Lord Moulton, who was the first chair of the MRC.[89] However, the main difference between vaccine therapy and other clinical laboratory work was that it was 'therapy', and as such directly encroached on the domain of the clinician.

Many of the issues surrounding the treatment, and its implications for medicine in general, were aired at two sessions at the Royal Society of Medicine (RSM) in 1910 and 1913. At both meetings vaccine therapy was said to have been 'on trial'. Wright began the first meeting with an address attacking clinicians' ignorance of bacteriology and of medical science in general. He pointed out that the most celebrated achievement of modern medical science, antiseptic surgery, had been adopted as a mere technique and that most clinicians still failed to understand the underlying bacteriology. A similar fate had befallen diphtheria anti-toxin:

> The medical man who had divested himself of all bacteriological work found here, to his joy, that it would be practicable for him to get his diagnosis from one bacteriological laboratory and his remedy from another, and that there would still remain for himself a dignified role as a middleman between the patient and the bacteriologist... The medical man even dreamed of polyvalent sera that would make everything in the nature of minute bacteriological diagnosis superfluous.[90]

These superficialities, he argued, could be tolerated no longer, for with vaccine therapy the laboratory had overtaken the clinic. Clinical methods of diagnosis were said to be *post hoc* and treatment 'hit and miss'. Against this, the new vaccinist-bacteriologist could offer exact diagnosis, even of incipient disease, and the precise targeting and monitoring of treatment. He went on to complain about the subordinate service position of the bacteriologist when his knowledge was now superior to that of the clinician. 'It is not at this hour of the day, arguable that the verdict of a bacteriologist stands in need of confirmation from a clinician. In blunt language, "the boot is on the other foot".[91]

Wright argued that every medical practitioner would have to become a clinician-bacteriologist; only then would they understand the true causes of disease, be able to interpret laboratory findings critically, and use the new therapeutics effectively. This ideal may have been achieved at St Mary's, but at Guy's and elsewhere vaccine therapy had been absorbed into existing clinical practices; a radical innovation had been turned into an adaptive one. Wright's calls for the restructuring of medical practice around the laboratory thus already rang somewhat hollow.

Apart from Wright's opening address, the discussion at the RSM in 1910 was not on relations between bench and bedside, but on clinical concerns – the mechanics of the treatment, and did it work? Wright and his supporters defended opsonic indices as reliable and necessary, though it was accepted that most practitioners used temperature to monitor effects. They also stressed the need for laboratory diagnosis and the manufacture of autogenous vaccines. Again it was well known that the choice of vaccine was often based on clinical signs and that stock vaccines were widely used. To the charge that the treatment worked only for a few chronic and relatively trivial diseases, the St Mary's workers replied with a long list of conditions said to benefit: pneumonia, typhoid fever, rheumatism, endocarditis, tubercular infections, dermatitis, tooth abscesses, hay fever, urinary calculi, food poisoning, epilepsy and the complications of cancer, gonorrhea, influenza, whooping cough, catarrh and glycosuria.[92] Most supporters gave detailed and optimistic case reports, whilst the relatively few critics present commented on the dearth of published papers, the absence of animal experiments and the ambiguities in those studies that had been published.

One clear problem was the absence of any agreed basis for evaluating new therapies. The absence of animal studies at St Mary's was commented upon, but nothing more. Wright, like almost all of his contemporaries, looked to accumulating clinical evidence, not to clinical trials or statistics.[93] It should not be forgotten either that the ethical implications of the treatment had been aired in *The Doctor's Dilemma*. The cautious endorsement of vaccine therapy by Rufus Cole, one of America's leading biomedical scientists is instructive.

> Clinical experience, however, is fallacious and critical judgement must be brought to bear upon the results ... One needs not only to review the

German literature of 1891, following the publications of Koch on tuberculin, to see how even medical men may at first be deceived as to the value of a remedy. We believe, however, that sufficient clinical evidence has been presented to make it more than justifiable to employ this method, but we should guard against too hasty conclusions.[94]

Part of the problem was that new treatments tended to be tried only when other, more established, practices had failed. The kind of chronic local infections on which vaccine therapy concentrated tended to wax and wane spontaneously, adding a further complication to assessments of efficacy. It was also said that there was patient demand for vaccines and hence that they had been misused.[95] It was suggested, therefore, that disappointments were in part due to promiscuous use of vaccines, both as regards technique and in unsuitable cases. Many doctors had been critical of Wright's association with an industrial company, both because of their dislike of 'trade' and because it was felt to compromise his objectivity. A common observation was that vaccine therapy, like many other therapeutic innovations, had been seized upon as a panacea and used indiscriminately, but that its real value, in skilful hands, was now emerging. In all, vaccine therapists themselves thought the 'verdict' at the RSM had gone in their favour.

Whilst Wright's rhetoric stressed the high scientific credentials of vaccine therapy, its practice suggested a different pedigree. The problem was summed up in two contemporary labels given to Wright: was he the 'British Pasteur' or a 'Celtic Siren'?[96] This ambivalence is evident in the attitude of the staff of St Mary's; they recognised that he brought world-renown to their hospital, with what they called his 'Find-a-bug-then-kill-it' Department, but apparently few bothered to find out what went on there. Also, the treatment was attractive to the medical fringe. The fact that it treated like-with-like and used minute doses interested homoeopaths; in 1907, the *British Homoeopathic Review* reported favourable results at the London Homoeopathic Hospital.[97] Furthermore the treatment did not offend anti-vivisectionist sentiment; unlike serum therapy, it did not use animal products and its research involved little or no animal experimentation. Wright was also a friend of George Bernard Shaw, who was also an outspoken critic of the medical profession.

The debate at the RSM had revealed other tensions within the medical community. There were suggestions that the treatment was seen by some as offering physicians the means to claim back from surgeons certain diseases which the latter had taken over in recent decades.[98] Vaccine therapy implied that infections had to be combatted by systemic treatments and Wright argued that neither the knife nor antiseptics could ever remove all infective material.[99] However, leading surgeons, like Arbuthnot Lane (Guy's) and C. B. Lockwood (St Bartholomew's) were enthusiasts and thought anti-streptococcal and anti-staphylococcal vaccines might strengthen the battle against wound infection by providing internal asepsis. Vaccine therapy was predominantly metropolitan and there are suggestions of resentment amongst provincial practitioners about patients wasting their money on expensive and fashionable 'London cures'.[100] It was clear, too, that vaccines could be seen as challenging public health campaigns, like that against tuberculosis. One campaigner noted sarcastically: 'No need now for fresh air in the treatment of tuberculosis: London slums would serve, provided tuberculin and the opsonic index were available'.[101]

Another tension highlighted was generational, between younger doctors trained in the new laboratory techniques and their older, more clinically-oriented, colleagues.[102] Bacteriology as a laboratory subject had been introduced as a formal part of the medical curriculum only in the late 1890s, so it was only from the early years of the century that Britain had graduates conversant with bacteriological techniques. The fact that Wright attracted so many young medical practitioners to his department, producing 'enthusiasts', did nothing to dampen the scepticism of older medical men.

The RSM met to discuss vaccine therapy again in 1913. This time a different verdict was delivered,

> Failures ... are more common than successes, and though there is no doubt about the efficacy at times of vaccine treatment, still in the mass the results are disappointing. This seems to be damning with faint praise; but we may put it less harshly, and say that the early enthusiasm and hopes excited have been tempered by experience.[103]

What had brought about this revision? There had been no sudden increase in critical reports of vaccine therapy. These had always existed – what had changed is that more notice was now taken of them. In part, it was because there were different speakers and a different audience. A complaint in 1910 was that the opponents of vaccine therapy had not turned out, three years later they did. The formidable personality of Wright no longer had to be faced; he had ceased actively to promote vaccine therapy, being occupied with his pneumonia work in South Africa, the introduction to which was being drafted as an ambitious discourse on 'The Logic of Medicine'.[104] Opponents may have gained confidence from the fact that the uses of therapeutic vaccines had become more circumscribed and that medicine had not been reshaped. A more modest estimation was being reflected in the literature, and it has been shown earlier that there had already been a levelling off and decline in the sales of vaccines after 1910. Certainly, the indeterminate effects of vaccines contrasted with the obvious effects, intended and unintended, of the new chemotherapeutic agents, like Atoxyl and Salvarsan. These shifts reversed the roles at the RSM by 1913, vaccine therapists were now on the defensive; significantly, though, the critics spoke out against the style of vaccine therapy as much as its results.

At the 1913 meeting, Thomas Horder, himself the author of a textbook on clinical pathology, admitted to having been 'hustled by those who at one time accused him on gross neglect' in not using vaccines.[105] Others spoke of this '"cocksure" method of treatment [giving] little or no benefit' and being 'futile and sometimes even disastrous'.[106] Batty Shaw, in a bitter attack on vaccine therapy, delivered at Norwich in October 1913, pointed to the widely-held belief that vaccines had 'improved the market value of the physician', imputing motives and assessments. He went on that the new hospital vaccinist was a officer who had:

> without exception very small clinical knowledge, in many cases he contributes nothing to the study of pure bacteriology, and he spends the greater part of his time putting to confusion the remainder of his colleagues, by a wholesale submission of them to 'terminological inexactitudes' about raising the power of resistance.[107]

This echoed an earlier remark of Horder's, warning of:

> The pseudo-bacteriologist [who], too often bearing the hall-mark of correct training, with his glib and parrot-like repetition of all the master's phrases. This imitation was pardonable; but he spoke of 'streps.' and 'staphs.' and 'gonos.' and of 'coli infections', as though, holding such a sway over life and death, there was now no time for ordinary language, nor need for any grammar; knowing the innermost secrets of disease whilst as yet his contemporaries were perusing a temperature chart or undertaking a blood-count. Here was a spectacle little likely to impress the clinician.[108]

It is significant here that Horder terms the vaccine therapist – 'the pseudo-bacteriologist', differentiating them from other bacteriologists and clinical pathologists – who were the acceptable face of laboratory medicine.[109]

CONCLUSION

There is little doubt that vaccine therapy was one of the most widely debated areas of medicine in the years before the First World War. Although it has not been possible to estimate accurately the extent of its practice in hospitals and in general practice, it would seem that this was not as large as claimed by Colebrook, Foster and Dyke. The attention it received seems to have been out of proportion to the scale of the innovation, as measured by numbers of practitioners, patients and the diseases treated. Thus, it has been argued that it was not the innovation as such that was significant, but the kind of innovation it was, and the way it was promoted. It has been argued that vaccine therapy was a potent symbol of the new system of laboratory medicine and its culture, representing an ideological as well as a technical discontinuity. It challenged the existing order of medicine in a way not done by previous laboratory-based innovations. From this perspective, it may be suggested that any conflict between bedside and bench in Edwardian clinical pathology was over bacterial therapeutics, not diagnosis.

That the long decline of vaccine therapy began as early as 1910 can make it difficult to take seriously the claim of the innovators of vaccine therapy that it represented 'the future of medicine'.[110] Equally, the much-maligned Almroth Wright now seems an unlikely figurehead for twentieth century British medical science. Yet, the history of vaccine therapy shows that both were possible trajectories for a short period around 1910. The tendency of medical historians to chronicle only 'successful' innovations and ignore those now deemed 'failures', has meant that this important innovation in Edwardian medicine has been overlooked. More generally, I have argued that innovations cannot be understood simply as changes in techniques or ideas. The adoption, adaptation or rejection of radical innovations, which is what vaccine therapy was initially, causes social conflicts and cultural change.

6
FROM THE TRENCHES TO THE HOSPITALS AT HOME: PHYSIOLOGISTS, CLINICIANS AND OXYGEN THERAPY, 1914–30

Steve Sturdy

Before the First World War, oxygen played only a limited role in medical treatment. It was widely recommended in the operating theatre as an adjunct to anaesthetics like nitrous oxide and chloroform, but it does not appear to have been so widely used in this way: more commonly, it was kept on hand for resuscitating patients who had been given too much anaesthetic. Outside the operating theatre, it was most frequently used as a means of relieving patients in the last stages of pneumonia. In none of these cases, however, was it clear that oxygen had any therapeutic value as such: it might help sustain life when anaesthetic gases were administered, especially when those gases themselves inhibited normal breathing; but there was little reason to suppose that it could actually cure disease. 'It is doubtful whether the inhalation of oxygen in pneumonia is really beneficial', declared William Osler in his authoritative *The Principles and Practice of Medicine*. 'Personally, when called in consultation to a case, if I see the oxygen cylinder at the bedside I feel the prognosis to be extremely grave'.[1]

If the place of oxygen at the bedside was uncertain, however, it was securely established in the medical laboratory, where research into the fundamental phenomena of breathing was avidly pursued by respiratory physiologists. Leading British scientists like J. S. Haldane, Leonard Hill, Joseph Barcroft and A. V. Hill had already done much to elucidate the rate of oxygen consumption, the regulation of breathing rate and the physiology of blood gas transport. But this work was largely confined to the laboratory. Physiologists usually worked in medical schools, but their research and teaching was generally restricted to the so-called 'pre-clinical' sciences; rarely were they given access to patients for research purposes, while the substantive relationship between the kind of systematic biological science they taught and the actual practice of medicine remained tenuous.[2]

Late nineteenth century physiologists generally seem to have colluded at this demarcation; independence from medical practice gave them the space they needed to further their research careers and to shape their professional identity as scientists. But by the early twentieth century, physiologists were beginning to look for a greater involvement in the advancement of medical practice. Doctors, jealous of their own professional identity, resisted such moves; by defining themselves as practitioners whose skills and interests differed from those of physiologists, they maintained the boundaries that separated the laboratory from the bedside.[3] Physiologists were left with little opportunity to influence the practice of medicine. Their work on oxygen exemplifies this impotence. Unlike practising doctors, respiratory physiologists clearly thought that the gas could be used to good effect in therapy. But without access to patients, they could do little to pursue their ideas beyond publishing the results of their research and their speculations.[4]

The situation changed markedly in the course of the First World War. By 1918, a number of physiologists had obtained access to patients with respiratory and other problems, and were looking specifically at the possibility of using oxygen as a therapeutic agent in a wide variety of diseases. Out of this work grew the Medical Research Council's Clinical Uses of Oxygen Committee, which eventually established the value of oxygen as a standard therapeutic agent in a wide range of respiratory disorders.

The story of these developments is worth telling in some detail. The growth of physiological research into oxygen therapy is of interest, not so much in its own right as because it exemplifies a much more general reorientation of medical science that took place during and after the war. The movement of laboratory physiologists into the sphere of clinical medicine at this time provided an opportunity to establish both an institutional base and a corpus of knowledge for a new scientific discipline called clinical science. A study of oxygen therapy throws light on some of the historical circumstances that made this possible. In particular, it shows how the war not only gave rise to novel medical problems, but also created a social and political milieu in which, for the first time, scientists could be seen to provide effective solutions to such problems.

That is not to say that the development of new methods and techniques of administering oxygen is irrelevant to this story. I will argue, however, that any attempt to write a purely technical account will fail to explain the career of this medical innovation. Even at the most basic level of evaluation, there were no simple technical criteria for judging the therapeutic efficacy of oxygen: such judgements depended so heavily upon previous experience of laboratory science and clinical practice, and upon expectations about how patients should be treated, that scientists and clinicians formed widely divergent views of the new therapy. In effect, the technique of oxygen therapy was incompatible with the needs and practices embodied in the prevailing culture of bedside medicine. On the other hand, it clearly commended itself to a very different culture – the culture of wartime medical administration – within which scientific medicine in general, and oxygen therapy in particular, were embraced for the solutions they offered to administrative problems.

The history of oxygen therapy, then, is inextricably bound up with the formation and consolidation of this scientific administrative culture. Insofar as this essay is a study of innovation, it is above all an account of social and cultural innovation. And while technical considerations undoubtedly have their place in the story, they are of less interest – and, indeed, of less historical importance – than changing attitudes towards science and technology, particularly in the hierarchies of British government.

PHYSIOLOGISTS AND GAS WARFARE

Before the advent of large-scale gas warfare in April 1915, physiologists had little direct involvement in war-related scientific research. But the devastating effects of the German attacks of 22 and 23 April quickly changed that: it was clear to those in authority that scientific knowledge of the effects of poison gases might be of value in developing suitable defences. J. S. Haldane was the first physiologist to become involved. On 24 April he was summoned to the War Office by Kitchener, and the following day left for France with Herbert Baker, Professor of Chemistry at Imperial College.

At the time, little could be done for gas victims other than providing general nursing care.[5] Besides, in view of the effectiveness of gas as a weapon, strategic priority was given to protecting those soldiers still capable of fighting. To this end, physiologists soon began studying the poisonous effects of gases and the development of suitable respirators. This was at least partly due to scientists' own efforts to become involved in war work. Shortly after the outbreak of hostilities, the Royal Society had established a War Committee to co-ordinate and represent scientific interests, and in November 1914 a Chemistry Sub-Committee was set up to discuss the production and supply of explosives and pharmaceuticals. On 30 April, in the wake of the gas attacks, this Sub-Committee requested that Haldane be added to their number, and on 12 May Haldane and another physiologist, E. H. Starling, were duly appointed.[6] By mid-June, the contributions of physiologists were beginning to gain official recognition; at the urging of Alfred Keogh, the scientifically-minded Director-General of the Army Medical Service, the Royal Society established a separate Physiology (War) Committee to co-ordinate the researches on a wide range of poison gases being performed in laboratories around the country.[7]

Meanwhile, a number of physiologists had been given official appointments in Army establishments: S. Lyle Cummins was appointed to the Gas Services Central Laboratory in St Omer;[8] C. G. Douglas studied gas victims in the same laboratory and in field hospitals, primarily to identify any new gases which might be used;[9] and late in 1915 Starling was appointed Director of Research at the Royal Army Medical Corps (RAMC) College at Millbank, to take charge of respirator development and further research into the nature of gas poisoning.[10] Later, as the British expanded their own chemical warfare capabilities, a team of physiologists was engaged to carry out both offensive and defensive research at the Experimental Station at Porton: in March 1917 Joseph Barcroft was given charge of the physiological laboratory, where he was subsequently joined by R. A. Peters, A. E. Boycott, J. Shaw Dunn and G. H. Hunt.[11]

Physiologists were frustrated by the subordination of medical to strategic views of the problems of gas poisoning.[12] But in spite of the constraints placed on them by the demands of war-fighting strategy, a number of scientists were able to perform at least preliminary investigations into possible ways of treating gas victims. While in France after the first attacks, Haldane briefly tried giving oxygen to gassed soldiers, and found that it relieved their condition. Subsequently, he urged that further studies should be made of the beneficial effects of oxygen.[13]

Other lines of therapeutic research were pursued under the auspices of the Medical Research Committee (later Council) (MRC), newly established in 1913. Medical scientists, especially physiologists, were strongly represented on the Committee, and were keen to promote clinical research into the causes and treatment of illness. So far, they had been unable to do much in this direction, but the medical problems of the war, in particular gas poisoning, provided just the kinds of opportunities they sought. From 9 to 17 May 1915, and again from 26 May to 26 June, Joseph Barcroft carried out physiological studies of gas victims at a base hospital in Boulogne, in an attempt to explain their symptoms and to suggest possible forms of treatment. In particular, he examined the chemical properties of their blood: it appears that he suspected a disruption of the acid-base balance, which could be treated with sodium bicarbonate injections, but his studies proved fruitless.[14] Despite this failure, Barcroft recognised the important precedent being set by his research. He expressed his excitement in a letter to his wife: 'This, I imagine, is one of the days which I will always remember. A European war is a rare enough event; the problem in this war of gas poisoning is unique; for a civilian to be at the official discussion of such a question is I suppose absolutely unknown'. But he confessed that he had been embarrassed at having to 'get on his pins and talk "gases"' to clinicians like Sir John Rose Bradford, Sir Wilmot Herringham, Sir John Atkins and Sir Almroth Wright, and admitted: 'I cut it pretty short!'[15]

In spite of his physiological interest in gases, Barcroft does not seem to have considered the possible therapeutic use of oxygen. The next step in that direction was taken by Leonard Hill, now full-time Director of the MRC's Department of Applied Physiology. Hill, like Haldane, believed that oxygen might be beneficial to gas victims, and by late 1915 had designed a mask through which it might be effectively administered.[16] Though Hill visited France to demonstrate the new apparatus,[17] however, his design did not find its way into regular practice. Instead, the impetus which eventually led to the development of a practical form of oxygen therapy came from a quite different quarter.

Some time in 1916, Haldane heard from an old friend, John Cadman,[18] Professor of Mining at Birmingham University. Cadman was involved in the development of new gas weapons and wanted Haldane to design an oxygen apparatus for the treatment of munitions workers poisoned in accidents at the factories. Haldane set to work with J. G. Priestley, who was working on a similar apparatus for use by pilots at high altitudes. Their previous experience with breathing apparatus proved invaluable, and by early 1917 they had developed an oxygen mask with a system of valves which provided an economical supply of oxygen and, eventually, permitted the treatment of up to four patients from a single oxygen cylinder.[19] This apparatus would come to be seen in certain quarters as one of the most important medical developments to come out of the war; and physiologists, in particular, would promote its use in the treatment of far more than just gas poisoning.

THE RECEPTION OF OXYGEN APPARATUS IN THE FIELD

The earliest use of the new oxygen apparatus with gassed munitions workers appeared to be successful,[20] and Haldane was keen that it should be tried out on gas victims in France. To this end, he collaborated closely with C. G. Douglas, who had access to various field hospitals and to the Gas Services Central Laboratory at St Omer, and did what he could to have the apparatus tested on gassed soldiers. The first trials were not encouraging. Initially, Douglas and his colleagues found the equipment difficult to use: the oxygen appeared to provide little benefit, and in some cases even seemed to increase the discomfort.[21] Nevertheless, they persevered, modifying the apparatus and adjusting the rate of oxygen flow until they found they could offer effective relief to many of their patients.[22]

In spite of Douglas's successful trials, however, the response from doctors in the field was disappointing. It was not that they were unwilling to adopt new therapies: even before Haldane and Priestley's apparatus became available, medical officers had attempted to treat gas victims with oxygen, but with little success. '[W]hen tried by medical officers', Douglas recalled, 'oxygen seemed to do but little good to acute casualties'.[23] In retrospect, these results seem surprising: oxygen is now widely accepted as a treatment in such cases, and we would expect it to have been obviously beneficial. It is consequently tempting to try and account for the failure of these early trials in terms which presuppose the therapeutic efficacy of oxygen. It may be, for instance, that the first cases on whom oxygen was first used had simply been too badly poisoned to benefit; the damage to their lungs was too extensive to permit oxygenation even from enriched air. However, such speculations divert us from recognising a more important and general reason why doctors should not have found any benefit from the new treatment. Oxygen therapy, as developed by Haldane and his colleagues, was not simply a matter of introducing a new form of apparatus: it also involved a very different conception of illness and its treatment, and a very different method of practice, from that with which most doctors were at that time acquainted.

Consider the theory and treatment of pneumonia, on which the clinical picture of gas poisoning was largely modelled. Pneumonia was generally regarded as a severe infection, not just of the lungs but of the body as a whole. The most dangerous effect of this infection was understood to be the acute inflammation set up in the lungs, which impeded the passage of blood through the pulmonary circulation and placed considerable strain on the heart: death from pneumonia was attributed primarily to heart failure. In such cases, the symptoms included cyanosis (a blue rather than pink tinge to the skin resulting from the poor oxygenation of the blood), but this was seen to be a secondary result of the real problem of circulatory failure.[24] The therapeutic value of oxygen was consequently limited. Osler, for instance, agreed that it 'does sometimes seem to give transitory relief and to diminish the cyanosis'.[25] But this was merely a matter of relieving the symptoms: the physical disruption of the circulation was in no way corrected. The only way in which the circulation might be directly assisted was by reducing the quantity of blood to be pumped round the body: in other words, by bleeding, a treatment which had been advocated for acute pneumonia cases in the years immediately before the war.[26] The same treatment was quickly adopted for the inflammation set up by lung-irritant war gases.[27]

If oxygen was understood to have any genuinely beneficial effects in such conditions, it was by acting as a stimulant to the heart and to the pulmonary circulation, along with other stimulants like strychnine, belladonna, atropine and warm mustard baths.[28] If it was to be used in this way, however, it should be supplied sparingly: in high concentrations oxygen was itself known to cause inflammation of the lungs, which could only exacerbate the congestion already present. It was therefore widely recommended that oxygen should be administered through an inverted funnel, held at some distance from the patient's face so that the gas would be well diluted with air. As soon as a reduction in cyanosis showed that oxygen had been absorbed into the blood, administration should be stopped.[29] This, Douglas recalled, was the 'customary old-fashioned way' of administering oxygen that medical officers had initially tried on gassed soldiers; when this failed, they had tried other means of delivering the stimulant – 'under the skin, by the rectum, or even by intravenous saline infusion enriched with oxygen'.[30]

It was clear to Douglas why medical officers should have had so little success with oxygen therapy: 'The patient ... benefited but little by such fragmentary supplies of oxygen ... Obviously none of these means could supply the quantity required'.[31] But while this may have been obvious to Douglas, it was based upon a very different understanding of gas poisoning from the accepted clinical conception of lung-irritant diseases. In the years before the war, Douglas and other respiratory physiologists had focused on the very precise regulation of the human breathing rate in response to the need for oxygen,[32] and so had an acute sense of how important a continual supply of the gas was for the maintenance of life. Haldane's work on carbon monoxide poisoning in coal mine explosions had reinforced this view: oxygen shortage could be harmful even without inflammation or damage to the lungs.[33] Consequently, he and his colleagues were inclined to see gas poisoning primarily as a state of oxygen deprivation[34] and not, as doctors saw it, as an inflammation. As a frustrated Douglas told Haldane: 'Eventually I suppose people will begin to appreciate that deficiency of oxygen is rather a potent thing, but personally I am sick of trying to persuade the unbelievers'.[35] Haldane took the issue up in the *British Medical Journal*, warning of the dire physiological consequences of a diminished respiratory exchange: 'this is a lesson which I wish to emphasise as strongly as I can – partial anoxaemia means not a mere slowing down of life, but progressive and perhaps irreparable damage to living structure.'[36]

This concentration on the progressive changes that accompanied oxygen deprivation, rather than the mechanical damage to the lungs, led respiratory physiologists to emphasise aspects of the pathology of gas poisoning which were not apparent to medical officers. In particular, Haldane and his colleagues regarded asphyxia as a disease process rather than a symptom, and recognised two distinct clinical stages through which the victim declined towards death. The first stage, easily recognised, was characterised by obvious cyanosis, with plum colouring of the face, intense venous congestion, and a fast strong pulse. Subsequently, however, oxygen deprivation led to weakening of various organs, including the heart muscle and the respiratory centre in the brain. The pulse weakened, becoming rapid and thready, while the obvious blue colouring of the face and the engorged neck veins subsided. The patient became pallid, only the blue colouring of the ears and lips betraying the cyanosis of asphyxia. Douglas called this 'the most dangerous condition of grey cyanosis and collapse'.[37] It was a condition new to clinical medicine, and Douglas made much play of it in his official publications on the subject.[38]

But the very novelty of this condition, and of the physiological conceptions by which it was understood, militated against its widespread recognition by medical officers. 'The majority of men who are working on the problem do not seem to me to be sufficiently acquainted with modern physiology', declared Douglas. 'I suspect that they do not see want of oxygen, unless the case happens to have a bright blue colour. I have been told that men refrained from giving oxygen to leaden-coloured, collapsed cases of gas poisoning, on the grounds that the case could not be short of oxygen because he was not blue!'[39] Haldane concurred, though he preferred to blame not the medical officers themselves, but the kind of education they had received. 'I don't wonder at MOs failing to recognise the leaden-coloured cyanosis', he counselled. 'The man from Harley Street never seems to recognise it, and consequently never points it out [when teaching] in the wards'.[40] As Haldane was aware, the adoption of oxygen therapy in the field was limited not by the scientific failings of individual doctors, but by the deep differences of training and experience that separated professional physiologists from medical practitioners.

These differences were manifest, not just in divergent theories about the pathology of gas poisoning, but also in the details of therapeutic practice: the new technique of oxygen therapy employed different methods of delivery, and different interpretations of clinical signs, from those in common use at the time. On the accepted view, the fact that oxygen was a potentially dangerous irritant implied that it should be used sparingly. Cyanosis was the crucial sign by which treatment was regulated: the disappearance of cyanosis showed that oxygen had been successfully absorbed into the body, and that administration should now cease. In contrast, respiratory physiologists regarded oxygen not as an occasional stimulant or palliative, but as a fundamental requirement of life, to be supplied in relatively large quantities until the lungs had healed themselves.[41] From this perspective, cyanosis had a quite different significance: since the primary aim of therapy was the continuous relief of oxygen deficiency, the disappearance of cyanosis indicated that oxygen administration was having a beneficial effect and therefore should be continued. It is hardly surprising that doctors were disinclined to accept such an indication, which ran contrary both to their theoretical understanding and to their practical experience of therapeutics.[42]

Before medical officers could regard oxygen therapy as beneficial, then, they would have first to reject much of what they already knew about the theory and practice of medicine; in its place, they would have to accept the very different views of laboratory scientists. They would also have to invest time and energy in learning a new technique, the therapeutic efficacy of which remained uncertain. Douglas had persisted with the oxygen administration apparatus, in spite of initially poor results, because he was convinced from the beginning that it would prove effective. Medical officers had no such convictions to sustain them, while the added difficulties of war-time practice – in particular the rapid passage of injured men through clearing stations and field hospitals – meant that they were unlikely to observe any beneficial effects of long-term oxygen therapy for themselves. Ultimately, the only way that medical officers could decide whether or not to use oxygen was on the basis of authority: on the one hand, the authority of their clinical teachers and of their own medical experience; on the other, the authority of a group of scientists whose experience and standing in clinical practice was known to be limited. In general, they chose the former. By the end of the war, the Army Medical Service had provided some 4000 oxygen cylinders for treating gas victims, but their deployment was limited by doctors' lack of interest in the new method. According to the official historian of the war-time medical services, 'the demands from the medical profession were not sufficient to stimulate the industrial effort for production and distribution [of oxygen] on a large scale.[43]

The point I want to emphasise is that oxygen therapy was as much a social as a technical innovation, made possible by the novel circumstances of the war. Under these circumstances, new medical problems like strategic gas poisoning created unprecedented opportunities for laboratory physiologists to breach the social barriers that normally kept them from the bedside: patients, once under the sole control of doctors, now became available to research scientists as well. But ironically, while they could now investigate clinical problems, physiologists could still do little to influence clinical practice. Again, this was as much a social as a technical limitation: despite their new place at the bedside, they had not yet acquired the status and authority within the clinical hierarchy that they needed in order to persuade doctors to adopt their therapeutic techniques and theories.

AN ALTERNATIVE CONSTITUENCY

If the new apparatus excited little interest among medical officers in the field, it was nevertheless hailed in certain quarters as one of the major medical successes of the war. The value of oxygen therapy in the field may have been limited, but it offered obvious benefits to a very different constituency from front-line medical officers. The most vociferous support for the therapeutic value of oxygen came from the MRC, which seized eagerly on any opportunities the war offered to demonstrate the value of the research performed under its aegis. The Committee's first two annual reports, for the years 1914–15 and 1915–16, had stressed the importance of laboratory-based pathological science in combatting the medical problems of trench warfare. But the third, for the year 1916–17, reserved pride of place for physiology; specifically, it hailed the new methods of oxygen therapy as a triumph of scientific medicine over older forms of practice: 'The application here of physiological principles – long ago established by English work but too little applied in England – will now, we may believe, bring permanent benefit back from the trenches to the hospitals and sick rooms at home.[44]

The emphasis on physiological principles was a strategic one. Broadly, the remit of the MRC included 'research in connexion with any disease to which insured persons [under the National Health Insurance Act of 1911] may be liable'.[45] During the establishment of the Committee, however, problems had arisen in defining precisely what kind of research this should include, a question that would be raised again in the immediate post-war years.[46] These difficulties were exacerbated when a physiologist, Walter Morley Fletcher, was appointed Secretary in 1914. Under Fletcher's leadership, the Committee inclined strongly towards research, not just in sciences like pathology, the medical applications of which were fairly obvious, but also in more abstract and 'fundamental' forms of laboratory science like physiology. The development of oxygen therapy, vaunted as a successful application of physiological principles, created a valuable opportunity to advertise the practical benefits of such a programme.

It was an advertisement that found a sympathetic audience in the War Office. In September 1917 the Royal Society's Physiology (War) Committee produced a pamphlet of *Notes on the effects of pulmonary irritant gases*, intended for circulation to Army Medical Officers, which included recommendations for the use of oxygen. Alfred Keogh clearly regarded this as a vindication of his scientific foresight in recommending that the Committee be set up. In January 1918, with the approval of the Chemical Warfare Department of the War Office, Keogh appointed practically the entire membership of the Royal Society committee to a new Chemical Warfare Medical Committee, established under the aegis of the MRC for the co-ordination of further researches into gas poisoning. The Royal Society pamphlet was adopted as the new Committee's first publication, and three months later Keogh issued a memorandum to Medical Officers substantially based on this report.[47] The appointment of the Medical Committee was an important development in the War Office's preparation for gas warfare, which had previously revolved around the production of weapons and of ways of protecting uninjured troops. The adoption of a programme of medical research revealed a new concern for the treatment of soldiers incapacitated by gas, combined with a growing faith that scientific innovations like oxygen therapy promised a solution to such problems.

This official interest in therapeutics was itself stimulated by a new kind of medical problem created by the war. By 1917, the medical services were coping well with the transport, hospitalisation and treatment of war casualties. But concern was mounting at the growing numbers of chronically ill soldiers who tied up hospital beds that were needed for acute casualties, and threatened a post-war burden of invalidity pensions. Gas poisoning added to this concern: from mid-1916, doctors began to recognise that many gas victims who survived the acute stages of poisoning remained chronically disabled. Effective therapeutic measures were needed to relieve and rehabilitate such cases, and it was this demand that medical scientists now sought to meet. They were encouraged by their recognition that the chronic symptoms of gas poisoning were remarkably similar to other long-term disabilities; in particular, it resembled a condition known as disordered action of the heart (DAH), which was a major cause of invalidity among troops. and was already being investigated in military hospitals by MRC physiologists.[48] A similar clinical picture was presented by shell-shock, which physiologists were likewise encouraged to study.[49] These researchers were strongly inclined to believe that the similarities between chronic gas poisoning, DAH and shell-shock were more than just symptomatic: indeed, they suspected that they were all manifestations of the same underlying phenomena.[50]

So far, physiological investigations had yielded no dramatic new therapies for DAH or shell-shock: the only treatment scientists had been able to recommend was a programme of carefully graded exercises intended to encourage recovery.[51] Therapeutic hopes were raised by their experiences with oxygen. If oxygen could assist the recovery of acute gas cases, perhaps it would also prove successful in combatting the longer-term disability that followed. And if that was the case, perhaps the same therapy would be beneficial in other related conditions. If a respiratory element could be adduced not just in chronic gas cases, but in DAH and shell-shock patients, perhaps oxygen would prove effective in treating all these kinds of disability, solving at one blow an enormous medical and organisational problem.

One of the first to investigate this possibility was J. C. Meakins,[52] a young Canadian researcher from McGill University, who had been studying DAH with other physiologists at the Army Heart Hospital at Hampstead. In May 1917 he joined Haldane at the Kitchener Hospital in Brighton, where they began working on the effects of oxygen in DAH cases. Two months later, they moved to the Duchess of Connaught's Canadian Hospital at Taplow, where the MRC set up a well-equipped laboratory for more extensive physiological studies of patients.[53] When Keogh set up the Chemical Warfare Medical Committee at the start of 1918, it was given responsibility for the Taplow work. This display of official interest soon led to further expansion, when Haldane and Meakins were joined by J. G. Priestley and H. W. Davies. By the end of the war their work had grown into a major clinical research project which combined experimental work on respiratory function in DAH, shell-shock and gas cases with studies of the effects of oxygen and other kinds of therapy.[54] Meanwhile, in Cambridge, the same Committee had constructed an oxygen chamber in which Barcroft and his colleagues investigated the effects of longer-term oxygen treatment on chronic gas and DAH cases.[55] The War Office even set up a special hospital for gas victims in Cambridge, to facilitate further clinical studies, but the war ended before it could be brought into use.[56]

For all this investment in research, oxygen found little use in the treatment of chronic gas poisoning and other war-related conditions: the clinical results were unencouraging; the gas was expensive, especially when used in large amounts and for long periods in oxygen chambers; and, with the end of the war, these conditions began to pose less of a problem. That is not to say that physiological investigations had no impact at all on the treatment of such cases, however: on the contrary, scientists' findings would prove highly influential in deciding how to deal with them. By developing a new understanding of the role played by respiration and other vital functions, physiologists helped to redefine chronic illness in ways that implied radically different forms of medical management. Before the war, doctors had attributed the symptoms of DAH, gas poisoning and shell-shock to intractable physical lesions of the heart, lungs and nervous system respectively. In the course of their investigations at Taplow and elsewhere, however, physiologists came to see these apparently different kinds of disability as essentially the same condition: they were to be understood, not as localised physical injuries, but rather in terms of a general disruption of nervous activity that resulted from the unusually intense physical, chemical and psychological disturbances of warfare, and led to a breakdown in the co-ordination of normal physiological activities like breathing and circulation.[57]

This new model of disease was seen to have important therapeutic consequences. Compared with anatomical damage, physiological disturbances might be expected to respond relatively well to therapy: given appropriate conditions and a certain amount of medical assistance, the regulatory processes that normally co-ordinate the various bodily functions should quickly reassert themselves. All that was needed, said the physiologists, was a suitably brisk course of rehabilitation, backed up by a calm reassurance that there was nothing physically the matter. Thus the Chemical Warfare Medical Committe advised doctors treating 'the late effects of gas poisoning' that, as with DAH and other 'nervous symptoms of a "functional" character', prolonged hospitalisation was 'Particularly apt, in these cases, to exaggerate the nervous conditions'.[58] As a result, the burden of disability on the military was substantially reduced: according to the MRC, the new approach to rehabilitation soon saved the War Office tens of thousands of pounds in heart disease cases alone.[59]

Of course, physiologists were disappointed that oxygen therapy did not help to speed up the process of recuperation and rehabilitation as they had hoped it would. But, in the long run, that failure was less important than the highly successful therapeutic principles they had helped to establish. The early therapeutic promise of oxygen had stimulated official interest in clinical research, as a result of which physiologists had become directly involved in questions of medical policy. They had gone on to demonstrate that, given access to patients, they could generate new knowledge of enormous value in medical administration if not in more traditional forms of clinical practice. This would stand them in good stead in the post-war years, when government support enabled them to continue their work in clinical medicine. Once again, oxygen therapy would occupy a special place in their programme: though doctors had not been impressed by oxygen in the field, physiologists clearly hoped that further research and further education would demonstrate its value to medical practice in peacetime.

SCIENTIFIC DEVELOPMENTS IN PEACE-TIME

In the course of the war, scientific research had been elevated to a new status in the state machinery. This was not just the case in medicine alone. In 1915 a Privy Council Committee for Scientific and Industrial Research, soon to become the Department of Scientific and Industrial Research (DSIR), was established with extensive powers to solicit and promote scientific investigations. Similar powers and resources were granted to the MRC immediately after the war. In 1919 the old Medical Research Committee of the National Health Insurance Commission was dissolved, to be reconstituted as the Medical Research Council; following the example of the DSIR, the new MRC was established not in the new Ministry of Health, as many had expected, but as a committee of the Privy Council. The choice of location had important implications for the role that medical research was expected to fulfil in the management of the state. Free of the demands of particular administrative departments, the scientists of the MRC should be as independent as possible in their choice of programme, and thus capable of taking a leading rather than a subordinate role in the development of scientific policy.[60] In effect, this was an official endorsement of the role that the physiologists of the Royal Society and the MRC had adopted in the development of gas warfare.

Both the DSIR and the MRC were interested in oxygen. In the case of the former body, which established a special Committee on the Uses of Oxygen, this interest was particularly in the development of new industrial supplies and uses for the gas. But these were understood to extend to medical uses also, and the DSIR funded some research in this area.[61] The

MRC stayed 'in direct touch' with this work,[62] and in 1919 set up its own Committee on Clinical Uses of Oxygen to correspond with and complement the DSIR investigation. Included in the membership of the MRC Committee were J. C. Meakins and G. H. Hunt, both of whom had carried out research into oxygen therapy during the war. This continuity of personnel from war-time to peace-time reflected a more general continuity of interest, especially in the development of new therapeutic responses to administrative problems of disability. As the Committee declared shortly after its foundation, it would promote research 'with a view not only to the needs of ordinary hospital practice, but also to the treatment of lung affections caused in industrial work by poisonous gases or other pulmonary irritants'.[63] During the war, the MRC had set up a clinical study of the effects of long-term oxygen therapy on poisoned soldiers and munitions workers, conducted in two specially constructed oxygen chambers at the North Staffordshire Infirmary in Stoke-on-Trent.[64] The new committee now took charge of this study, which was extended during the post-war influenza epidemic to include work on pneumonia patients.[65]

Research into the clinical use of oxygen was greatly facilitated by another official initiative to bridge the professional divide between pre-clinical science and medical practice, not just in institutions like the North Staffordshire Infirmary, but in some of the country's leading teaching hospitals. Before the war, all clinical teaching posts in medical schools had been honorary appointments, occupied by leading hospital doctors on a part-time basis, and jealously protected from the encroachment of full-time scientists. But moves were already afoot to create a new cadre of full-time salaried professors, who would not only take control of clinical teaching, but would also conduct scientific research; ideally, they would be provided not just with the usual complement of hospital beds, but with full clinical 'units' incorporating properly equipped and staffed laboratories. The original proposals had been laid down in 1913 in the Final Report of the Haldane Commission on the University of London.[66] They were subsequently pursued, against outspoken opposition from sections of the medical profession, by a departmental committee of the Board of Education.[67] The war delayed progress, but in 1918, with plans for reconstruction being laid, the Board renewed its campaign.[68] In 1919 the new University Grants Committee (UGC) adopted the clinical unit scheme as one of its main projects, and by the following year had provided funds for full-time clinical professors in medical schools in London, Sheffield and Edinburgh.[69] The successful exploitation of these new clinical research facilities owed much to the co-operation of the MRC, who augmented UGC funds with grants for additional research personnel and equipment.

This combination of UGC and MRC sponsorship gave medical scientists the opportunities they needed to develop their methods of oxygen therapy at the bedside. Much of the funding came from the MRC Committee on Clinical Uses of Oxygen, which provided grants to set up a co-ordinated programme of clinical investigations in teaching hospitals around the country.[70] A large part of the research was carried out by a team established under J. C. Meakins, who in 1919 had been appointed full-time professor in Edinburgh University's new clinical unit.[71] Other projects were set up in medical schools which had not yet adopted the clinical unit scheme, but which were nevertheless keen to establish clinical science facilities. Guy's Hospital Medical School, in particular, had chosen to continue its own well-established programme of medical research and teaching without the involvement of the UGC.[72] Nevertheless, the MRC established one of its major clinical research centres there when it appointed two assistant physicians, E. P. Poulton[73] and G. H. Hunt,[74] to the Clinical Uses of Oxygen Committee. With funds from the Committee, they constructed an oxygen chamber on one of the wards, where they carried out intensive studies of respiratory pathology and long-term oxygen therapy.[75] The other major grant went to Glasgow University, where D. K. Adams, assistant to the Regius Professor of Medicine and an assistant physician in the out-patient department of the Western Infirmary, was given access to patients for studies of oxygen therapy.[76]

Over the next five years, the Clinical Uses of Oxygen Committee built up an extensive and closely co-ordinated programme of research in these three main centres. Initially, the various workers collaborated in a broad survey of 'the occurrence of cyanosis in various diseases [including heart and kidney conditions], determining the gaseous content of arterial blood in different stages, testing the efficiency of different methods of alleviation by oxygen administration, and watching for evidence of consequent clinical improvement'.[77] In little over a year they were able to report that 'clear proof of improvement [had been] obtained in cases of pneumonia and broncho-pneumonia', but cautiously decided that it was 'too soon to attempt a general report on the subject'.[78] By 1922, studies carried out on patients at Guy's had given somewhat greater definition to this picture: 'The clinical data already collected lead the Committee to the opinion that patients suffering primarily from damage to the lung receive benefit from treatment with oxygen, but that those with circulatory disorders receive little if any'.[79] This established, the work of the Oxygen Committee came to focus chiefly on developing more efficient methods of delivering oxygen to the patient, particularly through various forms of the Haldane mask.

Meanwhile, as the work on oxygen therapy had begun to produce favourable results, it had spread from the three centres initially involved to a number of other medical schools. The Committee was joined in 1921 or 1922 by F. R. Fraser,[80] full-time Professor of Medicine and director of the clinical unit at St Bartholomew's, though it would be four years before Fraser initiated research in his own department into methods of delivering oxygen to hospital patients.[81] Meanwhile, in Manchester the Professor of Physiology, H. S. Raper,[82] was able to secure clinical facilities for two assistants to investigate the effects of oxygen on infants suffering from unexplained nutritional disorder (marasmus) who had failed to respond to dietetic treatment: the Clinical Uses of Oxygen Committee paid the costs of this research.[83] The last grant in this series was made in 1927 to H. W. Davies, who had moved on from Meakins's Edinburgh laboratory to the post of lecturer in physiology and pharmacology at Leeds, for clinical trials of a modified form of oxygen apparatus.[84] By this time the Oxygen Committee had ceased to exist as a separate sub-committee of the MRC. The interesting clinical investigations had been carried out, and no further need was seen for a concerted programme of research in this area.

OXYGEN THERAPY AND THE PRACTICE OF MEDICINE, 1920-30

During its lifetime, the MRC Committee on Clinical Uses of Oxygen generated a considerable body of information about the beneficial therapeutic effects of the gas in a wide range of medical conditions. These findings were widely promoted by clinical physiologists, particularly through the textbooks that they wrote. Shortly after the war, for instance, Poulton came to edit *Taylor's Practice of Medicine*, in which he recommended Haldane's apparatus for a wide variety of lung conditions, including pneumonic influenza. 'In cases of cyanosis it is essential to give oxygen practically continuously', he reminded his readers.[85] Fifteen years later, in his own popular textbook on *The Practice of Medicine*, Meakins reiterated that oxygen should always be used whenever there is 'the slightest cyanosis and continued as consistently as possible until all the cyanosis is permanently removed'.[86]

Nevertheless, in spite of these admonitions, many leading clinicians remained unconvinced of the physiological benefits of oxygen therapy, and continued to teach pre-war ideas and practices to their students. Thus, when Thomas McCrae edited the 9th edition of Osler's *Principles and Practice of Medicine*, he retained the warning that 'The value of the administration of oxygen is doubtful' in cases of pneumonia.[87] Ten years later the editor of the 13th edition, Henry A. Christian, noted the various new techniques of oxygen administration through masks, in chambers and in tents. But he could still argue that 'The value of such oxygen therapy, except for the comfort it may give, probably is over-rated'.[88] It seems that this critical attitude continued to be based on the old view that the chief danger in pneumonia and other diseases was the physical embarrassment of the heart, rather than the functional problem of asphyxia: insofar as non-physiologists recommended the use of oxygen, it was in the same terms as before the war, as one among a number of cardiac stimulants.[89]

Between these two extremes fell a third body of clinical opinion, which accepted the general physiological arguments in favour of long-term oxygen administration, but was unimpressed by the methods of delivering the gas that clinical physiologists had so far devised. Conflicting views on the practical value of the Haldane mask, in particular, reveal some of the divergent attitudes and priorities that shaped the debate over oxygen therapy. For many doctors who saw private practice as their main income, the first concern was the comfort of their individual patients. These doctors considered the Haldane mask to be 'quite unsuitable … as it produces a subjective feeling of suffocation, which more than counterbalances the value of the oxygen.'[90] Instead, they favoured the more comfortable method of delivering the gas through a nasal catheter.[91] The catheter was more wasteful of oxygen than the mask, but when wealthy individuals paid handsomely for the best treatment, their comfort was worth more to the physician than the extra oxygen expended.

In comparison, clinical scientists were not so inclined to indulge their patients, and instead regarded the need for administrative efficiency as their first priority. For that reason they favoured the Haldane mask, and shared Poulton's view that 'any objections on the part of the patient to wearing the mask must be overcome.'[92] This attitude reflects the professional identity of the new medical researchers, which differed significantly from that preferred by more traditional clinicians. Instead of private practice and the goodwill of individual patients, they looked chiefly to the state for remuneration and professional advancement. They consequently addressed themselves to problems of a sort that were likely to win them the continued interest and support of government officials. As the earlier sections of this essay have shown, the widespread and expensive disability suffered by workers in industry and in the military was one such problem. Oxygen, it appeared, might provide a solution, but only if the costs could be kept low enough: in the early years, at least, long-term oxygen therapy remained too expensive for routine use in institutions like the North Staffordshire Infirmary.[93] It was this that led the MRC and its researchers to favour the development of more economical over more comfortable means of administering the gas.[94]

These divergent attitudes towards oxygen reflect alternative views on the nature and organisation of medical practice in inter-war Britain. And I would suggest that the eventual adoption of oxygen as a standard therapeutic agent should be understood in the light of the compromise that was eventually achieved between the care of wealthy individuals and the treatment of the industrial working class. Hopes that oxygen therapy would provide an economical answer to the problems of industrial lung disease were never realised. But equally, it did not remain the privilege of the wealthy sick and the élite practitioner. Rather, it passed into routine practice in hospital medicine, for reasons that appear to have been as much due to do with improvements in hospital finance as with new methods of administering the gas or better knowledge of its efficacy. In this respect, it is perhaps significant that the method of oxygen administration that found greatest favour during the inter-war years was the oxygen tent developed by Poulton, which reconciled the interests of scientists and clinicians by combining comfort with comparatively economical use of the gas.[95] But in spite of such economy, oxygen remained a relatively costly form of treatment which appears to have become widely accessible only as hospital budgets were augmented by various forms of state support.

If this much is speculation, it is nevertheless clearly consistent with the main historical argument that I have pursued throughout this essay: namely, that the development and deployment of oxygen therapy cannot be understood simply in terms of the introduction and diffusion of new techniques or new theories, but must be seen in relation to a much wider programme of innovation aimed at transforming the social organisation of bedside medicine. Science, including technical innovations like oxygen therapy, undoubtedly played an important role in furthering this programme. But the motivation and impetus for the expansion of 'scientific medicine', as this programme might conveniently be called, came not from the technical developments themselves, but rather from social and political developments within and without the medical profession.

Thus I have argued that oxygen therapy was the property of a small body of full-time scientists who were looking for greater access to clinical medicine. Even before the war, their claims were endorsed by important members of the government and civil service who supported, in particular, the 1913 Report of the Haldane Commission, with its proposals for full-time chairs of medicine. Under the peculiar social and political conditions that prevailed during the First World War, this alliance was consolidated and strengthened when scientists demonstrated that, given the opportunity to conduct research in clinical medicine, they could develop new forms of therapy.

But the case of oxygen therapy shows that we must distinguish carefully between those sections of the medical community that could put the new knowledge and techniques to practical use, and those sections that could not. The élite clinicians who currently dominated the world of bedside medicine, especially in the teaching hospitals, favoured a kind of practice that was structured primarily by the close personal relationships they formed with their private patients. Oxygen therapy, and the scientific medicine that it was seen to represent, could not at first be easily accommodated to the body of knowledge and technique that constituted that kind of practice, and so had little impact on bedside medicine. Instead, it found favour chiefly among government officials, who saw it in terms of a very different form of practice: scientific medicine offered a solution to the problems of disease and disability as manifested, not in the personal discomfort of wealthy individuals, but in the administrative difficulties of organising large groups of workers in an army or an industry.

That some civil servants had been willing to embrace scientific medicine, even before it proved its value in the management of injured soldiers, suggests that the scientists' claims may have appealed in very general terms to the administrative mind. Such speculations will have to be discussed more fully elsewhere, however; for the time being, it is sufficient to observe that the connections between scientific medicine and the bureaucratic reorganisation of medical practice continued to be strengthened in the years following the First World War, when university medical schools, with their experts and their laboratories, came to be seen as central to the construction of a more efficient system of state medical care.[96] The development of oxygen therapy, far from being a discrete or isolated medical innovation, was part of the wider programme of scientific medicine which both contributed to and benefited from this process of social reconstruction.

7
X-RAY TECHNOLOGY IN OBSTETRICS: MEASURING PELVES AT THE YALE SCHOOL OF MEDICINE[*]

Anja Hiddinga

INTRODUCTION

X-ray technology is one of the most important technological innovations of modern medicine. Although it was not specifically conceived to be used in medicine, the medical possibilities were recognised within days of Roentgen's announcement of the discovery in January 1896. Apparatus for producing the rays was readily available in any physics laboratory and manufacturers responded rapidly by producing and selling roentgen tubes and other peripheral equipment. Clinicians in a number of specialties were intrigued by the possibilities of 'roentgenography' for their own clinical practices. There was of course as yet no specialty of 'radiology'. Within each field in which clinicians sought to apply roentgen radiation techniques, research was carried out to define and re-define problems in such a way that they could be tackled with the new technology.

The present essay analyses the introduction of a particular radiographical technique – x-ray pelvimetry – into a particular medical specialism – obstetrics. It provides an opportunity to explore the mutual adaptation of technique and intellectual problems within a developing specialty. The form of adaptation is, of course, partly determined by the form of practice and the goals of practitioners. Diagnosing, treating, and curing are tasks for the clinician, who may be at the same time a researcher: an investigator in a double sense.

The development of new techniques in clinical medicine depends on a wide range of factors, which vary over time and between countries. They vary with the organisation of goals of the specialism concerned, with the system of payment for medicine, with the organisation of hospitals, and with the relationship between medicine and commercial interests, including the firms who manufacture equipment. Further, for any given specialty, these relationships may vary from one technique to another, depending in part on the costs involved and the means for recovering these costs. Whether a technique is used in doctors' surgeries or in specialist hospitals, whether it is used for urgent or non-urgent problems, will all help determine the reciprocal accommodation of technique and professional goals.[1]

To study the introduction of a new technique, we need to understand the practices, priorities and aspirations in terms of which the innovation is presented to the professional community concerned. The new technique must fit the appropriate intellectual framework and the conditions of clinical practice. Here we analyse how clinical research in obstetrics provided both resources and constraints for a 'technological research programme' in the use of x-rays.

Radiation, for obstetricians, had several, contrasting uses. X-rays seemed to offer a means of sterilising women without reducing sexual function; they also offered a means of inducing abortion for medical reasons. At the same time, x-ray imaging techniques could be applied to the pelvis of mothers, so as to visualise bone structure. From our present perspective, where x-rays are seen as threatening to foetal development, it is perhaps hard to understand the engagement of obstetricians in roentgen technology. Since the 1950s the use of x-rays has been minimised, but in the inter-war years they were developed for routine use on pregnant women. The significance of this programme lies deep in the history of the field.

HISTORICAL BACKGROUND OF PELVIMETRY

One long-established concern of obstetric science is the determination of the pelvic size and shape in order to make predictions about possible difficulties in the passage of a baby during the delivery. The starting point of this concern can be located in the very early eighteenth century, before which time it was commonly believed that the baby used its own strength to crawl out. In this older view, long-lasting or difficult labour was due to a lack of strength; an impossible birth could be due only to intra-uterine death. Midwives and doctors believed that in the birth-process the pelvic bones would separate to permit passage of the child. In 1701 the Dutch teacher of midwives Hendrik van Deventer published a handbook of obstetrics[2] devoted, in large part, to the precise description of the pelvis. He showed that pelves which were too wide, too small or too narrow were responsible for characteristic difficulties in delivery. He observed that the size of a woman was not necessarily indicative of the shape of her pelvis, and so he insisted that the midwife was to examine each woman at the start of labour by inserting the hand and palpating the inner pelvis.

Van Deventer's book was translated into Latin, French, English and German, and was instrumental in focusing medical attention on the pelvis. Baudelocque, the famous French 'accoucheur', whose textbook for midwives appeared in 1753, also related the size of the pelvis to that of the child's head. But it was the equally well known British 'man-midwife', William Smellie, who first described a clinical method of measurement in a series of cases published in 1754, and who mentioned that absolute measurements of pelvic size might be a criterion for anticipating a difficult birth and so calling in a doctor rather than a midwife.[3]

The significance of 'malproportion' (in the relation of the child's head-size to pelvic measurement) was now established for obstetrics. In succeeding years scientifically-minded obstetricians tended to concentrate on the means of dealing with malproportion in practice, rather than on its nature. It was the anatomists who now began to make elaborate descriptions – measurements and categorisations – of the pelvic bones of skeletons. Later, pathologists picked up this work and sought to establish the causes of pelvic malformations, and the effects of diseases such as rickets and osteomalacia. During the eighteenth century the practice of the obstetricians gradually provided them with a sense of what should count as a 'narrow pelvis' in relation to childbirth. Concern with the individual patient and the possible complications of her delivery were transformed into the attempt to establish generally valid *norms*. This tendency became clearly visible in the work of G. A. Michaelis in the mid-nineteenth century. His posthumous treatise, *Das Enge Becken*, was completed and published in 1851 by C. C. Th. Litzmann, his collaborator and successor as professor at the University of Kiel. It is generally taken as one of the first main products of scientific obstetrics.[4] For Michaelis, seen as the founder of 'Beckenlehre' (theory of the pelvis), the essence of the question lay in the quantitative prediction of complications. He stressed the need for *precise* measurements (not estimations), and for *appropriate* measurements, that would not be affected by unrelated properties of the individual. For that reason, internal measurements came to be preferred over external body measurements, where thickness of bones and quantity of fat could distort the results.

Of the inner dimensions of the pelvis, the so-called 'conjugata vera' or diameter of the pelvic inlet is of particular importance (see Figures 7.1 and 7.2). The most frequent narrowing, in the flat rachitic pelvis, consists of a shortening of this diameter. But it was not possible to measure this directly in the living woman. Smellie, a hundred years previously, had proposed instead taking the 'conjugata diagonalis' (the diagonal distance between the promontorium and the pubic bone) as an indication of the length of the conjugata vera (see Figure 7.1). Michaelis and Litzmann proposed a categorisation of female pelves in which they distinguished between the normal narrow pelvis (deviating only in size from the anatomist's notion of the regularly shaped average pelvis) and the pathological narrow pelvis, differing also in shape. For each category a norm was established and at least on those measures that could easily be taken, general agreement on normal size was rapidly reached. Subsequently, measures became more precise, 'fingerwidths' giving way to inches, inches to millimetres. Soon, also, a variety of instruments came to be developed to standardise and objectify the act of measurement. New means of measuring angles with protractors allowed more precise estimations of the conjugata vera, which had not been accessible to direct measurement. The transverse diameter of the pelvis, too, was recognised as of importance, and though here internal measurement remained problematic, instruments were developed in an attempt to perform the task. Some of these 'pelvimeters', with or without

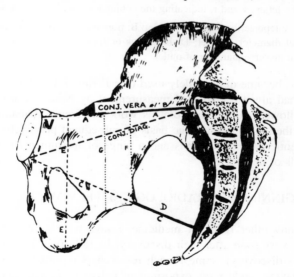

Figure 7.1   The bony pelvis seen from the side, published in an article by Herbert Thoms. The conjugata vera (cv) and conjugata diagonalis (cd), have been added.

Note: A: anteposterior diameter of inlet; B: posterior sagittal diameter of inlet; C: anteposterior diameter of midplane; D: posterior sagittal diameter of midplane; E: pubotuberous diameter; F: distance establishing level of ischial spine; G: distance establishing level of widest transverse diameter of outlet.

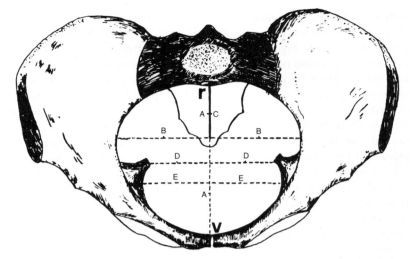

Figure 7.2 The bony pelvis seen from above, published in an article by Herbert Thoms. The letters v and r, indicating the conjugata vera, have been added.

Note: A: anteposterior diameter of inlet; B: transverse diameter of inlet; C: posterior sagittal diameter of inlet; D: bispinous or transverse diameter of midplane; E: widest transverse diameter of outlet.

protractor, became quite widely used (see Figure 7.3). Though Michaelis himself had always advocated measurement by hand, the use of instruments followed so directly from his work and from Litzmann's that we may see them as providing the basis of subsequent pelvimetry. Until the early twentieth century, hands and a range of instruments remained the basis tools of pelvimetry.

THE BEGINNINGS OF RADIOLOGICAL PELVIMETRY

As in many other fields of medicine, x-rays were applied to obstetric problems very soon after their discovery. In 1896 (that is, one year after Roetgen's discovery), some French researchers reported in an obstetrics journal having made x-ray exposures of pelvès in animals and in human corpses.[5] They concluded that an accurate picture was possible and that one could distinguish between symmetrical and asymmetrical pelves. Exposures on a living woman were not tried because the exposure-time was three *hours* (she might have moved). But the first examples of x-ray pictures of this kind followed very quickly. In 1897, at the international medical congress in Moscow, Pinard and Varnier reported on their re-

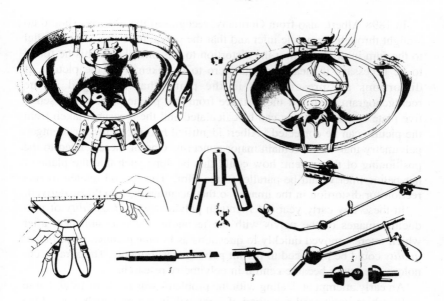

Figure 7.3 Nineteeth-century pelvimetry instruments, from a monograph on pelvimetry by the German obstetrician F. Skutsch. Note that although a bony pelvis is illustrated, the instruments were meant to measure pelvic diameters in the living woman.

Source: Skutsch (1887)

search on living women; exposure-time here is said to have been two minutes. They made the first attempt at measurement, by comparing an x-ray of a dried pelvis of known dimensions with an x-ray of the pelvis in a living woman.[6]

At the same congress, Kézmarsky showed his x-ray pictures of a woman with a contracted pelvis, his exposure time being 1¼ hours. In 1897 the first extensive treatise on roentgen pelvigrams appeared in Germany, in which attempts were made to deduce the true measures of the pelvis by mathematical calculations from the much distorted pictures. Average exposure-time here was two–five minutes.[7] The exposure-time constituted a particular problem in obstetrics. Safety was not yet a consideration and the inconvenience of the patient who would have to sit still for a long time did not seem to count for much either. The point here was that the equipment at that time could not produce rays strong enough to penetrate through the enlarged abdomen of a pregnant woman.

In 1898 Albert, also from Germany, recognised that the rays should go straight through the pelvic inlet and that the pelvic inlet should be parallel to the film in order to limit the distortion to this one plane. This he found to be necessary if measurements were to be taken from the picture. If distortions are present only in the plane that is imaged in the roentgenogram, then – if the distance from that pelvic plane to the sensitive plate is known – they can be calculated and the measures taken from the picture can be corrected.[8] Albert identified two problems for roentgen pelvimetry that would remain major concerns. One of these related to the positioning of the patient: how could this be done such that the plane of her pelvic inlet would be parallel to the film? The second related to correction for distortion in the image, so that accurate data could be derived.

In these few early years of roentgen pelvimetry the *feasibility* of producing images of the pelvis with the technology was established, after which interest moved quickly to questions as to *how* pictures of *sufficient quality* could be produced to permit pelvic measurements. The x-ray technology therewith became central in pelvimetry research.

An early attempt at dealing with the problems was made in 1899. Fabré and Fochier announced a method of x-ray pelvimetry in which they added to the regular x-ray equipment a square metal frame with an index.[9] This device had to be placed in a particular way round the pelvis of the woman so that the frame plus scale-index appeared on the picture in the same plane as the pelvic inlet, thus providing a means of assessing the distortion. So originated a line of research in which *measurement* of the pelvis was the objective; correctly positioning the plane of the pelvic inlet and constructing an index in this same plane to assess distortion were seen as ways to achieve accurate measurement.

It is a particularly productive and influential programme of research in this line, conducted in the 1920s and 1930s at the Yale School of Medicine in the USA that I want to discuss here. The context is the emergence of American obstetrics as a specialty.

AMERICAN OBSTETRICS IN THE FIRST DECADES OF THE TWENTIETH CENTURY

It was in these first decades of the twentieth century that obstetrics in America tried to upgrade itself, professionally as well as intellectually. In 1912 John Whitridge Williams, the well-known professor of obstetrics at Johns Hopkins, had written an alarming article on the standards of admission and instruction in obstetric education. He made a strong plea for grounding the practice of obstetrics on a firmly scientific basis, and for recognising and teaching obstetrics as a fundamental part of scientific medicine.[10] Williams's article seems to echo the very influential report of Abraham Flexner of 1910,[11] in which poor standards of American medicine were traced back to poor medical education and to lack of scientific spirit in the teaching of clinical subjects. One of the proposals for improvement was the employment of full-time specialists in university medical posts. Flexner pleaded for these arrangements so that physicians could devote maximum time and energy to research, without the heavy distractions of private practice. Freed from the need for extra, private income, they could achieve high quality in clinical research, for it was by the quality of research that university medical schools would now be judged.

Concern about the quality of obstetric standards was also stimulated by alarming data on maternal deaths. After the First World War, statistics indicated that maternity was the second highest killer of women aged 15–45.[12] This situation led to political concern and in 1921 the Sheppard–Towner Act was passed, providing for pre- and post-natal care, especially for poor women. According to Rosemary Stevens, the 1920s were characterised by rapid and profound changes in medical institutions, which created great uncertainty about the status and organisation of the profession. At precisely this time, obstetrics began to grow as a specialty. As Stevens argues: 'The movement for specialty regulation in this field was part scientific and part professional, but part also of the maternal and child welfare movement, a social rather than a professional reform process'.[13]

This movement, encouraging higher standards of medical obstetrics and gynaecology and the elimination of practitioners untrained in this specialty, led to the establishment in 1930 of the third American specialty board, the American Board of Obstetrics and Gynecology. The Board stressed the need for clear specialist criteria and tried to prevent the mushrooming of specialists.

The attempt at improving the quality of obstetrics was not limited to institutional reform. The content of obstetric practice and the nature of the obstetrician's relation with the birth-process were also discussed. Thus, Joseph B. DeLee of Chicago argued in his by now renowned article in the first issue of the *American Journal of Obstetrics and Gynecology*,[14] that delivery had to be seen as a pathological process that needed technical intervention and standardisation. This view, which became widely accepted, contributed to the increasing tendency of women to deliver in hospital, where specialists had access to technology and where well-trained staff were available to manage the delivery.[15] Technology thus became a central element in the discussions on safe maternity.[16]

But there was yet another purpose that technology could serve: it could help improve the profession's image, both with the public and with other medical specialisms. Obstetrics had never been a specialty high on the medical status ladder. Obstetricians complained about the mockery to which they were exposed,[17] and worries within the profession about the difficulties of recruiting sufficiently talented people have persisted into more recent times.[18] So one might expect a certain readiness on the side of the obstetricians to accept technological innovations in their practice.[19]

Contrary to expectations, however, the introduction of technology in the management of birth did not lead to a decrease in mortality. This became very clear from the figures on maternal deaths that were published in the beginning of the 1930s in two influential reports, by the White House Conference on Child Health and Protection and by the New York Academy of Medicine,[20] which demanded improvement of pre- and post-natal care and careful, well-considered procedures in delivery. According to the latter report, increased operative interference and use of instruments was a major cause of maternal deaths. The report therewith confirmed the opinion of many doctors and medical organisations who had been calling for a reduction in interference with normal labour and delivery.[21] So here the profession was confronted with a dilemma: on the one hand, they had to establish the confidence of the public in specialist obstetrical practices[22] of which scientific expertise and advanced technology were the key characteristics. On the other hand, these same characteristics seemed to be at the root of high mortality rates and had therefore to be closely scrutinised.

Technology provided means of quantification and thereby standardisation; it was also very important in setting boundaries between obstetricians and general physicians and midwives, who were not able or not allowed to use expensive complicated instruments and procedures. Moreover, hospital-based technology, symbolising the advanced modern qualities of medicine, was instrumental in attracting women to use the clinic for the delivery of their babies.[23] Technology was central, but interference seemed to be counter-productive. Confronted with this dilemma, it seemed that the interests of the American specialist obstetricians in the first decades of the twentieth century would best be served by the deployment of *non-interventionist* technology.

A RESEARCH PROGRAMME IN X-RAY PELVIMETRY

The research in x-ray pelvimetry that was carried out at Yale from the

1920s to the 1940s was largely the work of one man, the obstetrician Herbert Thoms. The start of Thoms's research work on pelvimetry dates from 1915,[24] but it was in 1922 that he published a landmark paper on a method of measuring by means of x-rays. The technical problems of such a procedure were still very considerable. As described above, spread of the rays resulted in distorted pictures which did not lend themselves to the precise measurement which obstetric pelvimetry needed. Thoms claimed to have overcome these problems in such a way that his method would be easily applicable in obstetric practice. But since measuring pelves by hand was an accepted procedure for obstetricians, the use of x-ray machines needed some justification. Thoms wrote:

> The methods of pelvimetry now in general use, particularly the measuring of the diagonal and external conjugate diameters give at best only an approximate estimate of but one diameter. If it then should become possible in cases where a contraction or deformity is suspected, to outline graphically by a simple method a diagram in its true proportions, I believe that it would constitute a very valuable aid to obstetric procedure. I feel that we are able to do this with the method described.[25]

'The question of mensuration', Thoms claimed, 'becomes one simple proportion'.[26] The described procedure was easy and accurate, but in the discussion of his paper in the annual meeting of the New York Obstetrical Society, Thoms found his method criticised. Doctors with experience of other methods of x-ray pelvimetry claimed that clinical data were much more important than x-ray data, and that the difficulties of labour depended not only on the size of the pelvis but on the size of the baby's head as well. Thoms's answer to these criticisms gives us some insights into the ways in which he perceived his method to be useful for others.

> I agree that in the *individual* case the best pelvimeter is the baby's head, but I also believe that this x-ray method has its value not only for exact mensuration of certain dimensions but certainly also for the *classification* of contracted pelves. As a routine method I do not recommend or use it. It is simply for the case in which it is felt that more accurate knowledge is wanted. Lateral pictures are a great aid, particularly in the diagnosis of rickets.[27]

Thoms saw his technique as an important *diagnostic* instrument but also as valuable in drawing up *classifications*. This double purpose would remain Thoms's usual argument when he sought to convince the medical community of his point of view. He appealed not only to practising physicians confronted with problems of predicting difficult deliveries, but also to anatomists or scientists working in so-called anthropometric research, who were more interested in the variety and regularities of pelvic size and shape.

After 1922 Thoms concentrated on further establishing the usefulness of roentgen technology as a *diagnostic* instrument for obstetric practice. Establishing the value and meaning of new dimensions that could be measured with this technology was of subsidiary importance. It is striking that this work shows none of the concern with improvement of quantitative norms for normal or abnormal pelves, which had been a major question for the nineteenth century pelvimetrists. These norms Thoms assumed to be known and accepted by the obstetricians.

There was a constant flow of papers in various journals, aimed both at obstetricians and at a general medical audience, offering slight improvements of the technique. In a 1927 paper, five years after the first publication of his method, Thoms was still dealing with problems concerning positioning the patient, assuming that this constituted the major reason why the procedure had not been more widely adopted outside the Yale clinic. In these papers, mainly concerned with the technical perfection of the method, hardly any clinical cases were reported. The illustrations showed a correctly positioned bit of body with a pelvis drawn in, and there were photographs of the perforated lead plate that was used to overcome the misreadings due to distortion in the picture. Mostly, one or two x-ray pelvigrams were also shown as examples of what a good picture looked like (see Figures 7.4 and 7.5). In trying to convince his colleagues to adopt the technique, Thoms was referring to existing norms of obstetric practice, norms that were known and accepted.

Safety, mentioned once at the beginning of the research, was established by ascertaining the healthy condition of the baby after the delivery of the x-rayed woman.[28] Replicability and accuracy were technically improved, and usefulness was assessed by comparison with traditional hand methods. But the method provided yet more opportunities for tackling novel problems. By the time the technical problems had been overcome, the original purpose of the technique, the determination of the precise length of the conjugata vera, was becoming less relevant. Pressures were building up which would shift the rationale for the use of pelvimetry.

First, the contracted pelvis that Thoms saw as the object for measurement became less common and the abnormalities less severe. This was because rickets was becoming less prevalent, not least in the white middle-class population of New Haven, which began to visit the pre-natal clinic

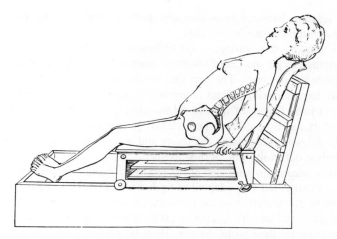

Figure 7.4 Making a pelvigram: the semi-recumbent position with back arched, published in an article by Herbert Thoms. In this position, the plane of the pelvic inlet is made horizontal and parallel to the sensitive plate beneath.

Source: *American Journal of Obstetrics and Gynecology*, 12 (1926) : 545

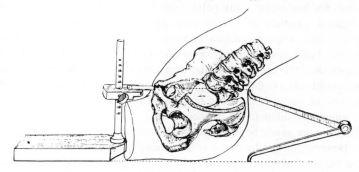

Figure 7.5 Making a pelvigram: making the superior strait horizontal and parallel to the sensitive plate beneath, published in an article by Herbert Thoms.

Source: *American Journal of Obstetrics and Gynecology*, 12 (1926) : 545.

of the hospital in growing numbers.[29] The diagnosis of rickets, which in any case was rather easy using hand methods, was becoming less of a problem.

Secondly, obstetricians increasingly began to pay attention to factors other than the size of the pelvis in their predictions of the problems of delivery. The force of uterine contractions, the maturity of the baby, and the relation between head-size and pelvis, were growing concerns and research topics for obstetricians.

Thoms tried in various ways to link his now perfected method to the changing circumstances of obstetric practice and to the emerging concerns of obstetric research. The fact that rickets was quite easily diagnosed by hand, together with its less common occurrence, was now turned into an argument in favour of the *lateral* imaging technique: 'lacking the experience offered by a clinic where rachitis is frequently seen, one may have difficulties in recognising the lesser and occasionally the moderate grades of pelvic deformity due to this disease'.[30]

New meanings were also attributed to *horizontal* pelvimetry. In the first place the method was adapted in such a way that the maturity of the baby could be calculated from the pictures. On the basis of data from the literature on foetal maturity, Thoms claimed maturity to be related to foetal *weight* rather than foetal *length*. Then he related foetal weight to a particular diameter of the baby's head, on the basis again of data from the literature and the Yale hospital *statistics*. This diameter of the foetal head, of course, was to be measured with Thoms's x-ray method.[31]

Secondly, apart from this cephalometry, Thoms also sought new meanings for horizontal x-ray pelvigrams by trying to establish the value for obstetric practice of pelvic dimensions other than the traditional *conjugata vera*. These were measurements that were never used in obstetric practice because they could not be taken by hand. But since anatomists had established the average, normal measures of several pelvic diameters, Thoms could rely on their conceptions of normal lengths, and in a 1930 publication he wrote of the 'long term recognised and obvious importance of the transverse diameter of the pelvic inlet'.[32]

Obstetricians would be interested in this measure in cases of *flat* pelves, where flattening might be partly compensated by enough transverse room. On the basis of observations on only six flat pelves, Thoms stated that shortening of the transverse diameter was related to a particular, deviant position of the foetus during labour. More evidence was needed better to establish this relationship. In subsequent work, now studying not flat pelves but deviant foetal positions, he had recourse to an even more unknown and abstract dimension. Shortening of the transverse diameter *relative* to the conjugata vera was now proposed as the new and important pelvic dimension. This proportion could be connected with a practical concern of obstetricians: deviant foetal position.

By 1934, Thoms had completed 12 years of research to establish his technique. He had constructed new, clinically relevant, pelvic dimensions, which he claimed as criteria for the involvement of the doctor in the birth-process. He was ready to make bolder claims, directed to a variety of groups:

> I realize that one is prone to make exaggerated statements regarding studies in which one is particularly interested. Nevertheless, I am thoroughly convinced of our woeful ignorance regarding the diameters of the superior strait of the bony pelvis, as gathered from the usual methods of pelvimetry. I have pointed out before that until the advent of x-ray methods in pelvimetry, there was no noteworthy advance in our methods of diagnosis of superior strait contractions since the time that William Smellie first described the estimation of the diagonal conjugate. The role which ignorance plays in our alarming obstetric statistics will certainly be less when men who deliver women possess accurate knowledge of the body pelvis of their patients. Every pregnant woman has a right to expect every scientific aid that can be given to safeguard her own and her baby's life and as a final word, I repeat once more that *regardless of external measurements, x-ray pelvimetry should be an essential part of the prenatal examination of every primiparous woman* (emphasis mine).[33]

Here Thoms marked the beginning of measurement of normal pelves. The strategies he had followed so far had resulted in arguments supported by a variety of interested groups, which maintained the importance of x-ray pelvimetry despite the declining number of contracted pelves. Performing measurements on every primiparous woman would for the first time give Thoms the possibility of doing research on large groups. He thereby created the opportunity to pursue a second goal, already identified twelve years before: x-ray pelvimetry for the classification of pelves. His 1934 article, 'What is a normal pelvis' can be seen as the starting point of this new research problem.[34]

In this section, I have presented Thoms's work as a succession of research goals that were tackled by particular strategies. The problems he addressed were shown to be rooted in the intellectual and technical development of pelvimetric research. What had come to be seen as the diameter containing crucial predictive information about possible complications was taken as a starting point for his research in developing x-ray technology, the technology and the problem of correctly measuring the conjugata vera mutually setting limits and creating possibilities for the ultimate goal of predicting malproportion. It was suggested that the broader context of American obstetric practice was also related to the development of the research at Yale. To understand the significance and saliency of this relationship, and the way in which the intellectual and material context of pelvimetry research fits in here, let us now turn to its institutional setting at Yale.

THE DISCIPLINE OF OBSTETRICS AT YALE MEDICAL SCHOOL

After a period of decline, financial as well as intellectual, Yale, around 1920, was given a large state fund to rebuild its shattered reputation. In the course of the 1920s, buildings were improved, technical equipment replaced and extended, and prominent foreign physicians employed.

For obstetrics there had been additional problems. Although a university chair of obstetrics had existed since the nineteenth century, women with diseases of the pelvic organs admitted to the hospital did not come under the care of this department. It was in 1915 that a special women's clinic was organised with a university resident staff, the first clinic of this type in the country.[35] Further reorganisation of the school, in line with Flexner's report, resulted in a department with a full-time professor and two staff members. In 1921, Arthur H. Morse became head of the department.

After these changes in the staffing of the women's clinic, Yale had too few patients for the staff members' research. In 1923 one of the staff members resigned, referring specifically to the untenable situation with respect to the number of patients.[36] Moreover, the need for patients became still more acute as the number of students grew enormously. It was felt that the education of the students in obstetrics did not provide adequate practical experience.[37] The situation became so pressing that in September 1924 Morse himself wrote a letter of resignation in which he referred to 'the inability of the department to retain first-class men in its organisation because adequate clinical opportunities were unavailable'.[38] The staff would have to be extended in order to train all students well and to produce scientific contributions of a high standard: this meant still more patients. In the negotiations with the Dean, the reasons for patient-shortage were seen to lie not only in inadequate facilities (e.g., too few beds, the scattered location of the different buildings) but also in an inadequate contact with the medical community of New Haven. The doctors of the town apparently did not refer many patients to the hospital.[39] As the Dean of the medical school subsequently reported to the Dean of the University:

> the recent survey ... indicated there was a large amount of obstetrical and gynecological material in New Haven, that could properly be expected to feed the clinic if the interest of this group could be reached and in the opinion of some ... this could be brought about through

prenatal clinics etc. I also pointed out to him [Morse] the desirability of entering into a close affiliation with at least two of the more prominent and better obstetricians and gynecologists in the community. Dr. Morse indicated that Dr. Lewis and Dr. Thoms would be the men he would pick out, and this would ... aid in the development of the Private Pavilion, and might aid Dr. Morse greatly from the standpoint of material for the wards and teaching.[40]

The compromise that was reached eventually provided a part-time appointment for Thoms and his colleague from New Haven in 1925, which was changed into a full-time appointment for Thoms in 1927.

This episode makes clear what was expected of Thoms's work at Yale. Naturally, in the first place it should be of high scientific quality. The capacity to produce such work he had already proved by numerous publications, including several papers on pelvimetry, and he would have been expected to continue this successful line of research.

Secondly, it should be mentioned that Thoms's method was also in accordance with the broader policy of Yale Medical School at the beginning of the 1930s to focus on preventive medicine; attention thus shifted from intrusive to preventive use of technology.[41]

Thirdly, he was expected to attract more patients to a clinic where only a limited number of beds was available. We can see how his attempts to develop a method of x-ray pelvimetry that would have a place in pre-natal examinations carried the possibilities of doing so.

Fourthly, contacts with the medical community of New Haven had to be improved. Coming from that community and remaining part of it with a private practice up to 1927, Thoms could easily help in this respect. However, it seems to me that his research on x-ray pelvimetry created special opportunities in this regard. Because the technology was hospital-based and thus out of the reach of most doctors, it could shape a practice which complemented rather than competed with those of the New Haven physicians. These doctors could refer patients to the hospital in case of a suspected pelvic contraction without fear of losing them. There was a clear division of labour that seemed to the benefit of all.

The measures soon had the result that was hoped for. By 1926 Morse could write that the work of the department had become widely known and appreciated in the community.[42] A 1927 analysis of the increase in the admissions to all the obstetric services of the department over the previous year showed that the main factor responsible for this growth was referral from outside sources. It was contentedly concluded that 'there appears to be a growing appreciation of the services rendered by the department on the part of the local profession'.[43] In the following year a report stated: 'At no period since the establishment of the Clinic upon a University basis have the ward beds been so consistently occupied.[44]

In the early 1930s, profiting from new financial incentives, the department moved to a brand new building where an out-patient clinic, wards and nursing rooms, newly-equipped laboratories, staff and teaching rooms were all located together. The annual report for 1935 then stated: 'In spite of the declining birth rate in New Haven ... clinics in this department have grown'. While in 1916 only 8 per cent of the total births in New Haven was cared for by the department, in 1934–5 this had gone up to 22 per cent. Of the pre-natal patients 221 were registered in 1915, against 700 in 1934–5.[45]

It was precisely in this period (when the number of beds in the clinic was *doubled*) that Thoms started his measurements on much larger groups: all primiparous women. In 1935 his 'widely recognised' method was explicitly mentioned in the Dean of Medicine's published report.[46] Nor was recognition restricted to medical men, to judge by a newspaper article headed: 'X-ray field widened by Yale doctor. Scientist develops apparatus to lessen dangers in maternity cases.'[47]

The (social) success of Thoms's method was very important in providing him with the motivation and legitimation to develop his own work in a situation that was becoming one of rivalry and competition. Two other successful obstetricians (W. Caldwell and H. Moloy), working at Columbia University, were developing an alternative method of x-ray pelvimetry and an alternative classification of female pelves. For the next decade these two and Thoms sought to convince fellow obstetricians of the value of their respective classifications.[48] Subsequently, when interest waned, Thoms radically shifted his focus: he became one of the first strong American advocates, in the USA, of natural childbirth.

DISCUSSION AND CONCLUSIONS

The general circumstances of American obstetrics in the 1920s favoured the adoption of technologies of a non-interventionist kind, like the x-ray pelvimeter.[49] This technique was also oriented to the prediction of possible complications and it was claimed that a thorough pelvimetric investigation could show whether a caesarian section was necessary. This emphasis on prediction fitted in well with a general concern to develop pre-natal care. It also fitted in with the specific wish to bring down the rate of caesarian section, since this reputedly much-abused operation was believed to contribute to the worryingly high mortality figures. Research to develop such a technology could reckon on an interested audience of practitioners and health care authorities. Researchers, too, had an interest, since the technology provided for the initiation and extension of research and for a scientification of existing practices. This does not, however, provide an adequate explanation of how a specific research programme like the one at Yale took shape, nor how the rationale for its development shifted.

It is suggested here that we have to take the local circumstances into account to be able to understand this. It seems clear that Thoms's research was not only in the spirit of the general development of American obstetrics at the time, but that it fitted in very well with local problems and concerns. The particular situation of obstetric practice at Yale called for a kind of research that was prestigious without being labour-intensive; ideally one with practical consequences that would help build up connections with local doctors and attract more women to the hospital. The development of Thoms's existing interest in x-ray pelvimetry fitted the bill very well. But whilst this might make it plausible that a research programme on roentgen pelvimetry could be started and continued for several years, it does not explain why Thoms addressed the problem of pelvic measurement in the way he did, or why he chose particular questions and approaches.

Of course one could claim that it was Thoms's own personality that made him persevere with the development of a technique that was complicated and difficult. In such a view the 'advocate' of a technology is seen as the essential driving force behind its development.[50] And indeed, Thoms's campaign-like struggle for recognition of x-ray pelvimetry seems to indicate a particular psychological make-up. He was described to me by people from Yale that knew and worked with him as 'a real Connecticut Yankee'.[51] Undoubtedly, this would have influenced his way of going about the research. On the other hand, if we want to learn something about the way in which technologies are developed, there is much to be said for focusing on the structural features which constrain or enable entrepreneurial drive.

We have seen that the validity of the technology for obstetric purposes and the feasibility of producing a meaningful image were established largely by research on dry pelves, using very little clinical material. In this period Thoms still worked as a private practitioner with the possibility of using the x-ray facilities of a New Haven hospital where he was a consultant. His other publications at that time were mostly case studies, programmatic statements about obstetric practice, and medical historical articles (a hobby on which he published throughout his life). The pelvimetry research was done in his spare time with borrowed equipment. When he was appointed part-time at Yale, the *clinical* application of the method was given all attention, and by 1927, with a full-time university appointment, he began to publish on applications other than pelvimetry. Obviously it was clinical practice that provided the opportunities to make this shift. At the same time, as was argued before, the direction of the shift accorded with the concerns of obstetricians and with concerns in the broader social context. We can thus conclude that Thoms's *clinical practice* focused his broader intellectual and social context onto specific research questions.

Finally, after 1930, when the number of pre-natal patients had grown (no doubt as a result of Thoms's work), the larger clinical research series were started from which he concluded that every primiparous woman should be x-rayed. Applying this recommendation in his own Yale practice, Thoms was enabled to address quite other research questions, for which a great number of cases was needed. In particular, he could work on the general characteristics of pelves and their classification. So not only was the changing practice at Yale shaping the research problems, the research, in turn, also shaped clinical practice.

If local concerns in Yale provided Thoms with an ideal context to make this shift possible and plausible, his objectives were not restricted to his own clinical practice or local context. In his publications, he addressed the wider community of obstetricians, not only with legitimations for his reported research, but also with general arguments for the introduction of x-ray procedures in obstetrics; this raises the question of how successful he proved in persuading the profession to adopt his perfected technology. How widely was Thoms's roentgen pelvimetry technique used?

The impression one gets from discussions of papers at local or national meetings of obstetricians, as they were reported in the *American Journal of Obstetrics and Gynecology* or in conference reports, is that x-ray pelvimetry was widely known and discussed, and certainly much published about. But it also seems that many objections were raised against its use and that a not inconsiderable number of obstetricians restricted themselves to using the older hand methods of pelvimetry. This impression is confirmed by the results of a 1947 survey among obstetricians in the South Atlantic region, from which we can estimate that in only 4 per cent of the cases were women x-rayed for pelvimetric purposes. For these cases, 50 per cent of the obstetricians used Thoms's method.[52] So, although Thoms seems to have been rather successful in establishing his method as the preferred one, roentgen pelvimetry, in general, was far from being a standard procedure in obstetric practice.

We would expect there to be major variations among institutions in early recourse to this (as to other) technology.[53] This does indeed seem to be the case. Associated with institutional variations, we find the technique more frequently used on private patients, who were more likely to be able to pay for the pictures, than on clinic patients. This may account for the low incidence of *routine* use of the technology.[54] Obstetricians were not over-enthusiastic. The 1947 survey showed that 'a third of the obstetricians find pelvic x-ray studies to be of real value, but only 50 percent of this group use x-rays as an important part of their obstetric armamentarium'.[55] For two-thirds of those interviewed, pelvic x-ray studies were of only occasional, or even questionable, value.

At this point a present-day reader might find herself or himself wishing to suggest an obvious explanation for the obstetricians' reluctance – i.e., the safety considerations involved. However, this does not seem to have been the case, although from the early years of the century it was known that x-ray radiation could burn the skin and produce deformations in animals and mutations in their offspring. Indeed, possible long-term damage to a baby that seemed perfectly healthy at birth was considered as early as 1926. The general argument of obstetricians and of radiologists working in roentgen pelvimetry was that the doses used for diagnostic purposes during pregnancy were far below the danger limit.[56] The well-known radiologist Paul Hodges and his colleague Jane Hamilton remarked about pelvic roentgenography in 1938:

> Even when one uses the maximum of ten films the amount of radiation absorbed by the skin of the mother is far below the erythema dose and that absorbed by the foetus is very small indeed. There is no reason at all to fear that this small amount of radiation will injure the foetus, and those who have expressed such fears presumably confuse the conditions of diagnostic raying with those of therapy.[57]

It was only after 1945 that radiation hazards became widely discussed in public media and only then did some pelvimetry workers become more careful about applying the technology to pregnant women. A 1956 statement by radiologists themselves that even the most minute radiation was hazardous led, according to one observer, to an 'almost hysterical radiation consciousness in the general public'.[58] When in that same year a British paediatrician published an article in which she showed a correlation between the incidence of childhood leukemia and radiation of the mother during pregnancy, the climate was fertile for the acceptation of her claims.[59] Although the interpretation of her statistics was contested, the publication had an enormous impact and 'killed off' pelvimetry.[60] For the pre-war period, however, we cannot include safety considerations as determining the sparing use which obstetricians made of x-ray pelvimetry.

There seem to be three sorts of reasons for their reluctance. One had to do with the fear that caesarean section rates would *increase* as a result of using the technique. Information about pelvic capacity gained from the pictures could ease the choice of this operation over the more difficult forceps or breech extractions. This reason was given in the 1947 survey, and we also find Thoms addressing the point of presumed higher incidence of caesarean section as a result of the use of roentgen pelvimetry.[61]

A second reason could be that the use of roentgen technology was relatively costly, not only for the patients, but also for the hospital. Costs of roentgen technology was a recurrent theme, not only in papers on roentgen pelvimetry, but generally in radiological literature of the 1930s.[62]

Finally, a third, and maybe most important, reason seems to be the difficulty of the different measuring techniques and the fact that a radiologist was involved in making the picture.[63] Hence a fear that obstetricians' own clinical skills might become subsidiary to technologies that were handled and controlled by others. In fact, the initial enthusiasm of the obstetricians about the possibilities of roentgen pelvimetry seemed to decline with the rise of the new specialty of radiology. DeLee's position is illustrative. Commenting in the *Yearbook of Obstetrics and Gynecology* in 1933, he wrote: 'There should be an x-ray room on the birthroom floor of every large maternity [sic] – although we older obstetricians are a little jealous of the technician who makes unnecessary the tactile skill we have developed in our fingers for many decades'.[64] But by 1938 his tone had become quite different:

> Yes, by all means dear x-ray men, give us all the information you can about the size and the shape of the pelvis, the size and shape and position of the baby, etc. etc, but please don't feel hurt if we do not accept all you tell us (or all we see ourselves) and use it exclusively in our deliberations on the conduct of our obstetric cases. We have suffered much from shadow-boxing ... The shadows are not easy to interpret, and even the stereos may cause optical illusions. The real obstetrician uses his hands, his eyes and his brains first, and the roentgenogram as supplementary or corroborative evidence.[65]

X-ray pelvimeters apparently had taken on characteristics of a real Trojan horse for the average obstetrician: status-enhancing, but at the same time embodying a high-risk professional strategy, resulting in loss of patients, loss of authority and loss of clinical judgement.[66]

The argument against *routine* x-ray pelvimetry continuously brought forward in discussions at meetings and conferences, was that hand methods were accurate enough to deal with the great majority of cases. Only if disproportion was suspected could roentgen pelvimetry provide useful extra information. It seems that the disadvantages of applying roentgen pelvimetry were such that this argument could take on decisive meaning and thereby limit the role of the technological procedure in everyday obstetrical practice.

In the Yale clinic, x-ray pelvimetry as a routine pre-natal examination was discontinued around 1957.[67] It is interesting to see how Thoms, by then head of department, had come to understand his own practice. In a letter to a friend he defended his previous work, but in recounting his steps he relativised the importance of routine measurements:

> I recognise of course that x-ray is a lethal agent and I am sensitive to the fact that exposure to x-ray can be harmful and its use should be avoided whenever feasible except when it is indicated for definite clinical purposes and also when it is used the exposure should be minimised as far as possible ... I feel that this knowledge [gained by previous work in roentgen pelvimetry] is so great and so useful to clinical obstetrics that these studies were justified even though further knowledge has shown us that x-ray radiation can be more harmful than we supposed ... Obstetricians should realise that although the routine use of x-ray pelvimetry can add to our knowledge ... the information is chiefly of value as assurance to the physician and the patient *that pelvic abnormality does not exist*. In other words the percentage of women in which x-ray information has clinical importance is relatively small probably less than five percent (emphasis mine).[68]

Thoms tried everything to fit his technique for general obstetric use. But although of all the x-ray methods of pelvic measurement his was the most widely adopted, it never became routinely used, certainly not for every primiparous woman as he had hoped. Precise measurements of the pelvis, with complicated and expensive techniques, really interested only the research-oriented obstetricians, not their practice-oriented colleagues.[69] The x-ray pelvimeter seems to have remained more a tool for research than for practice. After its use for obstetric purposes had been legitimised and after it had been adapted to existing obstetric problems, the exploration of the possibilities inherent in the technology became an end in itself. Obstetric problems were then adapted accordingly.

8
THE POLITICS OF A SPATIAL INNOVATION: FRACTURE CLINICS IN INTER-WAR BRITAIN
Roger Cooter

By conventional standards, the subject of this essay would seem hardly to qualify as a 'medical innovation'. Fracture clinics did not appear suddenly or sensationally, nor were they heralded as a scientific or technological breakthrough. Put forward, for the most part, in unsentimental businesslike terms, they have no obvious historical resemblance to the introduction of, say, x-rays, antibiotics or heart transplants.

Yet looked at more broadly, the innovativeness of fracture clinics is difficult to deny. Before the First World War most broken limbs were treated by general practitioners in their surgeries, by medical officers in the 'casualty' departments of voluntary hospitals, or by house surgeons on the wards. With the advent of fracture clinics during the inter-war period, however, these injuries came to be treated by specialists in purpose-built accommodation within or attached to hospital 'accident' departments. In this new environment patients no longer had their fractures merely reduced and splinted before being sent home; instead, they were subjected to a specially organised routine involving systematic case-recording at reception, x-ray examination, plaster-of-Paris bandaging by a specially trained nurse, after-care physiotherapy, and post-operative monitoring – all under the careful watch of a fracture surgeon-in-charge.

First and foremost, therefore, fracture clinics were an achievement in organisation and management, involving a seizure of therapeutic space by a specialist group for the exercise of control over a particular patient population. More precisely, they were (in Britain, at least) the means by which orthopaedic surgeons sought to acquire authority and status within the structure of hospital medicine. As such, fracture clinics were essentially a political innovation with significant implications for the distribution of power and authority in medicine generally. The late Sir Harry Platt (1886–1986), the first person to establish a fracture clinic in a British hospital, and one of their major advocates, recalled:

In the years between the wars the field of orthopaedics was to those of us then young an expanding universe, and we fought the battle for the control of fractures with gusto ... Our opponents often accused us of adopting the attitude of the German philosopher Nietzche [*sic*] – 'that a good fight sanctifies a cause'.[1]

But the cause was not all, or only, political. Fracture clinics were neither simple nor cheap to establish and operate, and unlike most other specialty services they could expect to receive little custom from the monied middle-class. It was estimated in 1937 that the minimum staffing requirement for a clinic in a city hospital dealing with between 2000 and 2500 fractures a year would be a surgeon-in-charge, two house surgeons, a radiographer, a records clerk, and a stenographer. The estimated annual cost for this minimum staffing was £1600.[2] Where new buildings and equipment were required, the costs were far greater.

The proponents of fracture clinics thus had formidable market forces to conquer in addition to the competing interests of general practitioners and hospital general surgeons. In pursuing their cause, they would mobilise not only the British Medical Association (BMA), but also leaders of industry, insurance companies, the Trades Union Congress (TUC), and on up to government both local and national. Ultimately, such involvements were to render the treatment of fractures in inter-war Britain nearly as public and controversial as the contemporary issues of tuberculosis and maternity.

This essay addresses the question of how and why fracture clinics were proposed when they were, and how the innovation was sold. In so doing, it seeks to illuminate the wider social, ideological and economic contexts, as well as the immediate medico-political circumstances, which constrained and which facilitated this innovation. Fracture clinics, in some ways, were representative of general developments in early twentieth century medicine; they were part of a pattern of increasing specialisation. But, as this essay will show, this innovation was also highly dependent on particular historical contingencies.

BACKGROUND

Although it is commonly believed that modern orthopaedics derived in large part from traditional bone-setting, it was in fact only through the First World War that the treatment of fractures was brought within the scope of orthopaedics. Until then, orthopaedic surgeons had been mostly concerned with the correction of deformities among children; fracture treatment, by contrast, had remained in the hands of GPs and hospital general surgeons.[3] The only exception was the work of the Liverpool orthopaedist, Robert Jones, which was conducted in a private clinic, mostly for the treatment of industrial injuries.[4]

During the war, through the political and organisational efforts of Robert Jones, a generation of young orthopaedists appropriated the treatment of fractures. However, after the war the special war-time centres for this work were disbanded and Jones and 'his men' were unable to carry over into civilian hospital practice their military gains. To old-guard general surgeons (especially in London), Jones and his colleagues appeared as the most conspicuous of several insurgent specialist groups threatening to outmode general surgery. In July 1918, for example, a special committee of the Royal College of Surgeons, headed by the president of the College, sought to circumscribe orthopaedics, viewing with 'mistrust and disapprobation the movement in progress to remove the treatment of conditions always properly regarded as the main portion of the general surgeon's work from his hands, and place it in those of "Orthopaedic specialists"'.[5] Thus, after the war, most of the members of the British Orthopaedic Association (BOA) – established by Jones and his colleagues in 1917 – were more or less compelled to retreat to the therapeutic territory that had been theirs before the war – namely, crippled children. Their energies thus directed, by the late 1920s they had succeeded in developing an impressive national network of orthopaedic hospitals and clinics for the treatment of children suffering from rickets, poliomyelitis and, above all, tuberculosis of the bone and joints.[6] For the most part, it was from this accomplishment, made *outside* the major centres of British medicine, that orthopaedists returned to the issue of fractures.

THE PILOT FRACTURE CLINIC

To this general picture of the post-war loss of fracture treatment by orthopaedists, there were two notable exceptions – the work of Meurice Sinclair, conducted at the St James Poor Law Hospital, Balham, London, and that of Harry Platt at the Ancoats Hospital, Manchester. Sinclair's work derived directly from his pre-eminent war-time handling of fractures at a base hospital in France, and it was to have considerable bearing on the London County Council's thinking on the organisation of fracture services in the 1930s. But it was little noticed before then.[7] Platt's work at Ancoats, by contrast, was intended from the start as a demonstration of orthopaedic specialisation. Much influenced by Robert Jones, it was also greatly indebted to developments in American orthopaedics, especially in Boston, where Platt had gone to further his training in 1913. Boston was then the world's centre for all that was most progressive in orthopaedics, and although segregated fracture clinics were not developed there until the 1920s, orthopaedics had already extended its territory from the treatment of deformities among children to the treatment of acute locomotory injuries among adults.[8]

Ancoats Hospital, where Platt was appointed honorary consulting surgeon in April 1914, was a small voluntary hospital. A typical place for a junior appointment, it was also in many ways ideally suited to a surgeon with budding orthopaedic interests anxious to try out American-style reforms. It had a relatively liberal board of managers; both of the other honorary surgeons were also young, ambitious and willing to try out new ideas and techniques; and the situation of the hospital, in one of Manchester's densest industrial districts, rendered it essentially an accident hospital with a large turnover of fracture cases.[9] Within months of his appointment Platt was conspiring with his colleagues to do away with the traditional routine whereby the surgeon on duty took responsibility for all incoming cases. Instead, the cases, along with the hospital's 75 beds, were divided among the surgeons according to their areas of specialist interest. Thus Platt was able to establish therapeutic control over all incoming fracture (and other orthopaedic) cases and impose uniformity in their treatment.

The strategic potential of small hospitals like Ancoats for rising specialisms (at least in the Manchester region) was also illustrated at the Salford Royal Hospital, where Platt's friend and war-time colleague, Geoffrey Jefferson, was allowed to concentrate on neurosurgery. But it was much more difficult for orthopaedists to carve out specialist niches in the major teaching hospitals where their interests threatened the physical territory (beds) and the confidence claims of general surgeons. Some London hospitals had established orthopaedic departments just before the First World War, but these were largely for chronic cases, rather than the victims of accidents, and were usually under the control of general surgeons rather than aspiring specialists. As we shall see, one of the staunchest opponents of the orthopaedic specialists in the 1920s was George E. Gask, who in the years 1900–10 had himself been Chief Assistant to the Orthopaedic Department at St Bartholomew's Hospital.[10]

During the war Platt was deeply involved with Manchester's military orthopaedic centre at Grangethorpe; his plans for Ancoats were delayed a little, but his experience was enlarged and his commitment to 'progressive orthopaedics' strengthened.[11] After the war, between 1919 and 1921, he was able (at Ancoats) to establish the world's first segregated fracture service in a general hospital under the control of an orthopaedic surgeon. He carried over into civilian practice the main war-time lessons of effective fracture treatment: segregation, expert supervision, team work, continuity of treatment and appropriate follow-up or after-care. To these features he added detailed record-keeping on the social and medical condition of patients, a technique he had learned in Boston and which, applied generally, was basic to the scientific management of American hospitals.[12] Such were the essential principles of the 'ideal fracture service' reported by Platt in the *Lancet* in 1921;[13] they were to be reiterated in the barrage of papers and lectures issued by orthopaedists from the mid-1920s to the 1940s; and they were to receive the backing of the medical establishment in the BMA's highly influential 'Report on Fractures' (1935) – written mainly by orthopaedists at the time when Platt was president of the BOA.[14] Directly as a result of the BMA's report, a motion was passed in the House of Commons in April 1936 which led to the setting up of the Government's 'Inter-Departmental Committee on the Rehabilitation of Persons Injured by Accidents', under the chairmanship of Sir Malcolm Delevingne. Through the Delevingne Committee's *Interim* and *Final* reports of 1937 and 1939, official endorsement was given to the princi-ples first implemented by Platt.[15]

CAMPAIGNING FOR THE CAUSE

The long interval between the establishment of the fracture clinic at Ancoats and the publication of the above reports strongly suggests that example alone was not sufficient to make the hospital treatment of fractures a public issue. Nor can it be assumed that consciousness was raised, and the innovation sold, merely through the rhetoric of zealous orthopaedists. Social, political and economical factors were also to play a crucial role. Nevertheless, it is worth considering first the rhetoric of the orthopaedists concerned with fractures, not only because it preceded the rise of certain prominent socioeconomic issues and intersected wider political ones, but also because it reveals the importance of professional interests, as opposed to technological factors, in the history of this innovation. Fracture clinics of course made use of certain technologies, just as they relied on a variety of surgical skills, but they were not principally advocated on those grounds. Plaster-of-Paris might have been their emblem by the 1930s, but their essence was scientific management and an implied recasting of the networks of professional power and control.

Among the more important of the battles for the control of fractures was that precipitated by Robert Jones in a lecture on 'Crippling due to Fractures' delivered in Liverpool in May 1925 and subsequently published in the *British Medical Journal*.[16] This may in fact have been encouraged by Platt and by Platt's orthopaedic colleague, W. Rowley Bristow of St Thomas's Hospital (the person, according to Platt, who was largely responsible for spearheading 'the attack on the idealogical [sic] barricades of the London teaching hospitals').[17] In any event, the lecture was regarded by friend and foe alike as a 'slashing attack on the "methods by which fractures are dealt with at the big teaching hospitals" '. Opening with the observation that the existing treatment of fractures was 'a blot upon our surgical escutcheon', Jones went on to decry as 'hopelessly wrong' the situation in 'the big teaching hospitals' where out-patient ambulatory fractures were treated by unsupervised junior medical officers and where in-patient fractures – ostensibly of little interest to general surgeons who resented them for blocking beds – were left to the care of under-trained house surgeons. 'Even more pathetic', according to Jones, was the fate of fracture cases sent to Poor Law infirmaries, most of which were totally unfitted for the modern treatment of fractures. In short, he believed, it would be 'far better for our hospital authorities to say, "we are not prepared to treat fractures", than that they should take on responsibilities which they cannot meet ... We cannot disguise the fact that great numbers of adult cripples are manufactured by want of adequate provision'.

Like subsequent advocates of specialised fracture clinics, Jones had few statistics to offer in evidence of the 'bad old way' of treating fractures. The badness of general surgeons in this area, he maintained, simply '[did] not admit of argument'. Convinced that better results could be obtained by properly-trained fracture experts, he could refer only to the organisational principles of care worked out during the war, and cite the reputed 20 per cent reduction in mortality from fractures that had resulted.

In reality, Jones's 'plain speaking', as he claimed it, was a reiteration of many of the complaints against the treatment of fractures that had been heard intermittently since the middle of the nineteenth century.[18] Now, however, such arguments struck a different chord; though Jones might protest that it was 'not a mere matter of the encroachment of specialism, but a call to our sense of proportion and sense of duty', his polemics hinted at a return to the war-time division of medical labour, with organised clinics under specialist control.

It was at a BMA meeting in Bath in August 1925, during a special session on 'Surgery and Orthopaedics' arranged by Platt and his friends, that the expected confrontation occurred.[19] It was here that George Gask made his début as the defender of generalism over surgical specialism. Though Gask (by this time the Director of the Professorial Surgical Unit at St Bartholomew's Hospital) was hardly a conventional opponent of specialisation, it was almost impossible for him to emerge from the proceedings at Bath as other than a 'quixotic [defender] ... of a losing cause'.[20] To the 'modernists' in orthopaedics, who imagined 'progress' in explicitly Spencerian terms of greater efficiency through greater division of labour,[21] Gask retorted that the extended division of labour which served the needs of the munitions factory hardly applied to the manufacture of good general practitioners.[22] But the orthopaedic ideologues were not interested in the education of GPs, for which Gask mostly spoke. They linked bad fracture treatment with the fact, made clear at the meeting, 'that probably one half of the fractures in this country were treated at home or in cottage hospitals by general practitioners', and concluded that fractures needed to be treated under their own expert supervision.[23]

Although the incompetence of GPs in fracture treatment was to figure in some of the rhetoric for specialist fracture treatment in the 1930s, this line of argument was never to move to centre stage (nor were there ever statistics on either the extent of the treatment of fractures by GPs or on their clinical results). Since orthopaedists relied on GPs to refer crippled children to their clinics, there was reason not to antagonise them. But GPs, for their part, did not generally feel threatened by the advent of fracture specialists,[24] for they could refer a patient to an orthopaedic specialist without losing that patient from their patient list and, hence, without forfeiting National Insurance remuneration. Overworked GPs were also mindful of patients' desires for specialist care, of the complications involved in different fracture cases and, not least, that complicated cases entailed extra costs which might go unremunerated from local National Insurance Committees.[25] More idealistic GPs could construe the orthopaedists' campaign for fracture services as akin to their own interest in improving community medical services. Indeed, Sir Henry Brackenbury, who spoke for GPs and who served on the BMA's Fracture Committee, seems to have seen orthopaedists and GPs as allied underdogs, fighting together against the reactionary general surgeons who were powerful in hospitals.[26]

The orthopaedic advocates of fracture clinics thus continued to perceive the old-guard general surgeons as the main targets of their rhetoric. However, after the meeting at Bath, they toned down their discussions of specialisation (in part for the reasons discussed below), and concentrated on the practical means to achieve the best possible results in the treatment of fractures. This, indeed, was how Jones had pitched his reply to Gask in 1925 and by 1928, when the budding proponent of fracture services, Ernest Hey Groves, delivered his presidential address to the BOA 'On the Treatment of Fractures', it was this practical matter that was emphasised as 'the problem'.[27] The solution to the problem, it was increasingly stressed, was 'not new knowledge' – least of all new technology or new research into methods for the internal or external fixation of fractures – but rather, 'the organized application of the knowledge we already possess', namely, the 'segregation of cases, the training of team workers, and the systematic tabulation of results'. Left unspoken were the implications of this 'application of knowledge' for the organisation and structure of British hospitals. Left unspoken, too, it might be added, was the shrinking market for orthopaedists in the cure of crippled children.

THE CAUSE IN POLITICAL CONTEXT

The debate at Bath is a good illustration of how specialisation was negotiated in inter-war Britain. But it would be misleading to cast the rhetoric of the leading advocates of fracture clinics wholly in terms of specialist self-interest. Fundamentally, the campaign for fracture clinics was conceived in terms of the 'modernist' ideology of expertise and organisation-cum-rationalisation, rather than in terms, simply, of generalists versus specialists. Old-guard general surgeons, who claimed mastery of the field, were attacked by Jones and his orthopaedic colleagues primarily because they were the most visible and politically pertinent opponents of 'modernism'. Indeed, for the main advocates of fracture clinics, the issue was in part a political resource for their own status and authority as medical politicians.

These broader ideological interests cannot be fully understood outside the political and economic context; the early 1920s were, by and large, a period of retrenchment in health care, but the financial crisis that lay behind that retrenchment also provided an increasingly compelling argument for fundamental change. Although in 1921 the Cave Committee, in its *Report on Voluntary Hospitals*, sought to shore up the existing system, in the following year the TUC and Labour Party, in a pamphlet on *The Labour Movement and the Hospital Crisis: A Scheme for a Hospital Service*, put forward a sweeping programme of reform. The latter's social and political impact is open to question;[28] but obviously basic to the scheme were the concepts of 'continuity of treatment', 'team work', and the 'unification of control' over all hospital facilities and services. This similarity to the rhetoric of the orthopaedic reformers was hardly coincidental; it reflected the shared faith in rationalisation, or the application of 'system and uniformity' to problems of ever-greater complexity and waste. As is well known, this faith was inspired by American commerce and industry where the 'scientific' principles of bureaucratic management had proved themselves 'rational' from the point of view of economic efficiency.[29] For many, the experience of the war had cemented this faith.

But the campaign for the standardisation and co-ordination of hospital services in Britain involved more than merely the extension of the principles of rationalisation as they were beginning to be applied in post-war industry, transport and agriculture (as well as in trade unions). For the TUC and Labour Party, the call for the rationalisation of hospital services was part of a bid for an egalitarian distribution of medical power and provision in a state medical service. As such, their plans were different in orientation and in principle from those contained in the famous Dawson Report of 1920. Whereas in the Dawson Report the argument for rationalisation was concentrated on general practice with little reference to accident and emergency services, in the TUC and Labour Party's vision of a unified and classless medical system, hospitals were central, and accident services were held to be at the front line of medical care.[30] In promoting a uniform, co-ordinated accident service they criticised the existing unequal, haphazard and confused hospital system. In fact, here, as nowhere else in medicine, it could be made apparent that in order to meet the needs of patient populations it was vital and urgent – as vital and urgent as in the war – to have a regionally co-ordinated, fully rationalised system. Ideally, too, according to some, this would be a hospital system in which doctors were salaried (as in the war), since there could be little scope for private practice in the treatment of acute injury.

To no group in medicine was this line of argument more pertinent than to the orthopaedic advocates of fracture services. But there was little hope of implementing such schemes in the financial climate of the early 1920s. To understand how the rhetoric of the orthopaedic reformers came to have its impact in the 1930s, it is important to appreciate not just the political context, but also the more immediate economic pressures and resources for debate that came to surround accident services – in particular, the growing concern over the incidence and cost of motor vehicle accidents, and the growing interest in the Vienna Accident Hospital as a model for the economical treatment of industrial injuries.

MOTOR VEHICLE ACCIDENTS

The rise in the number of motor vehicle accidents, became a prominent issue in the mid-1920s. 'Before the era of the motor car', commented the *Lancet* in 1926,

> the receiving officer of a large general hospital might be embarrassed by a run of Pott's fractures on the first snowy evening of the year; today a fine Sunday evening may overwhelm any cottage hospital with a glut of complicated injuries.[31]

For England and Wales non-fatal street accidents involving motor vehicles increased fourfold between 1913 and 1932: from 38 000 to 162 000, with the most striking rise (unsurpassed until the 1950s) occurring between 1927 and 1934.[32] And, as the Ministry of Transport was acutely aware in 1928, 'There are few questions, as reference to the daily press will show, which excite more constant and widespread interest'.[33] Although the number of domestic and industrial accidents was greater, the socio-medical profile of road accidents was considerably higher, in part because these accidents were not confined to specific geographical areas.[34]

Allied to the troublesome spatial aspect of motor vehicle accidents was the financial spectre they raised: the problem of recouping from insurance companies the costs of treating the victims, especially those injured away from home. The BMA reckoned that doctors were paid in only one out of five cases; and in 1931 it was estimated that some 25 000 victims of motor accidents treated as in-patients in voluntary hospitals had cost nearly one-quarter of a million pounds.[35] The problem was partly solved by the Road Traffic Act of 1934, but the general issue of recovering costs for accident victims who were covered by insurance for *other than* hospital medical services remained contentious. Indeed, it was partly out of this issue that the widespread interest in the Vienna Accident Hospital arose.

BÖHLER'S CLINIC

Erected in 1925, at a cost of £10 000, the Vienna Accident Hospital was a fully equipped 125-bed establishment which was entirely maintained by the Austrian National Insurance Company (at a cost of around £16 000 p.a.). It was the brainchild of Lorenz Böhler, a general surgeon who had developed a special interest in the treatment of fractures during the First World War.[36] Like Platt and his colleagues, Böhler had become a propagandist of segregated fracture treatment. A visit to his 'mid-European fracture synod', as the *Lancet* once referred to it,[37] became a badge of progressivism among orthopaedists in the 1930s.[38] Like an earlier generation of visitors to Robert Jones's private clinic in Liverpool, visitors to Böhler's clinic were impressed not only by his various surgical and manipulative techniques (many of which were praised as 'revolutionary'), but even more by his meticulous organisation. Böhler himself regarded his organisation for the overall control of fracture treatment as the secret of his success, and specifically likened it to the progress of rationalisation that in recent years had rendered industry, transport and agriculture more efficient. It was by means of the sub-division of the labour process, he insisted, that the parts of medical treatment were simplified and greater efficiency resulted.[39]

But it was above all the demonstrable economic advantages of this efficiency that gained Böhler his fame; he boasted an impressive – and, for the time, unique – accumulation of statistics. As the *Lancet* hastened to point out in its editorial on 'Accidents and Hospitals' in 1926 not only had Böhler shown the clinical and social advantages of specialised fracture treatment in restoring patients to their full economic-earning capacity, but he had shown the insurance companies that surgical specialisation could render enormous savings. That the Vienna Accident Hospital had 'saved the insurance societies something in the neighbourhood of £18 000' was just the sort of 'astonishing' fact to stimulate interest in the whole question of accident services and, at the same time, to draw attention to the 'grave scandal' that in Britain the insurance companies 'contribute nothing towards the treatment of their injured clients in hospital'. 'Both equity and self-interest alike', the *Lancet* maintained, 'should lead the [insurance] companies to support an accident department in every hospital'.[40]

Not surprisingly, precisely this conclusion was taken up by the orthopaedic advocates of segregated fracture services. Ernest Hey Groves, who was more insistent than most about the need to convince British insurance companies 'that organisation of fracture treatment would effect so much saving in compensation as more than to cover the cost of the treatment', translated Böhler's work on fractures into English. In his preface he noted that Böhler had 'demonstrated that the proper treatment of fractures is not only a scientific problem or a philanthropic duty, but also a business proposition. In other words, it pays to treat fractures well!'[41]

But there were two obvious reasons why the British proponents of segregated fracture services needed to be highly selective in their use of Böhler's model. First, Böhler's hospital was intended not just for fractures, but for accident cases of all kinds. References to the Hospital did not therefore automatically serve the interests of those seeking to expand their professional space on the basis of fracture treatment alone. Although in certain industries and industrial regions in Britain fractures predominated over other injuries, in general only about 10 per cent of all cases of injury involved fractures.[42] Arguments for the segregation of accident cases, therefore, did not necessarily promote a commanding role for orthopaedists, or even necessitate their employment in accident services.

The other reason why orthopaedists had to be selective in their use of the Böhler model was that Böhler's hospital was a separate institution. As such, it hardly furthered the orthopaedists' main ambition of securing for themselves and their specialism a permanent niche *within* the major teaching hospitals. The last thing that British orthopaedists wanted (as one of them put it), was to end up like Böhler 'cut off entirely from general hospitals and cut off entirely from the general surgical and medical staffs [and students]'.[43]

Yet it was not by rejecting outright the idea of accident hospitals that those interested in segregated fracture treatment established the priority of their claims. Rather, by putting accident hospitals forward as one option among others they exploited the wider social and political issue of accident services, and also served their own interest in hospital fracture clinics. While explaining the difficulties involved in establishing separate accident hospitals, they revealed the economic viability of such clinics. It was only after the establishment of the Emergency Medical Service during the Second World War – that is, after government backing had already been secured for hospital fracture services under orthopaedic control – that BOA members began collectively to push for the national provision of what they at first called 'Orthopaedic and Accident Services'.[44] And only then was it to be said (as by Platt in 1950) that 'the Vienna experiment ... has proved that *within the framework of comprehensive orthopaedic schemes* there is a place for the accident hospital' (emphasis mine).[45]

INDUSTRY, STATE AND ORGANISED LABOUR

Throughout the 1930s orthopaedists continued to argue that the private insurance companies ought to be induced or be compelled by 'legal enactment' to support fracture services. Hopes for tangible results from this quarter steadily diminished, however. The insurance companies were simply not interested in arguments for the more efficient treatment of fractures: '[they] simply say that so far as they are concerned, it is a question of finance, and the premiums are so regulated that they cover even the most expensive case'.[46]

Nor could the interests of employers be aroused. Although the hope that large industry might provide financial and moral backing to the fracture cause had existed from the beginning of the movement, and had been nourished in the 1920s by American examples of industry-based accident and rehabilitation services,[47] it was not until the early 1930s that the exponents of fracture clinics in Britain began in earnest to make their pitch to industrialists. A major source of motivation was the work of H. E. Moore at the London Midland and Scottish (LMS) Railway yards at Crewe. Begun on a small scale in the late 1920s at the suggestion of Gwynne Maitland (medical officer to the Cunard Steamship Company), Moore's rehabilitation of fracture cases became widely known in government and medical circles in the early 1930s. Although British orthopaedists were not interested in industrial medicine as such, Moore's work provided them with useful socioeconomic ammunition in the cause of fracture clinics. They could argue that it was economically advantageous (for large self-insured companies, at least) to have industrial injuries – and fractures in particular – treated by experts in order to avoid the unnecessary legal bother and expense of workmen's compensation claims and to avoid, as well, purported malingering.[48]

Very largely this was how the case was put in the appendix to the BMA's 'Report on Fractures' of 1935. Since Moore and Maitland were among those who made up the committee that produced the Report (along with Bristow and Hey Groves), the bias is hardly to be wondered at. Yet beyond alerting employers to the issue of fractures, the pitch had little effect. In a context of cheap labour, few employers were seriously concerned about labour 'wastage'. Moreover, few employers were like the Cunard Steamship Company or the LMS Railway in carrying their own insurance risks, and they therefore had as little incentive as the insurance companies to become directly involved with fracture treatment as a means to economy.

Increasingly, therefore, orthopaedists came to realise that only through state initiatives and through public funding could fracture clinics be effectively implemented. Through their work with crippled children (which involved remuneration from the Board of Education, Ministry of Health, and Local Authorities), they were already accustomed to state support. And, in fact, in the case of some of the municipal hospitals (those taken over from the Poor Law through the Local Government Act of 1929) local authority funded fracture services were already in operation by the early 1930s. Orthopaedists encouraged and collaborated in this development, but it failed to meet their ultimate objective of securing their place in the more prestigious voluntary hospitals. One of their proposals, therefore, was that the local authorities 'subsidize the voluntary hospitals for this service from public funds'.[49] At the very least, they argued, the voluntary hospitals ought to enter into collaborative fracture schemes with the municipal hospitals.

But the voluntary hospitals were reluctant to budge. Hard-pressed for funds, and without a legal basis for means-testing (unlike the municipal hospitals), most of them preferred to develop services for patients who could be expected to contribute. Only in exceptional hospitals, such as the one at Ancoats, did most of their income come from industrial workers. In voluntary hospitals such as this, the fund organisers had influence in determining services and the hospital had good reason to attend to workers' interests. But elsewhere in the voluntary hospital sector the outlook was bleak for advocates of fracture clinics. 'One almost fears that nothing short of some social cataclysm, such as Communism, bankruptcy, or war, will be strong enough to break old prejudices' complained Hey Groves in 1933 after a frustrating and futile attempt to secure the co-operation of the voluntary hospitals in Bristol.[50] Meanwhile, in London, some of the larger voluntary hospitals were arranging to send their fracture cases to those London County Council hospitals where segregated fracture services had been established.[51]

The recognition of the obstacles to fractures clinics in voluntary hospitals led orthopaedists in the early 1930s to occupy a middle ground between voluntarism and statism: they sought 'State aid, which does not involve State control'.[52] Like most hospital consultants, the orthopaedic advocates of fracture clinics feared bureaucratic controls, but they also wanted to be free of lay managers who could still use 'any excuse or justification for behaving as though they controlled a proprietary institution'.[53] State aid was thus seen as the means to bring the voluntary hospitals further under the control and direction of 'medical experts', as well as the means to new equipment and buildings.

The occupation of this middle ground was not unique to orthopaedists, yet it was their interests above all that were served by the argument for state aid, since they had so little scope for private practice. As pointed out during the discussions for a fracture service in Dundee in 1944, the financial situation of orthopaedists was almost unique among specialists: 'the Orthopaedic Specialist has a far higher percentage of charity work than any other clinician – i.e., any clinical lecturers may pick up a fair amount of private work in general surgery but your orthopaedic colleague gets very very little as operations on long stay cases simply cannot be paid in private and consultations are scant'.[54] Thus, for the political architects of orthopaedics to establish and maintain the sought-after niche for the specialism in the voluntary hospitals, they needed to secure salaries, on at least a part-time basis, for those of their number who were going to act as the surgeons-in-charge of the fracture clinics. But before the mid-1930s the idea of salaried consultants was virtually unthinkable within the voluntary sector, and even in the London County Council's hospitals it was not until the mid-1930s that the practice was begun on a small scale. It is hardly surprising, therefore, that Hey Groves and other orthopaedic spokesmen in the early 1930s did not explicitly refer to the state remuneration of fracture 'experts'. To have done so would have weakened their case in certain quarters and generally have played up the fact (appreciated by a sub-committee of the Voluntary Hospitals Committee for London in a report of 1939 on *Organised Fracture Services for London*), that 'the voluntary hospitals are being called upon to play a prominent part in what is virtually a new type of service and that payment of whole-time directors or registrars will inevitably involve additional expense'.[55]

But by the late 1930s, the situation had changed. In a context of falling unemployment, there was greater interest in rehabilitating injured workers. According to the then Minister of Health, the Conservative Edward Hilton Young, the BMA's 'Report on Fractures' – so well received in the press – had thrown into 'lamentable relief' the lack of facilities for treating fractures,[56] the time was now ripe for encouraging the voluntary hospitals to put their house in order. The establishment of the Inter-Departmental Committee under Delevingne was the Government's response, the pressure for which was intensified by organised labours' favourable response to the fracture cause.[57] Labour leaders took up the issue not only because of the serious effect of fractures on workers' wages and job retention, but also because it was an expedient means by which to re-state the accident-service rationale for a unified and 'classless' hospital service, centrally co-ordinated and regionally administered and financed. Thus the TUC and Labour Party made fracture services a part of their political platform on health care. Moreover, jointly with the orthopaedists (officially through the BMA), they presented a memorandum on rehabilitation and industrial injury to the Delevingne Committee in December 1937.[58] Drafted and promoted by the outspoken Liverpool orthopaedic surgeon and BOA activist, Reginald Watson-Jones,[59] the memorandum called for the deployment of experts in orthopaedic and traumatic surgery. It was subsequently presented to the (Hetherington) *Royal Commission on Workmen's Compensation* at its sittings of 1939-40, and, in general, its point of view and its emphasis on rehabilitation were taken up in the Beveridge Report and in other war-time and post-war legislation.[60]

Thus, by the late 1930s, orthopaedists were willing and able to go much further in their campaign for fracture clinics in the voluntary hospitals. With TUC and Labour Party backing, and within a generally more favourable social, political and economic climate for their cause, they were to become, as Frank Honigsbaum has noted, 'the "radicals" of the medical profession', pressing for the reorganisation of the health services as a whole. Indeed, the orthopaedic advocates of fracture clinics were to emerge 'almost alone among consultants [in welcoming] ... proposals for the state control of the voluntary sector'.[61]

CONCLUSIONS

If measured by the number of hospital fracture clinics established before the Second World War, the orthopaedists' campaign for fracture services would have to be reckoned, at best, only a partial success. The Government's (Delevingne) 'Inter-Departmental Committee on the Rehabilitation of Persons Injured by Accidents', after announcing in their Interim Report of 1937 that there were 'many indications that a widespread movement for the establishment of fracture clinics has begun and is likely to make rapid headway', was forced to confess in its Final Report of 1939 that

> progress in the general application of them has not been as rapid as we hoped. The matter had been taken up in a number of places ... [but] the hope that a *general* movement had been started and would be carried through by the hospitals themselves, both voluntary and municipal, individually or in co-operation, has not been realised (emphasis in original).[62]

By 1939 only four out of twelve London teaching hospitals had fully developed fracture clinics, and there were only 74 such clinics in the country as a whole (seventeen of which were in industrial Lancashire where funding often came from the Miners' Welfare Commission).[63] Elsewhere, partly 'organised' clinics were the norm, full implementation being held back by financial constraints and continuing disparities between hospitals, by foot-dragging on the part of the Ministry of Health and some local authorities, and by the shortage of qualified experts willing to undertake the work. To some extent, too, the idea of fracture clinics was overtaken by the ascending ideal of specialised industrial rehabilitation centres where, it was sometimes argued, greater continuity of treatment could be expected than in most of the voluntary hospitals.[64]

Historically more significant than the findings of the Delevingne Committee, however, was the fact of the Committee's existence, and the fact that, despite its official brief, it chose to concentrate exclusively on the organisation of efficient fracture services. Like the BMA's 'Report on Fractures', the Delevingne reports, by reiterating the orthopaedists' social, therapeutic and economic rationales for efficient fracture treatment, legitimated the professional self-interests behind them. Not only did the Delevingne Committee justify the control of fractures in the hands of 'fracture experts'[65] but, by complying with the notion that the voluntary hospitals were the most appropriate place for fracture clinics, it also granted to orthopaedists the status and authority they had been seeking, and the basis for the reproduction of that authority through access to undergraduate teaching. Further, by recommending 'departure from ordinary practice in respect of . . . remuneration' in the form of honoraria of between £300 and £500 per annum to the surgeon-in-charge of a fracture clinic, the Committee came close to accepting a salaried service for orthopaedists within the voluntary sector.

But the Delevingne reports have an importance beyond the professional interests of orthopaedists. They stand, with measures such as the Cancer Act of 1939, as evidence of government commitment to an organised, statutory health service which included medical specialists and their work in voluntary hospitals. Previously, where government had been involved in health-care activities (such as tuberculosis schemes and those for maternity and child welfare), the concentration was on 'public health' conceived largely in terms of preventive-cum-'personal health services'. By the 1930s, however, the focus of development for central government and for many Medical Officers of Health lay with curative services, including the development of municipal hospitals, their staffing with consultants, and their relations with the voluntary hospitals. In these discussions, which lasted through the Second World War, 'medical rationalisers' played a key role – some were Medical Officers of Health, some were medical academics, several of the most active were specialists who needed hospital rationalisation to develop more widely available services. Among the specialists (as Honigsbaum has observed) orthopaedists were conspicuous – an obvious and important example being Harry Platt, who was active on Manchester's Joint Hospitals Advisory Board before becoming involved, nationally, with the Nuffield Provincial Hospitals Trust.

That one of the first acts of the Joint Board in Manchester was the implementation of a city-wide fracture scheme, and that the first report (1939) of the Nuffield Trust outlined a scheme for a unified accident service[66] is not simply an indication of the influence and interests of Platt, however. His career was typical of several contemporary would-be specialists who similarly moved through the small voluntary hospitals to regional hospital centres to involvement with the organisation of the National Health Service.[67] In considering these parallel careers, what emerges clearly is not the particular, but the general, importance of the orthopaedic case for fracture clinics. Accidents, like cancer, were on a new frontier of 'public health' – a frontier that was part and parcel of the argument for rationalisation and specialisation. From this wider perspective, the reports by the BMA and the Delevingne committees appear less as orthopaedic reports in government dress than as particular instances of the reformist strategems of consultant specialists.

Clearly, the campaign for segregated hospital fracture clinics under the control of orthopaedic specialists was about much more than meeting the need in inter-war Britain for improved fracture treatment. As an emotive social issue which could draw in major interest groups from outside medicine,[68] the fracture issue provided one of the most visible and compelling of the arguments for technical expertise – an argument that could be, and was, used to legitimate the reform of hospitals and the medical services as a whole. In this sense, the fracture movement did indeed 'sanctify a cause'.

But what of fracture clinics for the study of medical innovation? From the evidence presented here, it would be difficult to argue that fracture clinics were inevitable, or merely compelled into being by the force of circumstances. Just as there is no inherent reason (but several historical ones) why today accident departments in the majority of British hospitals should be headed by orthopaedists,[69] so there is no inherent reason why fracture treatment or fracture clinics should have come under orthopaedic control. General surgeons might have done the work as well, with (or possibly without) special fracture clinics. (In fact, in some hospitals general surgeons continue to carry out this work.) Innovation and its legitimacy in this case was the consequence of political effort on the part of a professional interest group. Hospital provision for accidents, in other words, was neither 'natural' nor merely 'accidental'.

Yet neither political will and effort, nor even solitary examples on their own were enough to carry the innovation through, or to mobilise public opinion for its general adoption; had it not been for contingent historical circumstances, such as the rise of motor vehicle accidents, the financial crisis in hospitals, politico-economic interests on the part of trade unions, etc., it is unlikely that the innovation would have occurred. But the historical contingencies were not only of a material or social or economic nature. A significant feature of this example is the light it sheds on the role of ideology in informing and enabling the innovation. Underpinning fracture clinics, as well as the wider programme of hospital reform of which they were a part, was an abiding belief in the wisdom and necessity of 'rational' or 'scientific' organisation – i.e., of hierarchical divisions of labour with 'experts' (or teams of experts) in control at the top. That the need for fracture clinics became 'self-evident' when this economically-rooted ideology came to prevail is perhaps best attested by the sparsity in these debates of any quantitative evidence. Comparative statistics on the results of 'organised' and 'unorganised' clinics were seldom to hand, and rarely ever called for. It would therefore seem that in the history of this medical innovation, as in the history of medicine more generally, professional interests and underlying ideologies could count for more.

9
CORTISONE AND THE POLITICS OF DRAMA, 1949-55*
David Cantor

1949 saw a bumper crop of rheumatism cures and treatments. Rheumatics flocked to Golders Green to try Dr Michael Good's 'referred pain' treatment.[1] Mrs Joan Owen, the Mayor of Surbiton, and Gwyn Rees from Glamorgan recommended the use of bee stings, though Dr C. G. Butler, Head of the Bee Research Department at Rothampstead, was more cautious, advocating only more research.[2] Denying he had any connection with the blanket trade, E. J. Penton recommended two thicknesses of woollen blankets in the summer and three to four in winter.[3] A 42-year-old radio engineer and homoeopath, Eric King, cured a mother and son from West Norwood;[4] 'Fiery Jack' was advertised as a new rubbing ointment;[5] Dr Louis Moss developed an adrenalin vanishing cream;[6] and from Crewe came the Calmic lotion 'anti-rheumatism' mixture.[7]

Abroad, the South Africans announced a molasses treatment for rheumatism.[8] From Sweden came the hot-house treatment,[9] from Norway the exercise treatment,[10] and from Ireland the nettle-sting treatment.[11] In Paris, thieves stole what the papers called the elixir of life (a rejuvenating serum) from the Pasteur Institute, which also happened to relieve the pain of arthritis.[12] The Belgians announced a treatment based on derivatives of copper salts.[13] In the USA, Boston doctors helped rheumatic patients to walk again by inserting nylon strips into the knee joint.[14] The general manager of the Miami Quarterdeck Club told the world of the wonders of the wolf-fat cure.[15] The Americans also announced that a derivative of ox-bile might cure the disease – though 200 000 cattle would be needed to produce a mere 2700 grams of the substance.[16]

It was the derivative of ox-bile – cortisone – that attracted most attention. Yet amid the welter of cures and treatments available for arthritis and rheumatism, cortisone could too easily have been dismissed as just another miracle cure. As the *British Medical Journal* noted, 'the rheumatic diseases are notorious for the frequency with which dramatic successes have been claimed for various forms of treatment'.[17] To many doctors, dramatic portrayals of therapeutic innovations were fraught with ambiguity, at times aiding, at times impeding professional advance. Using a case study of cortisone and rheumatology, this article examines the politics of such portrayals. Drama provided a powerful means of advertising medical innovations. However, its associations with surprise, sensationalism, exaggeration, even enthusiasm and optimism could undermine the value to medicine of such advertisements. First, rheumatologists feared that sensationalist or exaggerated claims about the efficacy of cortisone could reduce public or official confidence in the drug and in the medical profession. Second, rheumatologists also felt that those who made such claims displayed feelings or emotions more appropriate to commercial advertisers than to scientists or doctors. In other words, those who produced dramatic portrayals of innovations exhibited an uncontrolled enthusiasm or optimism similar to that of the quacks or the purveyors of patent medicines. The key word is 'uncontrolled' for, as we shall see, most accepted that an element of enthusiasm or optimism was necessary for the advance of medicine. Who then defined the limits of acceptable, controlled, enthusiasm?

Cortisone, a hormone of the adrenal cortex, had first been isolated and identified in the mid-1930s.[18] Until 1948 when a practical method for producing it was developed there was little clinical interest in the drug, but all this changed when the physician Philip Hench and his colleagues at the Mayo Clinic demonstrated the drug's effect on rheumatoid arthritis.[19] In October 1949 an unnamed Fellow of the Royal College of Physicians told BBC listeners that Hench and his colleagues had 'opened the way to a real cure'[20] of rheumatism, a 'Beach-head' he called it. '[E]poch-making'[21] sprang to the lips of a rheumatologist speaking to the British Rheumatic Association, a welfare charity set up to help people with rheumatism. The *British Medical Journal* lauded it as a 'landmark',[22] the *Lancet* as 'a therapeutic discovery of the first importance',[23] and the *Times* reckoned that it would be as important to arthritis as insulin was to the treatment of diabetes and liver extract to the treatment of pernicious anaemia.[24] Indeed, so significant was cortisone that in March 1950 (less than a year after Hench's report) the Minister of Health, Aneurin Bevan, cited cortisone in his defence of rising NHS expenditure. Constant new developments, he argued, would be needed to keep pace with research progress, and penicillin, streptomycin and cortisone were his examples of 'research progress'.[25]

Dramatic images of cortisone's beneficial effects on rheumatoid arthritis at first worked to the professional advantage of rheumatology. Yet the persistence of shortages of the hormone, and the redefinition of the drug as potentially very harmful would in time undermine the portrayal. In addition, rheumatologists came to fear that their continued adherence to the early imagery of cortisone could prove a professional liability; instead, they would turn the dangers of cortisone to their own professional advantage, arguing that such hazards required specialist use of the drug. However, it would remain unclear how far rheumatologists themselves were exaggerating the harmful side-effects of the hormone, to limit demand. To pharmaceutical companies involved in the production and development of cortisone – and, indeed, to many patients – official portrayals of the harmful effects of the hormone revealed how the British welfare state, aided by rheumatologists, was willing to sacrifice the health of its citizens to economic circumstance.

In formulating responses to the early imagery rheumatologists drew on a long history of responses to drama and spectacle. During the eighteenth century orthodox medicine and natural philosophy had distanced themselves from public spectacle in response to fears of its dangerous religious and political associations.[26] Subsequently, Victorian and Edwardian advertisers had used spectacle and display to magnify the importance of patent medicines. As Thomas Richards argues, advertisements portrayed the body as both a container for such commodities and a commodity in itself. In other words, patent medicine advertisements equated the making of the self with the consumption of commodities.[27] By the outbreak of the First World War scientific medicine was still at odds with the sophisticated marketing and packaging of proprietary medicines. Indeed, Richards argues that the post-war pharmaceutical industry adopted many of the conventions of the patent medicine advertisements. Moreover, the efforts of advertisers to turn the body into a commodity at times posed a direct challenge to medicine's own authority over its patients. The advertisers' equation of the body with the consumption of patent medicines was antithetical to many doctors.

I

The worries of rheumatologists about dramatic portrayals of cortisone were products of debates in the 1920s and 1930s over the role of enthusiasm in shaping their field. Most textbooks on rheumatism rejected enthusiasm, especially that for particular techniques, as inevitably corrupting. As the authors of one textbook put it, 'Inventors of a new form of treatment and their disciples are bound to become enthusiasts. It is therefore not very surprising to find how remarkably close the statistical records are of the benefit obtained from various remedies which do not resemble each other in the least'.[28] In other words, enthusiasm blinded the creator to the true value of his or her treatment. Indeed, the authors suggested, it might even turn the search for new treatments against the very patients they were supposed to help, for 'often in an enthusiastic search for a remedy against the disease the patient's reactions are overlooked'.[29] Even simple optimism might be the start of the slippery slope to enthusiasm:

> Rheumatologists as a class are optimistic; the chronic and progressive character of rheumatism makes it imperative that they should be so, otherwise they would often feel inclined to give up treatment in despair. In certain recent books on rheumatism this optimistic attitude has tended to colour the chapters on treatment, which deal with the various difficulties and doubts we have mentioned, with an enviable coolness.[30]

Fears of the corrupting power of 'enthusiasm' and 'optimism' formed part of a wider concern about the extent to which commercial culture acted as a social leveller in late nineteenth and early twentieth century Britain.[31] As D. L. LeMahieu has argued, cultural élites attacked the popular press for appealing to people's 'baser instincts', dehumanising, homogenising and manipulating people in ways the élite found threatening. So, too, writers of rheumatism textbooks in the 1920s and 1930s rounded on the press, fearing that its sensationalist reporting and advertising of remedies encouraged such 'baser instincts'. They were concerned that such exaggerated reports and advertisements manipulated people in a way that was antithetical to science and medicine. In their view, 'enthusiasts' and 'optimists' distorted the sober assessments of new knowledge and therapies provided by science and medicine. Such people, the textbook writers felt, were not to be trusted. They were all show and no substance. And, they played to the lowest common denominator, often motivated by the same money-grabbing designs as advertisers, or the other purveyors of spectacle, the theatre and musical hall. To doctors, the theatre and advertisements could easily blur into each other as the common enemy. Thus the Buxton physician, Charles Buckley, criticised the money 'poured out lavishly in providing palatial hotels, theatres, ballrooms, concert halls etc.' for 'attracting the "cure-guest" exactly as a modern department store seeks to attract customers'.[32] Each, he felt, had exactly the same ultimate end in view, the profit of its ratepayers or shareholders as the case might be. To Buckley medical authority could not rest upon the same commercial practices as theatres and departmental stores – the latter being one of the chief advertisers in the popular press in the early twentieth century.[33]

Doctors also feared that advertising would dissolve the differences between orthodoxy and quackery. One author of a rheumatism textbook for general practitioners, W. S. C. Copeman, criticised 'ideas disseminated through the halfpenny post daily by the enterprising manufacturers of drugs with long names'[34] for encouraging people to try patent medicines or quack cures rather than to turn to orthodox medicine. In this context, David Vincent has noted that many advocates of literacy saw it rolling back the frontiers of superstition and magic associated with oral culture.[35] But for Copeman, advertising seemed to revive the spectre of magic. 'The public', he wrote in the *Spectator* in 1937, has 'not yet entirely lost its faith in magic – or its modern equivalent, radio-activity'.[36] To Copeman, this faith explained the sensational claims that a propriety brand of electrical heating pad (of high radio-activity) had achieved 90 per cent success in the cure of rheumatic diseases. So worried was the profession about this gadget that the Empire Rheumatism Campaign (ERC) – the major interwar rheumatism charity – issued a press statement denying that a panacea existed for the treatment of rheumatic diseases. It also promised to test all proclaimed cures for rheumatism, some of which, it claimed, were wholly fraudulent.[37]

The difficulty for physicians with an interest in rheumatism was that many orthodox cures themselves smacked of so-called 'quackery', notably the physical techniques which most rheumatism textbooks advocated as the mainstay of treatment.[38] The Bath physician George Kersley noted that the 'number and variety of the [physical] methods in use may savour of quackery',[39] and Copeman noted that many doctors rejected physical techniques for fear of the taint of folk-lore: 'The modern doctor is, I think, "rather afeared of old wives' remedies," in case his reputation for Science should suffer'.[40] It was, he noted, not too long since electrotherapy was known as 'elec-trickery'.[41]

The dominant method which rheumatism physicians used for distancing physical techniques from quackery was to assert their scientific credentials. Until 'recent years treatment has been mainly empyrical [*sic*] and no doubt at times somewhat abused', noted Kersley, as did many other writers of rheumatism textbooks.[42] But, he continued, the 'methods used today, however, are all based on sound physiological principles, and if these are understood and properly applied, nothing but good should accrue'. But science was defined as much by what it was not as by what it was. To most writers it was not drama, spectacle, folk-lore, quackery, commercialism or mystification; to the extent that enthusiasm and optimism were identified with these practices, they had to be controlled. Scientific discovery might be exciting, but such excitement had to be tempered with a denial of emotion and a willingness to investigate coolly and patiently.

Control of enthusiasm was also important in avoiding professional conflict. Thus, Humphrey Rolleston (Professor of Physic at Cambridge) gently warned physical medicine practitioners not to be too enthusiastic about the value of their treatment, lest they alienate general practitioners and physicians. Appearances could be deceptive:

> While, to the physical medicine practitioner, physiotherapy appears to offer the most dramatic results with regard to the prophylaxis, symptomatic relief and cure of rheumatism, the value of other measures as adjuvants are well recognised. Physical medicine should thus be regarded as a very valuable, though not the only useful, method of treating rheumatism.[43]

Perhaps some of the 'enthusiasm' here stemmed from the marginal status of advocates of physical medicine. In advocating physical techniques in the treatment of rheumatic diseases, Copeman felt aggrieved that many other doctors 'profess not to believe that any special skill or experience in the prescription or application of such methods is really essential',[44] and so were prepared to attempt such techniques themselves or to by-pass medical specialists by going direct to lay physiotherapists. For this reason specialists in physical techniques such as actinotherapy, balneotherapy, medical hydrology, and electrotherapy, began to combine in the 1920s and 1930s in an attempt to improve their status within medicine. Yet loyalties to particular techniques remained strong within the agglomeration of physical medicine, the solidarity of which was further strained as different factions tried to exploit the growing concern about the costs of rheumatism to national, industrial and military efficiency. In 1929, and again in 1936, national campaigns were launched to draw attention to the problem. The various advocates of physical medicine could unite over such 'promotions', but they were only too likely to divide when advising on specific treatments. Enthusiasm for particular techniques thus threatened the fragile solidarity of physical medicine.

Dramatic portrayals of treatments also disturbed the doctor-patient relationship which was of particular concern to rheumatism specialists. Almost all medical writers of rheumatism textbooks before cortisone had emphasised the non-dramatic, long-term nature of treatment. As the Professor of Medicine at Edinburgh, L. S. P. Davidson, was to note in the 1st edition of Copeman's *Textbook of the Rheumatic Diseases* in 1948: 'It must be made clear to the patient that, although no dramatic results are to be expected, and treatment will inevitably be a matter of months or even years, much can be done to relieve symptoms and restore function'.[45] Clearly, claims of new and dramatic 'cures' for rheumatism could undermine these months or even years of treatment, and to Davidson the whole problem was compounded by the uncertain prognosis of rheumatoid arthritis. Mild cases apparently making good progress could relapse for no reason, with a marked deterioration of the patient's general health and the condition of his or her joints; while other patients, whose progress seemed very bad, might suddenly improve, again without warning or reason.

But if optimism could be a problem, so could pessimism. Copeman argued that patients undergoing long-term treatment were at risk of 'defeatism' from friends and family, and most seriously from the medical profession itself. If most doctors found rheumatism dull and uninteresting, as Copeman admitted, they might transmit their pessimism about the outcome of treatment to their patients, so inducing that 'defeatism' which in turn encouraged medicine to ignore the rheumatic diseases.[46] It is in discussions of this kind that patients began to be conceptualised as potential defaulters, much as David Armstrong has suggested.[47] So some medical enthusiasm was necessary, but it had to be contained. It was chiefly in the inner-sanctum of the doctor-patient encounter that Copeman allowed physicians to express this modicum of hope.[48] Only in this way could the doctor safely transmit his enthusiasm to the patient, and so encourage the co-operation which Copeman saw as essential to the successful treatment of rheumatism. Uncontrolled enthusiasm damaged patient and doctor alike.

But was such controlled enthusiasm compatible with the recognition of rheumatic diseases as a public health problem? The paradox of medical antipathy towards uncontrolled enthusiasm and drama for therapy was that some doctors complained that the rheumatic diseases themselves lacked pizzazz. James Fenton, the President of the Society of Medical Officers of Health, argued in a 1938 ERC publicity drive that people often regarded rheumatism as an inevitable accompaniment to old age. It lacked the terror of a killer disease. He stated: 'One great difficulty in bringing home to the public the serious nature of rheumatism lies in the fact that it is not such a "dramatic" disease as cancer. The prospect of contracting it does not strike terror into the heart of the average man'.[49] Indeed, on the contrary, Fenton suggested people were often as not proud of their 'touch of rheumatism'; a mentality, he felt, that worked against the recognition of the disease as a national problem. To Fenton the solution was to turn to advertising. Indeed, he argued that the enthusiasm of advertisers for their products was essential to the success of the Campaign.

> If any of you went to the recent Ideal Homes Exhibition in London, you would have seen and heard demonstrators from a number of firms extolling the merits of water softeners, and had any sufferer from rheumatism been buttonholed by one of them for three-quarters of an hour, as

someone I know was, he would have written out a cheque on the spot. It is enthusiasm of that quality which we must have as a first essential if we are to make any impression. For medical men and public health officers imbued with the spirit of service, it should not be difficult to become and remain keen enough to carry out a successful programme.[50]

'Service', for Fenton, made the difference between commercial and medical attitudes towards enthusiasm. It was the 'spirit of service' that would temper the corrupting influences of enthusiasm and drama. For this reason, no doubt, he felt entitled to dramatise rheumatism as a 'masked killer' – a reference to rheumatic heart disease, mortality from which he felt was largely unrecognised. Indeed, to Fenton advertising blurred into public education. The ERC's campaign, he told his audience, was to educate the public to obtain treatment in the early stages of the disease and to avoid 'proclaimed remedies, which are often fraudulent'. The availability of treatment facilities, he noted, should be advertised, but without encouraging false hopes or using sensational methods. Medicine's authority was opposed to such sensationalism: 'We have to maintain the dignity of authority and avoid all sensationalism in our methods'.[51]

Yet Fenton was not alone in fearing the corrupting power of sensationalism. Some commercial advertisers adopted similar rhetoric. Thus when Fenton argued that it was 'essential that we take as an inviolable principle, "Truth in Advertising", and avoid raising false hopes',[52] he was merely echoing phrases that had circulated among advertisers since before the First World War. According to E. S. Turner, the phrase 'Truth in Advertising' had been first adopted by the Associated Advertising Clubs of the World in 1911.[53] The phrase was revived during the 1930s, following concern about the effects of sensational advertising. A prominent figure in this revival was Gordon Selfridge, the Oxford Street retailer, who was later to join the ERC as a vice-president.[54] Like the doctors on the ERC, Selfridge was worried about enthusiasm. Selfridge saw sensationalism and spectacle as dangers to the good reputation of shopkeepers. In 1933, he offered £10 to anyone who could spot an error in his advertisements – an offer that three years later had cost his company £1320.[55] In 1935, Selfridge, together with a number of other proprietors of leading department stores, launched the Retail Standards Association with the avowed aim of protecting the public from misleading advertisements, descriptions and prices.[56] The liaison officer between the Association and the newspapers, Tom Clarke, formerly editor of the *News Chronicle*, saw this organisation as an attempt to improve the status of the shopkeeping profession and its relations with the public.[57] To the department store proprietors the sensational advertising of some companies undermined the importance of the very commodities they hoped to sell. And in their fear of sensationalism some department stores turned to education, including health education, as part of their advertising schemes. Selfridge's and John Lewis's employed a Robot Doctor which measured the height, weight and rate of heart-rate for the price of 3d.[58] Also as part of his attempt to create a new shopping experience, in 1935 Selfridge persuaded the Minister of Health to open a 'health and housing' exhibition in his Oxford Street store.[59] The department stores at which Dr Buckley had sneered were now to attract their own equivalent of the 'cure-guest'. As in the spas, customer and health-seeker would be one. And in blurring the distinction between customer and health-seeker, the department stores used the very rhetoric of 'truth in advertising' employed by the ERC (of which Buckley was a member). Indeed, the advertisers had supplied the rhetoric.

II

Concern about the dramatic portrayal of cortisone thus drew on a long tradition of conflict over the role of advertising and of enthusiasm and spectacle in shaping rheumatology. It occurred in the aftermath of the atomic bomb, when Hiroshima and Nagasaki emphasised the cultural ambiguity of traumatic events on the prestige of science. Reports of the Japanese explosions combined both fears of the end of civilisation and, paradoxically, hopes of unlimited sources of energy. They portrayed the reality of death and destruction, and the promise of cures for disease and disablement. Cancer and rheumatic patients stood to benefit from these 'cures'.[60] In medicine, the 'wonder-drugs' penicillin, aureomycin and streptomycin provided 'new wonders for the doctor's bag', and set the stage for cortisone's dramatic reception.[61] And in his original paper on cortisone, Philip Hench told the tale of a bed-ridden woman getting up and walking.[62] This was the stuff of miracles. Newspapers described the hormone as a 'miracle-drug' and a 'wonder-drug' in 1949, and again in 1950 when the first British trials of cortisone began.[63]

Cortisone came at a crucial time for British rheumatologists, who used the wonder-drug to distance themselves from physical medicine, and to ally themselves more closely with general medicine. Boundaries between specialist fields were fluid as the NHS took shape. Physical medicine had grown substantially since 1939 as a result of war-time man-power needs, and the post-war concern about disabled ex-servicemen. This growth threatened to subsume rheumatology into physical medicine. For example, Ministry of Health officials suggested that a rheumatism service could most practicably be created by expanding physiotherapy services.[64] Partly as a result of such threats, rheumatologists began to define themselves as a part of general medicine, and persuaded the Ministry of Health to include rheumatology under general medicine rather than physical medicine within the NHS. Nevertheless, rheumatology remained dogged by a shortage of recruits, and it seemed for a while as if their field might go by default to physical medicine specialists. Now, despite years of doubts about the association of the specialty with one form of remedy, identification with a particular drug became a way for rheumatologists to distance themselves from physical medicine. At last, rheumatology had a treatment that could supplant physical medicine as the core of rheumatological practice. Indeed, the links between rheumatology and general medicine became all the stronger when it became clear that cortisone could be of value in diseases other than rheumatoid arthritis. By August 1950, the Ministry of Health listed Addison's disease, acute rheumatic fever, acute gouty arthritis, acute inflammatory diseases of the eye, rheumatoid arthritis, psoriasis, certain acute diseases of the skin, certain acute and malignant diseases of the blood, disseminated sclerosis, osteoarthritis, pernicious anaemia, and hypertension among the diseases possibly benefiting from cortisone and the related remedy, pituitary adreno corticotrophic hormone (ACTH).[65]

British rheumatology was fortunate indeed in having a 'major scientific discovery' fall straight into its lap. It was a discovery that rheumatologists themselves used to transform their field from a specialty that had been 'as unscientific as it was unfashionable'[66] to one at the forefront of medical advance. The state-financed Medical Research Council (MRC), the bastion of scientific orthodoxy in Britain, immediately became interested in cortisone. As its secretary, Harold Himsworth, wrote 'when a subject becomes of burning importance (as rheumatism has since the discovery of Compound E [Cortisone]) we have to take up the question very actively'.[67] The MRC's interest in rheumatoid arthritis had been negligible, but in October 1949 it appointed an advisory committee to 'advise and assist in promoting research on rheumatism, with special reference to treatment with cortisone'.[68] This committee amalgamated with the Nuffield Foundation's clinical trials committee the following year when the American pharmaceutical company, Merck, gave the joint project a gift of 1 kg of cortisone for clinical trials.[69] This gift – the first substantial quantity to arrive in Britain – ensured that by 1951 the *British Medical Journal* could state that with 'the advent of cortisone there has been no shortage of medical and other scientists to work upon the new clues'.[70] Steroid chemists, biochemists, endocrinologists and a variety of clinicians moved into the field, and as the consultant in physical medicine J. H. Glyn notes, rheumatologists themselves broadened their interests to embrace the metabolic and endocrinological implications of cortisone therapy, some even acquiring a working knowledge of steroid biochemistry.[71] As it became clear that cortisone had more uses than in rheumatoid arthritis the MRC expanded its cortisone research programme, appointing panels for this purpose in rheumatic fever, ophthalmology, haematology, dermatology and industrial pulmonary disease.[72]

Finally, for rheumatologists the spectacle of cortisone meant that the newly-created NHS could be pressed for greater resources, effectively creating an instant lobby for the drug and hence (they hoped) for rheumatology itself.[73] Although rheumatologists had been incorporated into the state's planning bodies for the new Service during the Second World War, they had entered the NHS profoundly unhappy about what they felt was the poor public provision for rheumatology. Rheumatologists thus formed pressure groups through their professional societies and through the ERC to lobby government for greater resources. The dramatic publicity for cortisone provided a welcome boost to this campaign. A 1950 report in the *Daily Mail* on the dearth of rheumatological services in the NHS linked cortisone to demands for an extension of facilities.[74]

Such publicity gained even greater significance as the escalating costs of the NHS appeared to damage the possibility of cortisone's widespread use. Almost from the start of the Service it had seemed to many that NHS expenditure was out of control. Bevan had warned the Cabinet in December 1948 that the original estimate of £176 million for 1948/9 would turn out to be £225 million. A supplementary estimate of £59 million for 1948/9 was followed by one of £90 million for 1949/50. The growing drugs bill was a particular worry. In 1951 the Public Accounts Committee suspected negligence over determining fair levels of profit for suppliers of drugs: suspicions which led to suggestions that prices could be controlled by the state, and to Ministry of Health meetings with the Association of the British Pharmaceutical Industry concerning voluntary cost investigation.[75] As Harry Jephcott, the chairman and managing director of the British drugs company, Glaxo, remarked in July 1952, there would always be restraint placed upon the use of cortisone so long as it remained relatively expensive.[76]

Indeed, the whole problem was further complicated by problems in Britain's external account with the dollar area. Britain had ended the war as the world's leading debtor country, recording substantial deficits in its account with the dollar area. On average, the dollar deficit accounted for more than the total deficit on current transactions through the years 1946-51. Even in the years of overall surplus – 1948 and 1950 – Britain recorded substantial imbalances with the dollar area.[77] The discovery of cortisone's beneficial effects on rheumatoid arthritis followed shortly on the first major dollar crisis – the 1947 crisis in the convertability of sterling – and coincided with a sharp deterioration in the dollar account in the second quarter of 1949. Both these emergencies promoted a search for ways to save on any unnecessary dollar expenditure, and the point was emphasised by the outbreak of the Korean War in 1950, which sparked the third major payments crisis. In 1952 Jephcott noted that government desires to minimise dollar expenditure tended to overshadow what he described as 'a cautious, but not adverse, medical reception'[78] and felt that 'a more liberal policy towards cortisone might well, probably would, follow from any proposal which resulted in substantially reducing or eliminating the dollar cost'.[79]

But just as the dramatic imagery associated with cortisone began to work to the professional advantage of rheumatology, rheumatologists began to draw back from their early euphoria about the drug. They loudly blamed the press for over-dramatising the value of the hormone. 'Like many new treatments it was hailed by the Press, although not by the discoverers, as a "miracle cure"',[80] reflected W. S. C. Copeman in a 1954 popular handbook on rheumatism. For Copeman, the worry was not just that the drug might fail, but that the dramatic publicity might rebound to the discredit of rheumatology itself. He tried to limit the damage in two ways. First, he excused the early optimistic reports of cortisone by situating them in the context of a long history of failed 'cures'. Many doctors originally doubted even the 'obvious' beneficial effects of the hormone. Reports of such 'striking effects' sounded too much like those of any other cure. And it was only when physicians with a reputation for scepticism were persuaded by hands-on experience of their benefits that others came round. The focus of drama switched from the drug to the profession.

> It soon became obvious that these hormones had a striking effect in rheumatoid arthritis and the Mayo Clinic invited 5 physicians who were known to be extremely critical of any claims for the successful treatment of rheumatoid arthritis to see the investigations and each was given sufficient cortisone to treat 1 case under their own control. One of the most dramatic moments at the International Congress of Rheumatic Diseases in New York in June 1949 was when after Hench and Kendall had read their papers on this subject these 5 physicians one after another confirmed the effects of this drug in their own cases.[81]

But such comments risked implicating physicians in the original reports of a 'wonder-cure', even if it was an understandable reaction to the praise of the sceptical five. Copeman thus tried a second means to limit the damaging effects of the early optimism, by indicating that the profession itself no longer induced such 'striking effects' in its patients. Hench had used high doses of cortisone to enable his patients to walk again. By 1953, however, Copeman pointed out that such dramatic results were undesirable when cortisone was part of long-term therapy. Patients, he felt, were often disappointed that the improvement did not continue at the same rate when the dosage had to be reduced to a maintenance level, and there was a danger that they would strain muscles weakened by disease when attempting movements denied them for months or even years.[82] For this reason Copeman now avoided the spectacular response achieved by high initial dosages, using a low starting dose instead.

But the reports of dangers extended far beyond strained muscles. Responding to persistent rumours that side-effects were not being reported in the medical press, the MRC sent a letter to the *British Medical Journal* and to the *Lancet* in July 1951 asking users to report any untoward effects of the drug.[83] Diabetes mellitus, hypertension (both benign and malignant), peptic ulceration (including perforation of such ulcers), cardiac failure, psychotic changes and the exacerbation of tuberculosis were already rumoured, and soon replies came in confirming these and more.[84] Such dangers meant that far from being the panacea presented to the public in the heady days of 1949, cortisone could be used in only a limited number of cases. These dangers also strengthened the rheumatologists' argument that it should be used only in the hands of experts. The Ministry of Health told one arthritic who complained about the lack of cortisone available for treatment that 'increasing experience both in this country and in the United States has shown that serious and sometimes fatal side-effects are more common than had been supposed and experienced physicians are consequently using both Cortisone and A.C.T.H. with even greater caution than before'.[85] In May 1952, H. M. Walker, an official of the British pharmaceutical company, Glaxo, felt that 'cortisone will be a substance that should be used only with the greatest care and discrimination, and probably in the hands of those with real background and experience'.[86]

But to the pharmaceutical companies such claims themselves began to smack of sensationalism. By the end of 1952 Walker had told Merck that the dire side-effects of cortisone were being exaggerated in Britain to excuse the shortage of supplies.[87] Only a few arthritic patients, he felt, were being treated, and these just long enough to be relieved of their symptoms, only to suffer all the physical and mental anguish of abrupt withdrawal when the cortisone supply gave out. This view he claimed concurred with that of an official at the Department of Health for Scotland, Dr J. M. Johnston. According to a furious Merck official the basic reason for the shortage was that

> the British medical authorities, i.e. Medical Research Council, and the Ministry of Health are frightened to death of the almighty dollar, and are content on this account to disregard the welfare of a very considerable segment of the sick and disabled in their country. If I interpret Dr. Walker's and Dr. Johnston's remarks correctly, it is even suspected that these two august bodies are conniving to perpetuate or enlarge the specter of toxicity, so that rheumatic sufferers will continue to be bamboozled.[88]

The Merck official went on to argue that cortisone's costs were in fact comparable to those of aureomycin or terramycin therapy, and were probably not much more than streptomycin-PAS, or streptomycin-isoniazid therapy 'which the British have gone in for so promptly and gaily'.[89] The cost of cortisone therapy, he argued, was comparable to two persons having a moderate-sized dinner at a British hotel once a week. Such comments may not have packed much punch in austerity Britain, but the official concluded that to circumvent the MRC's 'feeble, backward and stupid'[90] policy towards cortisone Merck should target 'good clinicians', 'in the hope that eventually the British practitioner can teach the Medical Research Council a little medicine.'[91]

To Glaxo and Merck the exaggeration of the side-effects was also a product of the ascendancy of hospital medicine.[92] Indeed, Walker had heard it said 'that cortisone will never be a drug to place in the hands of the general practitioner',[93] and it was for this reason that he suggested that a clinical trial of cortisone with general practitioners should be organised. Walker argued that too little attention had been placed on suppressive as opposed to curative treatment. In particular, little work had been done on the long-term administration of relatively small doses, precisely the sort of use that general practitioners might make of cortisone. Walker further argued that the MRC should not be involved in this clinical trial as it was dominated by 'expert rheumatologists' whom he felt were not the best people to judge the use of the drug by general practitioners. It was these same experts, he felt, that were partly responsible for the view that cortisone was too dangerous for the ordinary practitioner. Yet, Walker claimed, they also were responsible for the fact that patients were being treated in hospital, only to be sent out to almost certain relapse.

A Merck official concurred with Walker's comments about the need for a clinical trial involving general practitioners. In a revealing internal memorandum on Merck's policy towards clinical trials in Britain, he highlighted the importance the company attached to encouraging enthusiasm for cortisone among British clinicians:

> the major achievement which MERCK & CO., Inc. wants in Great Britain is not so much the sale of cortisone there, as it is the enthusiastic acceptance of it by the best of their medical men. The reason for this is the powerful influence which good British clinicians and clinical investigators exert in establishing the climate of medical opinion, both throughout Europe and in America. We can say what we like about American superiority, but a thoughtful article in the LANCET or the QUARTERLY JOURNAL OF MEDICINE will in its influence often outweigh a dozen or two of American publications.[94]

But the acceptance was never as entire as Merck would have liked. In 1955 cortisone became available on prescription from general practitioners, but rheumatologists continued to warn that the final effects of long-term administration were yet unknown. Some even suggested that long-term administration could not be considered safe or practical in many cases.[95] Many newspapers echoed this warning. While most welcomed the 'end of rationing' of the 'wonder-drug', the overall reception was more muted than in 1949. 'Don't Pester Your Doctor – this new drug is useful but ...' read one headline.[96]

III

Rheumatologists had capitalised on both the beneficial and harmful effects of cortisone. But they were also much criticised, especially while cortisone remained in short supply. Decisions about its use could be hotly contested, not least when they revealed a system of priorities which patients did not share. Inevitably in a situation where shortages occurred, patients would be unhappy about being denied access to the 'wonder-drug'. But the narratives of those who wrote to the Ministry to complain about being denied cortisone often linked their personal problems to wider political issues. Implicit in many of these letters is a critique, not only of medical attempts to control patient reactions, but also of the close association between the state and medicine, the persistence of inequalities within the NHS, and conflicts between foreign policy and social welfare.

At the root of these criticisms was the dramatic transformation which cortisone wrought in the lives of many of those who wrote to the Ministry.[97] To some it was a matter of changing their day-to-day lives, restoring 'normal power and movement'[98] and banishing pain. It could improve household economies wrecked by the burden of caring for an arthritic, and a repeated theme was the extent to which cortisone would enable an individual to be a 'useful citizen'.[99] The obverse was a fear of dependency: 'Nobody wants to be tied to a cripple',[100] wrote one arthritic. Yet these very transformations would at times be turned against the patient. Some doctors would suggest that cortisone stimulated uncontrolled enthusiasm for the drug among their patients, which worked against their better interests. Cortisone could, they would argue, work against effective treatment. But effective treatment would often be defined in a way that negated the patients' political voice.

First, let me outline some of the patients' criticisms from the Ministry of Health's complaints file. For some of those complainants who wrote to the Ministry, the shortage of cortisone revealed continued inequalities of access to health care within a Service that was supposed to have done away with it. To IS, official suggestions that cortisone still needed further trials worked against the poor: 'is it so essential to be so cautious for the poor, whilst the fortunate & wealthy are so satisfied with the results they can buy for money[?]'[101] This writer felt that a black-market in cortisone existed, an allegation that had appeared in the newspapers in 1951.[102] Others noted that they had been advised by their doctors to try to purchase cortisone privately from the States, and some appear to have done just this.[103] While a few stated that they preferred to obtain cortisone privately rather than through the NHS, for others the expense stretched the domestic budget, or put cortisone beyond all reach.[104] 'I am told she [my mother-in-law] would require 75 mgm. daily to commence and am told this would cost in the region of £8 a day. The sum is far beyond my means',[105] wrote CL. This writer consoled himself with the thought that his mother-in-law was over 70 years old, and might have got the price wrong: but she was not far out. The cost of a 'hotel meal for two' was substantially higher on the 'black-market' than in the Merck price index.

Some correspondents interpreted the shortage to mean that the needs of the patient were being sacrificed to economic requirements. The export of cortisone was a particular concern: 'I know that this country must export in order to live but I feel it is very sad if something needed so much here must be exported'.[106] Others saw the lack of cortisone as yet another example of the waste of public resources – 'when one considers the millions of public money that has been wasted from Groundnut schemes to carpets for [the] R.A.F. is it any wonder that some of us are beginning to lose heart?'[107] Indignation replaced discouragement for IS, 'It is a moral wrong that sufferers should be denied this treatment while so much money is being wasted on unessential things'.[108] IS looked on in dismay as the shortage confirmed her belief in traditional biases against chronic illness,

> it takes the threat of death or contagion to spur financiers, politicians & doctors to action. The prospect of prolonged human suffering does not stir their imaginations, it only bores them into inaction, thus we have built up a heritage of rheumatism and are still doing so! How different it would be if we were contagious, then cortisone or A.C.T.H. would be pumped into all of us ad lib whether suitable or unsuitable[.][109]

Thus for IS the NHS appeared to perpetuate this traditional bias against the chronically sick. She saw herself as victim of government crime – 'You no doubt would shrink from flogging a gangster or a cosher[,] & Parliament will waste time & money on them. Yet you do not shrink from inflicting unnecessary torture on innocent victims of the scourge of arthritis'.[110] Medicine and 'finance' were conspirators in 'torture' by neglect.

Few other correspondents made such direct attacks on 'finance', but along with government (waste and export policy) medicine was a favourite target of attack. For some, cortisone revealed the depths of medicine's ignorance and its reluctance to take rheumatism seriously – 'The doctors I have been under know practically nothing of Arthritis, Cortisone or A.C.T.H.';[111] 'Ordinary doctors and the hospitals are not very interested and usually prescribe exercise and massage'.[112] This last writer contrasted British neglect with a more enlightened American attitude, and felt that the indifference of British medicine was compounded by commercial exploitation of the sick – 'It may not be generally know[n] that when a person has a complaint like this, they become fair game for every pill and lotion maker, every so called specialist, and a host of others who fatten on suffering'.[113]

Clinical trials were viewed suspiciously by a number of writers. I have noted that IS saw trials as perpetuating the division between rich and poor within the NHS. BM saw comparative trials of cortisone and aspirin as an attempt to claim that aspirin was as good as cortisone. Her experience showed otherwise:

> Please do believe me what this your humble constituent knows from its [*sic*] own experience and from the experience of other arthritics that no number of aspirins can compete with the miracles wrought by Cortisone.[114]

But experience could too easily be turned against the patient. BM's belief that cortisone was a gift from God could easily have consigned her to the crackpot fringe. Furthermore, doctors and officials could easily discount experiential justifications on the grounds that the arthritis or cortisone had unduly influenced their perceptions. To many doctors the patient was still little more than a potential defaulter, easily swayed by spectacle. As J. J. R. Duthie and L. S. P. Davidson put it in 1955: 'The dramatic suppression of symptoms which follows the administration of ACTH or cortisone has temporarily obscured the importance of the basic therapeutic régime.'[115] To the MRC, demands for cortisone were an assault on the scientific validity of clinical trials, as well as being practically impossible to meet, and for these reasons they petitioned Merck not to advertise its gift of 1 kg of cortisone. Merck agreed as they 'knew how very trying such appeals *ad misericordiam* could be'.[116] To the Ministry political commentaries were little more than special pleading. All had an interest in denying the experience of the patient, and reinforcing medical interpretations of that experience.

The case of IS who complained that clinical trials perpetuated inequalities in access to health care can further illuminate the ways in which medical interpretations were given priority over those of the patient, and used to deny the patient's political criticism. Even her own doctor used IS's experience of illness against her. When she wrote to the Ministry, IS complained that she had been given cortisone, had been relieved of pain, and then had had the hormone withdrawn.[117] The Ministry wrote to her consultant for more information, and the registrar who had administered the cortisone replied. He dismissed her complaint as an undesirable psychological effect of the withdrawal of cortisone. He claimed that IS had 'extreme faith in this remedy'[118] (an instance of medical distrust of enthusiasm?). In addition he stated, that she had pressured him into giving her cortisone against his better judgement from a small 'illicit'[119] supply of his own. Despite the fact that her orthopaedic surgeon had also asked for her to be put on cortisone, the registrar maintained that his original judgement had been the correct one. 'I felt that even although there was no objective benefit yet the feeling of well-being, so often noticed in the early stages, would be a disadvantage'.[120] Disadvantage to whom? It may have been that the further administration of cortisone would have been a disadvantage to the progress of the treatment, but it was also a disadvantage to the authority of the registrar in controlling the patient (and perhaps his attempts to get himself off the hook for using, by his own admission, an 'illicit' supply of the substance).

Note here the opposition of 'objective benefit' and 'feeling of well-being'. To the registrar the patient's own enthusiasm for cortisone and her feelings after its administration worked against what he defined as 'objective benefit'. This 'feeling of well-being' also worked against the Ministry's desire to cut back on drugs expenditure. Thus, despite its misgivings over this case, the Ministry supported the registrar, telling IS that 'specialists' decisions are made on purely medical grounds to ensure that the best use is made of their supplies, and the social position of the patient does not influence them.'[121] This was written despite the fact that the registrar, by his own admission, had given her cortisone for what they would have regarded as 'non-medical' reasons – i.e., the demands of the patient, albeit with the backing of her orthopaedic surgeon.

Finally, it was not true to say that the social position of patients did not influence doctors. Even if doctors allocated limited NHS supplies with no regard to their patients' social position, the Ministry could hardly have been unaware that doctors turned a blind eye to a loophole in the law which allowed those patients who could afford it to import cortisone privately.[122] The private importation of cortisone was illegal unless it was an 'unsolicited gift', but patients could pay vast sums for these 'gifts'. The *Sunday Mail* reported the comments of one doctor: 'if the patient came to him with the drug he tacitly assumed it had come to this country as an "unsolicited gift". He asked no questions, even when he had his doubts'.[123]

IV

Debates about drama were debates about power, prestige and profit. Therapeutic innovation provided a significant arena for such debates. In 1949, the dramatic portrayal of cortisone's beneficial effects on rheumatoid arthritis worked to the advantage of all: to rheumatologists hoping for professional advantage and effective treatment; to pharmaceutical companies hoping to supply the NHS and to exploit the international prestige of British medicine; to patients hoping for a cure and for the alleviation of pain; even to the Minister of Health hoping to stave off cut-backs in the health service.

But by 1952, because of the shortage of cortisone and the dangerous side-effects of the drug, dramatic portrayals had become much more problematic. Rheumatologists felt that sensational presentations of the beneficial effects of this innovation had begun to work against them. In their face-to-face contact with patients, these doctors found themselves besieged by demands for a drug that they could not supply, and frustrated by hopes of cures they could not deliver. Rheumatologists now saw enthusiasm, whether induced chemically or by desperate hope of a cure, as a danger to medical authority. This view was supported by the Ministry of Health, which hoped that doctors would help control the escalating costs of drugs as well as dollar expenditure. It was also supported by the MRC which saw patients' demands for cortisone as a challenge to the utility of clinical trials.

Whereas rheumatologists, the Ministry of Health, and the MRC saw the publicity regarding the dangers of cortisone as toning down the sensational claims about the curative effects of cortisone, pharmaceutical companies felt that reports of hazards went too far. In the view of Merck and Glaxo the harmful effects of cortisone were being exaggerated in response to fears of the 'almighty dollar', and the encroachment of general practitioners on hospital medicine. In this view they edged near to the position of some of the patients who complained that they were being denied cortisone for reasons of foreign and financial policy, as well as medicine's traditional neglect of the rheumatic diseases. Yet, for all their concern with 'bamboozled' patients, Merck and Glaxo had little interest in the political criticisms of those patients and others who complained to the Ministry of the shortage of cortisone. While they might agree with patients that official caution regarding benefits of cortisone was questionable, the pharmaceutical companies shared with the Ministry, the MRC and the rheumatologists the view of experts as 'gatekeepers' of knowledge and therapy. These companies were principally interested in encouraging sales of the drug and, as Merck noted, in using British medicine to advertise their wares throughout the world. This may not have been the conspiracy of industry, medicine and the state which some patients perceived. However, those patients correctly deduced that companies, doctors and the state all had interests in common which were not those of the arthritic or his or her family. Patients were aware of their own powerlessness against such groups and institutions. But not until the 1960s would the possibility exist of organised patient-consumer intervention in the management of innovation.[124]

10
PSYCHIATRY IN DISTRICT GENERAL HOSPITALS: HISTORY, CONTINGENCY AND LOCAL INNOVATION IN THE EARLY YEARS OF THE NATIONAL HEALTH SERVICE*

John V. Pickstone

INTRODUCTION

The history of the NHS, like that of other government services, is usually written in terms of national policy formation and then of policy implementation. Action appears to come from the centre; argument and representation may come from the periphery, but only via central decisions will pressure be turned into policy, only then will action result - unless, of course, the inertia of the system or the resistance of the periphery nullify central directives. The present essay explores a historical counter-example. I shall argue that the development of psychiatric units in district general hospitals was pioneered in the Manchester region around 1950, not in response to national policy but because of the way new *regional* decision-makers reacted to peculiar problems in the provision and staffing of mental health services.

Through this case study we can explore the importance of deep historical patterns in the development of mental health services, both generally and for the features peculiar to the Manchester region. We can elucidate the responses of the decision-makers at regional level by exploring their backgrounds and their attitudes to related questions of hospital development. We can show how the structure of the NHS brought mental health services to the attention of these medical policy-makers, none of whom had much previous experience of psychiatry. And in this way, we can construct an explanation for a policy which was out of line with those in other regions. The Manchester region's mental health services came to be planned, not on specifically psychiatric models, but according to a mode of hospital 'upgrading' which was widespread in non-psychiatric services.

PSYCHIATRY AND GENERAL MEDICINE

The history of British psychiatric provision since the Second World War might be summarised in two simple observations. From about 1960, after a century and a half of increasing asylum accommodation, the emphasis in psychiatric care shifted towards 'the community' – i.e., patients were increasingly accommodated outside institutions. Associated with this pattern one sees a shift in the relationship between psychiatry and somatic medicine. For most of the nineteenth century the particular role of the asylum doctor was considerably different from that of other 'hospital' doctors. In the present century psychiatric medicine has come closer to other medical specialisms, for example in its use of drugs. One aspect of this latter development in Britain has been the tendency, from the 1960s, to place psychiatric units in general hospitals.

On the roots of the policy of 'community care', there is a significant literature and much discussion, both supportive and critical, though the making of policy within the Ministry of Health still requires elucidation. About the tendency to place psychiatric units in general hospitals, we have very little historical information. By the early 1960s, when the NHS was first able to provide significant capital for hospital building, it had already been decided that large asylums would be run down, at least partially. That was the thrust of Enoch Powell's 'Watertower' speech in February 1961. It was also agreed that the district general hospitals which were to be developed throughout the land would contain psychiatric services, though not necessarily such as would relieve the district concerned from all dependence on the nearest asylum.[1]

There was never any public debate about the advisability of putting psychiatric units in general hospitals. As far as we can presently ascertain, there was no recorded policy debate within the Ministry of Health. The policy 'just growed', for three reasons. The use of general hospitals for out-patient and 'short stay' in-patient cases seemed a reasonable extension of the push towards 'community care'. The inclusion of psychiatry was in line with the notion of 'comprehensive service' which underlay planning for district general hospitals. And thirdly, the Ministry knew and approved of pioneering units developed since 1950 in the region around Manchester.[2] It is the origins of these pioneering units that this essay will analyse.

THE DEVELOPMENT OF MENTAL HOSPITAL SERVICES BEFORE THE SECOND WORLD WAR

That the institutions for the care of the mentally ill should be more like general hospitals is a claim which goes back to 1900 at least. The Maudsley hospital was set up in 1907 by the London County Council using a donation from the psychiatrist Henry Maudsley. It was to concentrate on early treatment and out-patient treatment of people who were not certified as lunatics. The same concerns were stressed by the Royal Commission in the 1920s, and to some extent underlay the changes in procedure and terminology introduced in 1930. Again, during and after the Second World War, the report undertaken by C. P. Blacker, largely on out-patient psychiatry, stressed the need for more early care, for out-patient clinics, for wards in general hospitals, etc.[3]

The introduction of new physical treatments in the inter-war years – insulin therapy, cardiazol, malarial therapy – helped psychiatrists in asylums to feel that they were engaged in activities paralleling those of other hospital specialists. No more would they be merely bureaucrats or custodians, or general practitioners to asylum populations (though when discussing doctors in asylums, we do well to remember the huge amount of *physical* illness, notably tuberculosis, which was to be found in institutions of 3000 patients or more).[4]

That some asylum doctors sought roles akin to those of specialists in non-mental conditions, or that some non-asylum psychiatrists undertook clinics in general hospitals is not surprising, given the general pattern of inter-war psychiatry in Britain. But there was a considerable gulf between such aspirations and the then common forms of psychiatric care. Physical therapies and acute units were marginal to most asylums; out-patient clinics in voluntary hospitals were marginal both to the voluntary hospitals and to the local asylums. If we are to understand the conditions which hindered or encouraged the general hospital model in psychiatry, then we need to consider the organisation of medical and social services in early twentieth century Britain, and how these changed with the introduction of the NHS.

PATTERNS OF SERVICES

In the eighteenth century, several asylums had been founded alongside general voluntary hospitals, but the bulk of the nineteenth century care was provided by legal authorities of counties. By the end of the century, responsibility lay with county councils and with county borough councils, for the local government of large towns (county boroughs) was independent of local government for the surrounding rural areas and smaller towns. Sometimes a given county and the county boroughs within its boundaries would jointly manage a single asylum (or single set of asylums). In other counties, the county council had one or more asylums, and each of the large towns had its own.

By the 1920s most of these counties or county boroughs were also responsible for considerable non-psychiatric medical services – isolation hospitals, TB sanatoria, clinics, health education, etc., so one might have imagined that there was scope for reducing the barriers between physical and mental medicine, or at least for reducing the barriers between asylums and community services. But this very rarely happened, partly because many asylums, even when they served a single town, were somewhat remote; mainly because the asylums, though owned by local councils, largely functioned as extensions of the Poor Law, the statutory system of relief for paupers which until the 1929 Local Government Act was quite separate from the rest of local government.

Until 1930, asylum occupants were admitted through Poor Law relieving officers, not by the Medical Officers of Health who worked for counties or county boroughs. That many of the patients came from the workhouses underlined the association of asylums with Poor Law rather than public health services. Asylum doctors might, if they were very enterprising, run a clinic at a local *voluntary* (i.e., charity) hospital, but local government had no legal power to finance such a clinic as part of the public health services.

The legislation of 1929–30 altered this situation a little. The Poor Law services were transferred to public assistance committees of the county and county borough councils. Those councils *could* transfer the workhouse hospital wards to the administration of their public health committees, while keeping the non-hospital sections under 'public assistance'. Many of the workhouses then contained 'mental wards' as well as hospital wards, but it was very rare for these mental wards to be recognised as a health committee responsibility; they remained a part of public assistance.[5] Theoretically, in as much as local authorities were now responsible for asylums and ex-workhouse hospitals as well as clinics, etc., and in as much as they could help fund 'voluntary' welfare activity and voluntary hospital clinics, there was a possibility for a considerable integration of mental services, but real developments were few and far between. Hugh Freeman has described an integrated service in the southern harbour town of Portsmouth. There were other cases – for example, Nottingham – but they were rare.[6] Some cities with very large municipal hospitals and public assistance institutions (e.g., Manchester) were able to concentrate their mental cases into a section of one institution; a former workhouse might thus serve as a kind of mental hospital, taking all the new cases from its area, passing the intractable ones to larger asylums.[7] But most workhouse mental wards remained essentially as a public assistance service, as did asylums. Overwhelmingly, the mental service continued to be provided through the large asylums, though some now had acute units and out-patient clinics, and others provided out-patient clinics at voluntary hospitals. There were a few psychiatrists in the big cities who concentrated on clinics and private practice, but the bulk of psychiatrists continued to be strongly identified with large public asylums.

THE INTRODUCTION OF THE NATIONAL HEALTH SERVICE

When the NHS was introduced in 1948, almost all the hospitals of the country, whether voluntary or municipal, general, specialist or asylums, came under the ownership and control of central government. The Ministry of Health exercised this control through Regional Hospital Boards (RHBs) based on the major cities with medical schools. The membership of the Manchester RHB, like the others, was decided by the Ministry after local consultation. All the Boards contained senior medical personnel, plus influential figures previously active in charity hospitals, and some local councillors (active as individuals). These RHBs in turn appointed Hospital Management Committees (HMCs) to serve each of the major towns in the region.[8]

In 1943 it had been announced that mental hospitals would *not* be included in the new post-war arrangements. In the 1944 White Paper, however, this intention was reversed because of pressure from psychiatrists and others who wanted mental care to be pulled much closer into the main stream of medicine. Yet in the arrangements for the new service, psychiatry was very much an add-on. Each of the new RHBs was supposed to appoint a regional psychiatrist to its administrative hierarchy, so that he or she could oversee the mental hospitals.[9] Where psychiatrists were so appointed, it was usually on a part-time basis. In some cases – for example, Manchester – the role was eventually taken by a (non-psychiatric) member of the Board's medical secretariat. Nor was there much integration at district level. In general, because the asylums were so large and remote, each was given its own HMC, rather than being covered by the committee based on the nearest town. The local administration of asylums thus remained separated from the local administration of ex-voluntary and ex-municipal hospitals. The whole of the hospital service (under RHBs and HMCs) was now substantially separate from county and county borough councils which continued to be responsible for public health measures and many clinics.

Under the NHS, if we may judge by the Manchester case, the regional policy documents drawn up for the asylums followed the familiar guidelines of the Board of Control, which had previously controlled national policy on mental hospitals and which continued to be influential. They stressed voluntary (i.e., non-certified) admissions, early acute care, and out-patient clinics. The Board of Control were keen to see that psychiatric services got their share of the (very thin) new cake. But the introduction of the NHS did not precipitate any general, widespread change in psychiatric services. For the most part, the asylums ran in their old grooves, now alongside but still largely independent of non-mental hospital services.

Indeed, the NHS set back some of the enterprises of the 1930s. As I have argued at length elsewhere, the fact that scarcely anyone before Aneurin Bevan had expected local authorities to lose control of their hospitals meant that no-one had stressed the deep and inevitable connections between hospitals, welfare, domiciliary services and public health. The functioning links, or the possibility of such links, were often disrupted when municipal hospitals passed with voluntary hospitals into state ownership, leaving other services with local authorities. That spilt damaged the Portsmouth municipal mental scheme as it, more generally, disrupted the services for tuberculosis, maternity, etc.[10] Any integration of community and mental hospital services would henceforth depend on the co-operation of agencies which were administratively separate.[11]

Overall, the planning for the NHS had not been dominated by concern for integrated services, or even for the day-to-day operation of general hospitals; rather, it was focused on the need for new services in all but the best hospitals. Municipal and non-teaching voluntary hospitals were to be upgraded; above all, they were to be staffed with consultants. The spreading of consultants from the regional medical capitals into ordinary 'peripheral' hospitals was one of the major aims and achievements of the early NHS. Note the underlying model: an improved hospital was a federation of units, each headed by a consultant. This model was central to the NHS. We shall show that it was crucial, at least in Lancashire, for the development of local psychiatric services based on district general hospitals. Here was the bridge by which psychiatry came to be seen in the same terms as other specialisms.

THE LANCASHIRE EXPERIMENT AND ITS EXPLANATION

By the 1960s the Manchester region had become recognised for the innovative psychiatric units which had been developed in general hospitals at Oldham, Blackpool, Bolton, Burnley and Blackburn, all towns with a hospital catchment population of about 100 000 – 200 000. This was not because the Regional Board was generally innovative: it was not. Indeed it was known rather for its economy with taxpayers' money, a feature perhaps relevant to this case study.[12] Why, then, had it pioneered in this field?

The key decisions were taken in 1949 and 1950, soon after the Manchester RHB had taken responsibility for all the hospitals in its region. In this early period the Board was collecting information about its properties and the immediate need for repairs, building works or new basic services. It was concerned with regularising staffing patterns, including the assignment of grades to existing staff, and with assessing immediate needs for recruitment of consultants and others to the various hospital districts. The Board's own committee structure was still novel and somewhat experimental. Questions of consultant staffing, for all specialisms, including psychiatry, were overseen by the Medical Advisory Panel.[13]

For psychiatric medicine, as for other specialisms, the RHB had established a Technical Advisory Panel (TAP), most members of which were consultants in that field. In the case of psychiatry, for reasons already outlined, the majority of the representatives were asylum doctors. When asked to plan the development of the service, the TAP, unsurprisingly, recommended extensions to asylum-based services, plus a corresponding increase in staffing. The panel was also concerned that asylum psychiatrists below the level of medical superintendent should be eligible for full consultant status.[14] If we may judge by events, other hospital regions seem to have accepted recommendations of this sort, so that services there showed no obvious short-term effects from the incorporation of mental hospitals within the new NHS bureaucracy.

But in the Manchester RHB, when the recommendation of the psychiatrists was passed upwards to the Medical Advisory Panel, it was rejected. An *ad hoc* group was set up to consider the question, partly because of the known difficulty in recruiting well-qualified psychiatrists.[15] When the *ad hoc* sub-committee produced a report at the end of March 1950, the question was again referred back.[16] At the end of April, however, a policy was agreed. Development would no longer be based on asylums, but instead new psychiatry posts would be attached to district general hospitals, at least as an experiment. In spite of attempts by the psychiatrists to push the claims of asylum doctors, this new policy was accepted by the Regional Board.

The report of the *ad hoc* sub-committee had been direct in its criticism of the TAP recommendations:

The idea of a service based on the four large mental hospitals with peripheral clinics in hospital centres staffed by mental hospital staff is alien to the principles of the national hospital and specialist service, and will only serve to divorce the diagnosis and treatment of mental disorders still further from the broad stream of general medicine.

The Sub-Committee suggests that consultant psychiatrists should eventually be appointed to all general Hospital Centres in the Region, either on a whole-time, but preferably on a part-time basis (minimum 9 sessions).[17]

In the summer of 1950, the Board decided to advertise for a consultant psychiatrist who would be primarily attached to a pair of general hospital districts, with but a secondary appointment to a large mental institution. The experiment worked, in that a good candidate was found, who made the most of the opportunity. As we shall describe, the first such appointment was at Oldham-Rochdale. Other such appointments followed, largely to Lancashire mill-towns north of Manchester. By 1960 the Lancashire experiment was well-known among psychiatrists and others who took an interest in mental health services.[18] But to outline the decision process is but the first stage in explaining it. We can begin our search for explanations by ruling out some of the possible reasons.

The policy did not depend on a psychiatric 'prophet', though it benefited greatly from the commitment and enterprise of two individuals. The first 'district psychiatrist' was Dr Arthur Pool, without whose energy and commitment the experiment might well have failed. That the first trial was followed up in several other towns owed much to Dr J. S. B. Mackay, the deputy senior administrative officer at the Manchester RHB. He was a doctor, though not a psychiatrist; he, more than anyone else at the RHB office was identified with the policy, but he seems to have become involved *after* it was initiated in 1950. It may be of some importance that Dr Mackay had a background in public health administration rather than in psychiatry.[19]

If no prophet was responsible for the conception, were there perhaps other advocates of the medicalisation of psychiatry, pushing for the Manchester Region to catch up on the indices of psychiatric progress? There is a little to be said for this view. The chairman of the Manchester

RHB, Sir John Stopford, was also Vice-Chancellor of the University, and a former Dean of Medicine. With some of his key medical associates (e.g., Sir Geoffrey Jefferson), he had been involved, just before the NHS, with schemes to get more and better psychiatry into the hospitals of Manchester city and its neighbouring authorities. There had been a report on these services, which was linked with the Blacker Report of 1946, and the University had recently appointed a new Professor of Psychiatry. It is also true that when their asylum psychiatrist recommendations were rejected by the Medical Advisory Panel, the *ad hoc* group which produced an alternative contained the new professor and the psychiatrist at the teaching hospital, the Manchester Royal Infirmary.[20] Here, then, was a source of specialist advice, or at least legitimation, which was directly substituted for that of the psychiatrists employed in the Region's own large asylums. Yet there is little or no evidence for any continuing connection between the university department (or the teaching hospital) and the development of the district services. When I asked the former medical administrator of the Regional Board (Dr F. N. Marshall) about the pre-NHS report on Manchester services, he denied all knowledge of it.[21] The Manchester RHB policy, it would appear, was not seen as an extension of city-centre patterns, or of war-time musings on psychiatric policy. To explain the new direction we need rather to consider the Regional Board's medical politicians and the physical condition of the services which had become their responsibility.

The Manchester RHB was peculiar in having a medical chairman, one whose experience and other positions gave him great authority both nationally and in the region. Sir John Stopford had been Vice-Chancellor of Manchester University since 1934, before that he had been a long-serving Dean of Medicine and Professor of Anatomy. When, in 1935, Manchester and its contiguous boroughs had established a voluntary liaison committee for hospital development, Stopford had been the Chairman, as we have noted. During the war he had been deputy chairman of the Goodenough Committee on medical education, an essential aspect of national planning for post-war medical services. As a keen supporter of the NHS, Stopford had been a natural choice as RHB chairman for the Manchester Region.[22] The Senior Administrative Medical Officer of the new Board, Dr F. N. Marshall, was a Manchester medical graduate who had served there during the war as a regional officer with the Ministry of Health. He was well known to Stopford, who had arranged for his appointment to the new NHS post.

Stopford's power was further increased by the appointment to the Board of several of his long-time associates and friends from the Manchester medical school. Sir Harry Platt and Sir Geoffrey Jefferson were his close contemporaries who had attained international standing by developing surgical specialisms: orthopaedics and neurosurgery respectively. Platt, especially, was a keen medical politician, closely involved with the Nuffield Provincial Hospitals Trust, which for a time during the war had made the running in planning for a post-war service. As his 'regional advisor in medicine', Stopford had appointed Dr Norman Kletts, whose brilliant career as a physician in Manchester had been curtailed by illness, but who had returned to serve the new Board on a part-time basis.[23] These men were the power centre of the new Board, and what we know of their general policy preferences goes some way to explaining the 1950 decision. Jefferson and Kletts were both on the *ad hoc* panel on psychiatry. Marshall has claimed that he had himself been unhappy with the report produced by the psychiatrists on the TAP, and that he and Kletts suggested the 'district' alternative.[24]

There was, however, another group of Board members who are important to this story. They helped represent the peripheral towns in the region. One was a teacher from Burnley, Mrs Edith Watson, who had long been dedicated to improving the services of that town; another was Mr Walter Briggs, the leading surgeon in the neighbouring mill-town of Blackburn, again a person of strong local commitment, determined to see improvements in ordinary hospital services.[25] But to understand the perceptions of the two groups we must understand, first the general policies of the Board, and second the peculiar circumstances of mental services in the Region.

As to medical politics, the key factor may well have been the attitude to medical specialisation shown by Stopford and his close colleagues. They had no objection to specialisation, *provided* the specialists were co-ordinated with each other in hospitals or educational structures which represented medicine as a whole. Platt and Jefferson were both pioneers of surgical specialisms, but within general hospitals. Stopford had aided their careers by appointing them to University positions; there and in the associated teaching hospital they formed part of a single élite group which was the apex of the region's medical hierarchy.[26]

But, in their eyes, medical specialisation could become a problem if specialists developed their own institutions outside the main hierarchy. Of this pattern, the best example in the Manchester region was the Christie Cancer Hospital and Holt Radium Institute, which had developed very rapidly during the 1930s under Dr Ralston Paterson, a radiologist of extraordinary tenacity and organisational skills. The 'Christie' was sited about three miles south of the main teaching hospital; by the Second World War it was the centre of a regional cancer service which covered most of the hospitals for which the RHB came to be responsible in 1948. Thus, a decade before the NHS, it had been one of the best examples in Britain of an organised regional service based on a centre of excellence – the very pattern regularly used to justify the post-war organisation of hospitals within a national service. But it was not a teaching hospital, and Paterson, for all his international eminence, was marginal to the Manchester medical hierarchy, to which he was not accustomed to defer.

Under the NHS, the Christie might have been made part of the teaching hospital complex, but Stopford was not keen. Paterson wanted to have his own hospital management committee, to maintain some degree of independence for his hospital, and he pushed the case with considerable energy and influence; but Stopford blocked him and the Christie hospital was linked with a general hospital, as if it were an ordinary institution serving only south Manchester. The outcome would almost certainly have been different if the Manchester RHB had had a lay chairman, less concerned with medical politics and less worried by the separate institutional development of particular medical specialisms.[27]

That the attitude so exemplified extended to other specialist services and to psychiatry is suggested by the early structure and other actions of the RHB. As we have seen, for some time the RHB tried to handle all specialist development through a single, wide-ranging committee, advised by technical panels. Only later did they set up a series of special committees for special subjects. When the Ministry asked each RHB to appoint a full-time Regional Psychiatrist, Manchester refused;[28] they appointed an institutional psychiatrist as a part-time adviser. Given such attitudes, they were not likely to leave psychiatry to the psychiatrists. In the Manchester region, more than elsewhere, psychiatrists were up against keen medical planners, widely experienced in medical education and general hospital services, but not experienced in psychiatry, nor sympathetic to the special claims of staff in large mental hospitals.

Even so, it seems unlikely that such men would have devised a new policy, had the old proved less troublesome. Here one recalls the concern over the RHB's inability to recruit psychiatrists to the Lancashire asylums.[29] To understand this problem and the other unattractive features of the mental hospital inheritance, requires a digression into the administrative history of asylums in Lancashire.

The geographical county, which included Liverpool and Manchester as well as many smaller, industrial towns, had a population of about 5 million. Between the wars its asylums had almost all been controlled by a peculiar federation called the Lancashire Mental Hospitals Board, on which the county council and all the county boroughs were represented.[30] This had been formed after 1888 when county councils and county borough councils were first given responsibility for asylums. It was then deemed impossible to allot the existing asylums to Lancashire County Council or to the many separate municipalities in this region. Instead, all these councils were to contribute to a federal board controlling all the region's existing and future asylums. By 1948, the vast majority of the mental beds were in huge institutions including Lancaster Moor (2800 beds), Whittingham near Preston (3200) and Prestwich, near Manchester (2800). Had Europe's asylums been ranked by size, Whittingham would have been second only to the asylum in Vienna. The Lancashire Mental Hospitals Board was widely regarded as an administrative failure and a model of what *not* to do; it was too far removed from local politics, its members showed little interest. Gargantuan responsibilities had produced gargantuan asylums, much criticised and little praised.

The Manchester RHB inherited the three institutions just listed, plus another large asylum in Cheshire. Altogether it controlled about 13 000 'mental' beds, excluding mental deficiency. 11 000 of these beds were in very large institutions and several of the buildings were in poor condition. There was a plan already prepared for the rebuilding of Prestwich hospital, which had been the subject of a public scandal in the 1920s.[31]

Faced with these all too solid facts, Stopford and his colleagues were very wary of plans for extending asylum services. The existing institutions were far too large to be added to; there was no real possibility of building substantial new mental hospitals. Indeed, the Manchester RHB, in the early years, came in for considerable criticism from the Board of Control for low capital investment in psychiatric services. They replied that the existing buildings could not be extended; they would build a new mental hospital if the Ministry would give special funds.[32] But it was not just a question of bricks and mortar. As we have seen, when the Manchester RHB advertised for new asylum-based psychiatrists, they could not find candidates of sufficient standard. That failure was one of the immediate reasons why the RHB rejected the expansion plan based on asylums.

What, then, was the alternative? Here we note that the 'solution', like the difficulty, stemmed in part from the peculiarities of Lancashire. For as well as having several monster asylums, the county also contained about eight large towns (up to 200 000 population) which were distant from asylums, and which had very sizeable 'mental-blocks' in their former workhouses. These blocks, of 100–300 beds, were largely custodial. But from 1948 they were in the care of local HMCs to which they had been transferred together with the municipal, ex-workhouse, hospital accommodation; hence the interest from local representatives on the Regional Board.

The only specialist care provided for patients in these 'mental-blocks' was from asylum-based psychiatrists who, typically, drove over for an afternoon session. They came, it was said, and handed out the pills, before rushing back to the asylum grounds for a game of tennis. That seemed even more unsatisfactory now that mental-blocks were formally included in the 'district hospital service'; if mental wards, like the rest of the municipal hospitals, were to be upgraded, shouldn't they, too, have consultants in control?[33] Marshall recollects that he had been interested in upgrading such blocks since he had surveyed them, at the end of the war, in connection with a lady from the Board of Control.[34]

Here we return to the major theme of hospital improvement around 1950, the appointment of consultants to upgrade services. That is what local representatives wanted for psychiatry in their areas, and this hope meshed directly with the main concerns of the key members of the Board, who then were engaged in appointing consultant surgeons and physicians especially to the peripheral districts, largely to upgrade the supervision of ex-Poor Law accommodation. Why not do the same for psychiatry? Marshall, Kletts and Stopford seem to have favoured such a move.[35]

The Board in 1950 advertised for a psychiatrist to work chiefly at Burnley or at Oldham. Mrs Watson was keen that it be Burnley; but Oldham had the largest of the mental-blocks.[36] The psychiatrist appointed chose Oldham for personal reasons. He took five sessions per week there, plus two at Rochdale and two at the nearest large asylum, Prestwich, just outside Manchester. The Board was fortunate to find this consultant – Dr Arthur Pool – for he was experienced in senior posts, including the medical superintendency of The Retreat at York. He was idiosyncratic but enthusiastic and inventive; he teamed up with the Medical Officer of Health in Oldham and together they built up a mental service which was largely self-contained. In Oldham, Pool worked closely with the 'Duly Authorised Officers', the psychiatric social workers who were officially under the Medical Officer of Health; in Rochdale, he did not receive this degree of co-operation, and his interest in Rochdale declined. Pool's activities in Oldham soon took up most of his time and he asked to be relieved of his asylum duties. Thus the model was established. By 1953 the Manchester Board regarded its psychiatric services as essentially based in general hospitals.[37]

The Oldham experiment was followed by similar developments in other mill-towns (and in Blackpool). The Regional Board was able to get good applicants for these posts, though they often failed to find satisfactory psychiatrists for their big asylums. The psychiatrists appointed to the districts vied with each other to 'burn their own smoke' – i.e., to operate without sending patients to asylums – and this became easier as drug therapy increased. Without the new drugs, the general hospital units would probably have continued to depend, partially, on the big institutions.[38] But, if I read the history correctly, that would not have prevented the extension of the general hospital system for the majority of cases. By the 1960s there was a significant professional literature on the Lancashire experiment. It would seem to have fitted the contemporary pressure for 'community care' because the new units operated with a considerably lower bed-population ratio than had been characteristic of asylum services.

When Enoch Powell, in the early 1960s, called forth a national plan for hospital building, he had already committed himself to 'community care' and the run-down of large asylums. New general hospital units were to figure in the new plans, new asylum buildings were not. At this stage Powell was still thinking of small acute units in general hospitals, for communities which would still send some patients to asylums. The Manchester RHB had to argue for larger units to be allowed in its own plans; in this, they were successful. By the end of the 1960s, it would seem, larger, more self-sufficient units were coming to be accepted as the national norm.

LESSONS?

This historical case study, though peculiar, would seem to be illuminating for those interested in innovations under the NHS. The new service had no money for new buildings, but it did have funds for new consultant staff. And it had a general philosophy, widely accepted from the inter-war years as the royal road to hospital improvement in *non-mental* fields: take hospital patients from the care of generalist practitioners or of visiting consultants, and put them under the care of specialist consultants based on the hospital concerned. The pattern could be applied to psychiatry, but was so applied only where there was some reason to depart from the existing, asylum-cum-clinic routines which usually dominated plans for mental health. In the Manchester Region the asylums were large and decrepit and the clinics were relatively few, so there was little chance of expanding the main-line services. On the other hand, the ex-workhouse mental-blocks, often large and distant from the asylums, afforded the opportunity for an alternative mode of development, one which proved to overcome staff recruitment difficulties and which involved no major capital costs.

This type of innovation might be known as coming from behind, or 'the last shall be first'. Where there was neither the opportunity nor the inclination to catch up on old trajectories, then new patterns might be tried. It was under these conditions that English psychiatry became most obviously approximated to general medicine. If the aspiration was general, the conditions for the departures in that direction proved to be rather specific.

I am not arguing here that the development of general hospital psychiatric units was necessarily beneficial. Indeed, some may read this case-study as an unfortunate example of orthodox medical models being extended into fields of care which should be organised quite differently. I make no judgement in this essay. But perhaps there is a general lesson here for analysts and for promoters of change in health services. The lesson would be about local concerns and local initiatives, about the possibilities of seizing difficulties as opportunities, about analysing local situations for their individual potential and not just for their conformity to national norms. Innovations do not all come from the centre, or through extensive policy reviews; some come, rather quietly, from the historical peculiarities of particular people and places.

NOTES AND REFERENCES

1 Introduction

1. 'Neo-technics' is Patrick Geddes's term, popularised by Lewis Mumford and used in contrast with the 'palaeotechnics' of the first industrial revolution. The palaeotechnic phase of machine civilisation was based on coal and iron, neo-technics on science, electricity and new materials. Palaeotechnics wasted nature, neo-technics tended to restore, conserve or indeed, to imitate nature (e.g., artificial sunlight). A good book could be written on medicine as neo-technics. See Lewis Mumford, *Technics and Civilisation* (New York, 1934).
2. But for a discussion of social process and S curves see James S. Coleman *et al.*, *Medical Innovation. A Diffusion Study* (Indianapolis, 1966).
3. For a convenient introduction, see Peter Temin (ed.), *New Economic History: Selected Readings* (London, 1973); Peter Mathias (ed.), *Technical Change: The United States and Britain in the Nineteenth Century* (London, 1970); Donald N. McCloskey, *Econometric History* (London, 1987); Nathan Rosenberg, *Perspectives on Technology* (Cambridge, 1976) and *Inside the Black Box, Technology and Economics* (Cambridge, 1982).
4. But see George Rosen, *The Structure of American Medical Practice 1875–1941*, ed. Charles Rosenberg (Philadelphia, 1983); J. Liebenau, *Medical Science and Medical Industry* (Basingstoke, 1987); Peter Temin, *Taking Your Medicine. Drug Regulation in the United States* (Cambridge, Mass., 1980); Anne Digby and Nick Bosanquet, 'Doctors and patients in an era of national health insurance and private practice', *Economic History Review*, second series, 41 (1988): 74–94.
5. John H. Warner, *The Therapeutic Perspective: Medical Practice, Knowledge and Identity in America, 1820-1885* (Cambridge, Mass., 1986); Martin Pernick, *A Calculus of Suffering: Pain, Professionalisation and Anaesthesia in Nineteenth Century America* (New York, 1985).
6. See, for example, A. J. Culyer, *Need and the National Health Service. Economics and Social Choice* (London, 1976); Brian Abel-Smith, *Value for Money in Health Services* (London, 1976).

7. See Christopher Freeman, *The Economics of Industrial Innovation*, 2nd edn (London, 1982).
8. John B. McKinlay, 'From "promising report" to "standard procedure": seven stages in the career of a medical innovation', *Millbank Fund Quarterly*, 59 (1981): 374–411; Mary M. Fenell and Richard B. Warnecke, *The Diffusion of Medical Innovations, an Applied Network Analysis* (New York, 1988).
9. Julius Roth and Sheryl B. Ruzek (eds), *Research in the Sociology of Health Care*, 4 (1986), especially pp. 1–32: Susan E. Bell, 'A new model of medical technology development: a case study of DES'. Also see S. J. Reiser, *Medicine and the Reign of Technology* (Cambridge, 1982) and S. J. Reiser and Michael Anbar (eds), *Medicine at the Bedside: Strategies for Using Technology in Patient Care* (Cambridge, 1984).
10. For a useful introduction and bibliography see Donald Mackenzie and Judy Wajcman (eds), *The Social Shaping of Technology. How the Refrigerator got its Hum* (Milton Keynes, 1985). For recent attempts to apply to technology insights from sociology of science see W. E. Bijker, T. P. Hughes and T. J. Pinch (eds), *The Social Construction of Technological Systems: New Directions in the Sociology and History of Technology* (Cambridge, Mass., 1987). In stressing the social construction of technological artefacts, this school tends to neglect pre-existing strong traditions in history of technology which have given more attention to economic and political structures. See, for example, David Noble, *America by Design: Science, Technology and the Rise of Corporate Capitalism* (New York, 1977), and his *Forces of Production. A Social History of Industrial Automation* (New York, 1984). Other major recent works include: Ruth Schwartz Cowan, *More Work for Mother: The Ironies of Household Technology from the Open Hearth to the Microwave* (New York, 1983); David A. Hounshell, *From the American System to Mass Production, 1800–1932* (Baltimore, 1984); David Hounshell and John K. Smith, *Science and Corporate Strategy; Du Pont R & D, 1902-1980* (New York, 1988); Thomas P. Hughes, *Networks of Power; Electrification in Western Society 1880-1930* (Baltimore, 1983); Judith A. McGaw, *Most Wonderful Machine: Mechanisation and Social Change in Berkshire, 1801–1885* (Princeton, 1987); Merrit Roe Smith, *Harpers Ferry Armory and the New Technology: the Challenge of Change* (Ithaca, 1977); and M. R. Smith (ed.), *Military Enterprise and Technological Change: Perspectives on the American Experience* (Cambridge, Mass., 1985). All these books are

by Americans, almost all are on American topics. For two recent British analyses of the meanings of key technologies in modern Britain, see Bill Luckin, *Questions of Power. Electricity and Environment in Inter-war Britain* (Manchester, 1990) and David Edgerton, *England and the Aeroplane*, published in this series (London, 1991).

11. *Technology and Culture* is the admirable journal of the (American) Society for the History of Technology, where much of the best recent work has been published. For a survey of the contents see John M. Staudenmaier, *Technology's Storytellers: Reweaving the Human Fabric* (Cambridge, Mass., 1985). For a recent work on the history of medicine sensitive to the economic and cultural reverberations of science and technology see, Charles Rosenberg, *The Care of Strangers: The Rise of America's Hospital System* (New York, 1987).

2 'On This Principle I Have Based a Practice': The Development and Reception of Antisepsis in Britain, 1867–1890

1. Joseph Lister, 'On the antiseptic principle in the practice of surgery', *Lancet* Pt 2 (1867): 353–6. Lister's first articles appeared in 1867, the main ones that year being: Joseph Lister, 'On the antiseptic principle in the practice of surgery', *Lancet*, Pt 2 (1867): 353–6; 'On the antiseptic principle in the practice of surgery', *British Medical Journal*, Pt 2 (1867): 246–8; he had already published case studies in March 1867 in the *British Medical Journal*. This article is based on my MA thesis, 'The Reception of Antisepsis in Britain, 1867–1880', Bryn Mawr College (1978) and a paper given at the American Association of the History of Medicine, Kansas City (1978).

2. Biographers of Lister have consistently emphasised that Lister revolutionised surgical practice. See, for example, Hector C. Cameron, *Joseph Lister, The Friend of Man* (London, 1948); William Watson Cheyne, *Lister and His Achievement* (London, 1925); Cuthbert Dukes, *Lord Lister* (London, 1924); Richard B. Fisher, *Joseph Lister, 1827–1912* (London, 1977); Rickman J. Godlee, *Lord Lister* (Oxford, 1924); Douglas Guthrie, *Lord Lister: His Life and Doctrine* (Edinburgh, 1949); A. Logan Turner, *Joseph, Baron Lister: Centenary Volume. 1827–1927* (Edinburgh, 1927); and Kenneth Walker, *Joseph Lister* (London, 1956). This view is also implicit in other

historians' work. See for example Frederick F. Cartwright, *The Development of Modern Surgery* (London, 1967) and John Woodward, *To Do the Sick No Harm: A Study of the British Voluntary Hospital System to 1875* (London, 1974): 82.

Almost the only dissenting voice has been that of David Hamilton, with his suggestion that an increase in infection at the mid-century was later countered by new defences of patients which related to better nutrition: this development went in parallel to the introduction of antisepsis, and Lister could not therefore be given all the credit. David Hamilton, 'The nineteenth century surgical revolution – antisepsis or better nutrition?', *Bulletin of the History of Medicine*, 56 (1982): 30–40.

A recent contribution is Nicholas J. Fox, 'Scientific theory choice and social structure: The case of Joseph Lister's antisepsis, humoral theory and asepsis', *History of Science*, 26 (1988): 367–97, which has a number of interesting insights but is marred by some inaccuracies and an apparent lack of awareness of the primary sources. For Lister's changing notion of the germ theory, see Richard Dixey and Christopher Lawrence, 'Practising on Principle', paper given to the Wellcome Institute (December 1989), based on Dixey's MSc thesis for the University of London on Lister and antisepsis.

3. There have been studies of reactions to Lister's work in other countries, including those of I. H. Upmalis, 'The introduction of Lister's treatment in Germany', *Bulletin of the History of Medicine*, 42 (1968): 221–40; C. G. Roland, 'Early years of antiseptic surgery in Canada', *Journal for the History of Medicine and Allied Sciences*, 22 (1967): 380–91; and Thomas P. Gariepy, 'The Acceptance of Antiseptic Surgery in the United States', MA thesis, Notre Dame (1976).

4. Lindsay Granshaw, 'Stepping to fame and fortune by means of bricks and mortar', in Lindsay Granshaw and Roy Porter (eds), *The Hospital in History* (London, 1989).

5. The long-standing interest in these questions can be seen, among others, in John Howard, *An Account of the Principal Lazarettos in Europe* (Warrington, 1789); see also Woodward, *To Do the Sick No Harm:* 98–107.

6. This level of mortality was considered normal, for instance, by J. H. James in his 'On the causes of mortality after amputation of the limbs', *Transactions of the Provincial Medical and Surgical Association*, 17 (1849): 49–102.

7. Thomas Bryant, 'Contributions to the subject of compound fracture,

being an analysis of 302 cases', *Medico-Chirurgical Transactions*, 26 (1861): 30.
8. Thomas Nunneley, in his *A Treatise on the Nature, Causes, and Treatment of Erysipelas* (London, 1841) gives a comprehension account of factors in the incidence of the disease.
9. Thomas Spencer Wells, 'Some causes of excessive mortality after surgical operations', *British Medical Journal*, Pt 2 (1864): 384–8. Wells was no doubt getting his own back on London surgeons who were so critical of his operation of ovariotomy, judged by them to be a recipe for disaster.
10. Sir James Y. Simpson, *Anaesthesia, Hospitalism, Hermaphrodism and A Proposal to Stamp out Small-pox and Other Contagious Diseases* (Edinburgh, 1871): 289–90.
11. Timothy Holmes, 'On the influence exerted by treatment in hospital upon the event of surgical operations and accidents', *British Medical Journal*, Pt 2 (1866): 687.
12. Lindsay Granshaw, 'St Thomas's Hospital, London', PhD thesis, Bryn Mawr (1981); Grace Goldin, 'Building a hospital of air: the Victorian pavilions of St Thomas's Hospital, London', *Bulletin of the History of Medicine*, 49 (1975): 512–35.
13. For Sir William Lawrence's use of nitric acid, see Holmes Coote, 'On the treatment of wounds', *St Bartholomew's Hospital Reports*, 6 (1870): 113-14. A range of other popular treatments is described in Nunneley, *A Treatise*; the problems and advantages of limb 'conservation' are discussed in Report, 'A mirror of the practice of medicine and surgery in the hospitals of London: St Bartholomew's Hospital', *Lancet*, Pt 2 (1857): 601.
14. Fisher, *Joseph Lister*; Joseph Lister, 'The early stages of inflammation', *Philosophical Transactions of the Royal Society of London*, 148 (1868): 645–702.
15. Joseph Lister, 'On the antiseptic principle in the practice of surgery': 353.
16. Lister, 'On the antiseptic principle in the practice of surgery': 353.
17. Joseph Lister, 'On a new method of treating compound fracture, Abscess, etc.', *Lancet*, Pt 1 (1867): 327.
18. Joseph Lister, 'On the antiseptic principle in the practice of surgery': 353.
19. Joseph Lister, 'On some points in the history of antiseptic surgery', *British Medical Journal*, Pt 1 (1908): 1557. Calvert argued strongly for antisepsis, promising a good supply of carbolic acid. Lister later experimented with other antiseptics and in methods of impregnating

gauze with antiseptic substances, dealing at times with Thomas Morson and Sons Ltd. See A. Clifford Morson and Basil C. Morson, 'Memories of Lord Lister', *Proceedings of the Royal Society of Medicine*, 64 (1971): 1060–3.
20. Joseph Lister, 'Illustrations of the antiseptic system of treatment in surgery', *Lancet*, Pt 2 (1867): 668.
21. Lister, 'Illustrations': 668.
22. Lister, 'Illustrations': 668.
23. Lister, 'Illustrations': 668.
24. An account of the general use and early enthusiasm for carbolic acid as a dressing can be found in: Editorial, 'Carbolic acid in surgical practice', *Medical Times and Gazette*, Pt 2 (1867): 355. Carbolic acid was applied to all kinds of wounds and sores, as well as being administered internally for various diseases. John Wilson, for example, used the acid in the treatment of spina bifida: 'Case of spina bifida treated antiseptically', *Glasgow Medical Journal*, 4 (1871–2): 1–6, while Henry Greenway was one of many who used it on syphilitic sores: 'Treatment of constitutional syphilis by carbolic acid', *British Medical Journal*, Pt 2 (1867): 635.
25. Joseph Lister, 'On the antiseptic principle in the practice of surgery', *Lancet*, Pt 2 (1867): 353–6; 'On the antiseptic principle in the practice of surgery', *British Medical Journal*, Pt 2 (1867): 246–8; he had already published case studies earlier in 1867 in the *Lancet*: Joseph Lister, 'On a new method of treating compound fracture', *Lancet*, Pt 1 (1867): 326–9, 357–9, 507–9, and *Lancet*, Pt 2: 95–6.
26. Sir James Y. Simpson, 'Carbolic acid and its compounds in surgery', *Lancet*, Pt 2 (1867): 546–9; Jules Lemaire, *De l'Acide Phénique: De son Action sur les Végétaux, les Animaux, les Ferments, les Vénins, les Virus, les Miasmes; et de ses Applications à l'Industrie, à L'Hygiéne, aux Sciences Anatomiques, et à la Thérapeutique* (Paris, 1863).
27. F. W. Ricketts, 'Letter to the Editor', *Lancet*, Pt 2 (1867): 614.
28. For example, John Dewar, 'Letter to the Editor', *Lancet*, Pt 2 (1867): 757; Editorial, 'Compound comminuted fracture of femur without a trace of suppuration', *Lancet*, Pt 2 (1868): 324; T. Pridgin Teale, 'Letter to the Editor, including letter to Joseph Lister', *British Medical Journal*, Pt 2 (1869): 256–7; and Robert Hamilton, 'Carbolic acid in surgery', *Liverpool Medical and Surgical Reports*, 3 (1869): 49–63. As an example of a medical practitioner who used carbolic acid, without reference to Lister or to antisepsis, see D. Lloyd Roberts, 'On the local application of carbolic acid in uterine

disease', *Manchester Medical and Surgical Reports*, 1 (1870): 168–73.
29. Editorial, 'Carbolic acid in the London hospitals', *Medical Times and Gazette*, Pt 2 (1867): 575–6; 'Reports of medical and surgical practice in the hospitals of Great Britain: St George's Hospital. Three cases treated by Lister's method with carbolic acid', *British Medical Journal*, Pt 2 (1867): 588–9; Editorial, 'St George's Hospital: the use of carbolic acid', *Lancet*, Pt 2 (1868): 634; Editorial, 'Middlesex Hospital: Employment of carbolic acid', *Lancet*, Pt 2 (1868): 695; Editorial, 'The use of carbolic acid', *Lancet*, Pt 2 (1868): 728; Editorial, 'King's College Hospital: carbolic-acid treatment of suppurating and sloughing wounds and sores', *Lancet*, Pt 2 (1868): 762–3; Editorial, 'Charing-Cross Hospital: the use of carbolic acid', *Lancet*, Pt 1 (1869): 47; Editorial, 'University College Hospital: the use of carbolic acid', *Lancet*, Pt 1 (1869): 86–7; Editorial, 'London Hospital: the use of carbolic acid', *Lancet*, Pt 1 (1869): 122; and Editorial, 'Great Northern Hospital: the use of carbolic acid', *Lancet*, Pt 1 (1869): 361–2.
30. Editorial, 'Carbolic acid in the London hospitals', *Medical Times and Gazette*, Pt 2 (1867): 576.
31. Editorial, 'The use of carbolic acid', *Lancet*, Pt 2 (1868): 728.
32. Editorial, 'Compound comminuted fracture of femur', *Lancet*, Pt 2 (1868): 324.
33. Report, 'Belfast General Hospital: wounds of joints treated on the antiseptic method by carbolic acid', *Medical Times and Gazette*, Pt 1 (1869): 142.
34. Editorial, 'Deodorisers, disinfectants, and antiseptics', *British Medical Journal*, Pt 2 (1867): 431.
35. Editorial, 'Deodorisers': 431.
36. J. R. Wolfe, 'The removal of cancer by caustic arrows and carbolic acid', *British Medical Journal*, Pt 2 (1868): 150; Henry Greenway, 'Treatment of constitutional syphilis by carbolic acid,' *British Medical Journal*, Pt 2 (1868): 635.
37. One early proponent of antisepsis who was apparently dissuaded from its use because of instances of carbolic acid burns among patients he had treated was Robert Lightfoot, 'Poisoning by local application of carbolic acid', *British Medical Journal*, Pt 2 (1870): 331–2.
38. Joseph Lister, 'An address on the antiseptic system of treatment in surgery', *British Medical Journal*, Pt 2 (1868): 53, 101, 461, 515; Pt 1 (1869): 301.

39. Thomas Nunneley, 'Address in surgery', *British Medical Journal*, Pt 2 (1869): 152.
40. Nunneley, 'Address in Surgery': 153.
41. D. Campbell Black, 'Letter to the editor', *British Medical Journal*, Pt 2 (1869): 281.
42. Joseph Lister, 'On the effects of the antiseptic system upon the salubrity of a surgical hospital', *Lancet*, Pt 1 (1870): 4.
43. As quoted in: Joseph Lister, 'Further evidence regarding the effects of the antiseptic system of treatment upon the salubrity of a surgical hospital', *Lancet*, Pt 2 (1870): 287–9.
44. Lister, 'On the effects of the antiseptic system of treatment upon the salubrity of a surgical hospital', *Lancet*, Pt 1 (1870): 4–6, 40–2; Joseph Lister, 'The Glasgow Infirmary and the antiseptic treatment', *Lancet*, Pt 1 (1870): 210; and 'Further evidence regarding the effects of the antiseptic system of treatment upon the salubrity of a surgical hospital', *Lancet*, Pt 2 (1870): 287–9.
45. Lister, 'Further evidence regarding the effects of the antiseptic system of treatment upon the salubrity of a surgical hospital': 288.
46. Joseph Lister, 'Address in surgery', *British Medical Journal*, Pt 2 (1871): 226.
47. Lister, 'Address in surgery': 225–33.
48. Lister, 'Address in surgery': 227.
49. Lister, 'Address in surgery': 225.
50. Lister, 'Address in surgery': 226.
51. Lister, 'Address in surgery': 227.
52. Hector Cameron, *Joseph Lister: The Friend of Man* (London, 1948): 88–9.
53. To trace the development of Hamilton's stand on antisepsis see Robert Hamilton, 'Two amputations, treated upon M. Maisonneuve's method of pneumatic aspiration', *Liverpool Medical and Surgical Reports*, 2 (1868): 124–35; 'Carbolic acid in surgery', *Liverpool Medical and Surgical Reports*, 3 (1869): 49–63; and 'The after-treatment of large amputations', *Liverpool and Manchester Medical and Surgical Reports*, 2 (1874): 85–92. For Bickersteth's view, see E. R. Bickersteth, 'Remarks on the antiseptic treatment of wounds', *Lancet*, Pt 1 (1869): 743–4, 811–12; and 'Notes of cases treated upon antiseptic principles', *Liverpool Medical and Surgical Reports*, 4 (1870): 99–103.
54. William H. Barlow, 'Two cases of deep abscess treated by carbolic acid', *Manchester Medical and Surgical Reports*, 1 (1870): 37–40. Lund's early publications on antiseptic surgery include Edward

Lund, 'On the use of antiseptic cere-cloth for covering wounds', *British Medical Journal*, Pt 2 (1869): 267; 'Observations on some of the more recent methods of treating wounds', *Manchester Medical and Surgical Reports*, 1 (1870): 123–60; 'On the Antisepticity in Surgery', *Manchester Medical and Surgical Reports*, 2 (1871): 20–34; 'Fallacies and failures in antiseptic surgery', *British Medical Journal*, Pt 2 (1873): 454–6. For reactions in Birmingham, see 'Review of *The Antiseptic System: A Treatise on Carbolic Acid and Its Compounds*, by A. E. Sansom', *Birmingham Medical Review*, 1 (1872): 96; and 'Review of *On the Treatment of Fractures of the Limbs*, by Sampson Gamgee', *Birmingham Medical Review*, 1 (1872): 105–6; A. E. Sansom, *The Antiseptic System: A Treatise on Carbolic Acid and Its Compounds; with Enquiries into the Germ Theories of Fermentation, Putrefaction and Infection; and the Practical Applications of Antiseptics, especially in Medicine and Surgery* (London, 1871).
55. Report, 'The Edinburgh Royal Infirmary: the antiseptic treatment of wounds', *British Medical Journal*, Pt 1 (1870): 285–6.
56. Report, 'Glasgow Royal Infirmary. The antiseptic treatment of wounds', *British Medical Journal*, Pt 1 (1870): 361.
57. John Wood, 'Address in surgery', *British Medical Journal*, Pt 2 (1873): 147.
58. Report, 'The surgical section of the forty-first meeting of the British Medical Association', *British Medical Journal*, Pt 2 (1873): 257.
59. J. Cooper Forster, 'Clinical records', *Guy's Hospital Reports*, 18 (1873): 100–1.
60. 'Clinical Society of London, proceedings', *British Medical Journal*, Pt 2 (1875): 556–8; Editorial, 'Antiseptic surgery', *Lancet*, Pt 1 (1875): 565; and Timothy Holmes, 'On the amputation book of St George's Hospital, no. II', *St George's Hospital Reports*, 8 (1874–6): 304.
61. As quoted in: Editorial, 'Lister's antiseptic dressings', *British Medical Journal*, Pt 1 (1873): 97.
62. Editorial, 'Edinburgh Royal Infirmary: the antiseptic method', *British Medical Journal*, Pt 1 (1873): 84.
63. Holmes Coote, 'On the treatment of wounds', *St Bartholomew's Hospital Reports*, 6 (1870): 113.
64. Coote, 'On the treatment of wounds': 113.
65. Coote, 'On the treatment of wounds': 115.
66. Coote, 'On the treatment of wounds': 116.
67. Coote, 'On the treatment of wounds': 116.

68. Coote, 'On the treatment of wounds': 113–17.
69. Robert Hamilton, 'The after-treatment of large amputations', *Liverpool and Manchester Medical and Surgical Reports*, 2 (1874): 85–92.
70. George W. Callender, 'The isolation and treatment of wounds', *British Medical Journal*, Pt 2 (1873): 424–5.
71. Callender, 'The isolation and treatment of wounds': 424–5; and George E. Callender, 'Two years of hospital practice', *St Bartholomew's Hospital Reports*, 9 (1873): 1–46.
72. Callender, 'The isolation and treatment of wounds': 424–5.
73. Callender, 'Two years of hospital practice': 19.
74. J. Cooper Forster, 'Clinical records', *Guy's Hospital Reports*, 18 (1873): 100–1.
75. Editorial, 'Antiseptic surgery', *Lancet*, Pt 2 (1875): 565.
76. Editorial, 'Antiseptic surgery', *Lancet*, Pt 2 (1875): 597.
77. Editorial, 'Antiseptic surgery': 597.
78. Editorial, 'Mr Wood's address in surgery', *Lancet*, Pt 2 (1873): 272.
79. Editorial, 'Antiseptic surgery', *Lancet*, Pt 2 (1875): 565.
80. Sampson Gamgee, 'The treatment of wounds on the antiseptic method', *Lancet*, Pt 1 (1874): 51.
81. George Thomson, 'Letter to the Editor', *Lancet*, Pt 2 (1873): 432.
82. George Thomson, 'The antiseptic system in surgery', *Medical Times and Gazette*, Pt 2 (1875): 517.
83. Thomson, 'The antiseptic system': 517.
84. Joseph Lister, 'An address on the effect of the antiseptic treatment upon the general salubrity of surgical hospitals', *British Medical Journal*, Pt 2 (1875): 769–71.
85. See, for example, Editorial, 'Antiseptic surgery', *Lancet*, Pt 2 (1875): 598.
86. Joseph Lister, 'On recent improvements in the details of antiseptic surgery', *Lancet*, Pt 2 (1875): 365.
87. Joseph Lister, 'Demonstrations of antiseptic surgery before members of the British Medical Association', *Edinburgh Medical Journal*, 21 (1875–6): 193, 481.
88. Lister, 'Demonstrations': 193, 481, and Lister, 'On recent improvements', *Lancet*, Pt 2 (1875): 365.
89. Editorial, 'The vacancy at King's College Hospital', *British Medical Journal*, Pt 1 (1877): 242.
90. Editorial, 'Professor Lister', *British Medical Journal*, Pt 1 (1877): 277.
91. Editorial, 'Professor Lister': 277.

92. Editorial, 'Prof. Lister and London clinical teaching', *Medical Press and Circular*, Pt 1 (1877): 192.
93. Editorial, 'Prof. Lister': 192.
94. Editorial, 'Professor Lister', *Lancet*, Pt 1 (1877): 361.
95. Sir St Clair Thompson, 'A house-surgeon's memories of Joseph Lister', *Annals of Medical History*, 2 (1919): 98–9.
96. Editorial, 'King's College Hospital and School', *British Medical Journal*, Pt 1 (1877): 657.
97. L. J. Hobson, 'Letter to the Editor', *Lancet*, Pt 1 (1878): 851. The editor of the *Lancet* congratulated those at the hospital on this decline in mortality, but cast doubts upon the credit given to antisepsis by remarking that the mortality of the earlier years was appallingly high and that this could not be considered a fair average of the mortality 'under the best non-antiseptic methods'. Editorial, 'Antiseptic surgery in large hospitals', *Lancet*, Pt 1 (1878): 851. For evidence that antisepsis was being lectured upon at Guy's Hospital, see Henry Davy, 'Five cases illustrative of antiseptic surgery and pyaemia', *Guy's Hospital Reports*, 23 (1878): 261–328.
98. Editorial, 'The antiseptic treatment in the Glasgow Royal Infirmary', *British Medical Journal*, Pt 2 (1879): 473.
99. Robert Druitt, *The Surgeon's Vade Mecum: A Manual of Modern Surgery* (London, 1870).
100. Robert Druitt, *The Surgeon's Vade Mecum: A Manual of Modern Surgery*, 11th edn (London, 1878): ix.
101. Henry Davy, 'Five cases illustrative of antiseptic surgery and pyaemia', *Guy's Hospital Reports*, 23 (1878): 261–328.
102. Report, 'Antiseptic surgery: report of the meeting of the South London District of the Metropolitan Counties Branch of the British Medical Association, at St Thomas's Hospital', *British Medical Journal*, Pt 2 (1879): 1002.
103. Report, 'Antiseptic surgery': 1001.
104. Report, 'Antiseptic surgery': 1002.
105. T. Messenger Bradley, 'Antiseptic surgery', *British Medical Journal*, Pt 1 (1878): 259.
106. Druitt, *The Surgeon's Vade Mecum*, 11th edn: ix.
107. Druitt, *The Surgeon's Vade Mecum*, 11th edn: 132.
108. Druitt, *The Surgeon's Vade Mecum*, 11th edn: ix.
109. Bradley, 'Antiseptic surgery': 258.
110. Bradley, 'Antiseptic surgery': 258.
111. Editorial, 'Professor Lister at Amsterdam', *British Medical Journal*, Pt 2 (1879): 454.

112. William MacCormac, 'Antiseptic Surgery: an address delivered at St Thomas's Hospital, with the subsequent debate', *British Medical Journal*, Pt 1 (1880): 927.
113. MacCormac, 'Antiseptic Surgery': 927.
114. MacCormac, 'Antiseptic Surgery': 927.
115. MacCormac, 'Antiseptic Surgery': 927.
116. 'Adjourned discussion on Mr Stokes's communication on cases of ovariotomy treated antiseptically', *British Medical Journal*, Pt 1 (1880): 594.
117. Joseph Lister, 'Ten years' surgery in the Kilmarnock Infirmary', *British Medical Journal*, Pt 2 (1880): 340.
118. Lister, 'Ten years' surgery': 342.
119. Lister, 'Ten years' surgery': 342.
120. 'The antiseptic theory tested by the statistics of ovariotomy', *Lancet*, Pt 1 (1880): 250.
121. Joseph Lister, 'An address on the treatment of wounds', Lancet, Pt 2 (1881): 863.
122. Lister, 'An address on the treatment of wounds': 863.
123. Lister, 'An address on the treatment of wounds': 863.
124. Lister, 'An address on the treatment of wounds': 863.
125. Lister, 'An address on the treatment of wounds': 864.
126. Lister, 'An address on the treatment of wounds': 864.
127. Joseph Lister, 'An address on the present position of antiseptic surgery', *British Medical Journal*, Pt 2 (1890): 378.
128. Lister, 'An address on the present position': 378.
129. 'Military Medical Society, Woolwich', *British Medical Journal*, Pt 1 (1884): 364.
130. 'Military Medical Society': 363.
131. Edgar M. Crookshank, 'Remarks on the antiseptic treatment of the wounded on the battle-field', *Lancet*, Pt 1 (1884): 422.
132. 'Antiseptic surgery in war', *Lancet*, Pt 1 (1884): 352.
133. C. H. Godwin, 'Abstract of an address on antiseptic surgery in its application to field service', *British Medical Journal*, Pt 1 (1884): 350.
134. 'Influence of the antiseptic method on medical jurisprudence', *British Medical Journal*, Pt 1 (1881): 61.
135. 'Rules of antiseptic surgery in Germany and Austria', *Lancet*, Pt 2 (1884): 978.
136. 'Rules of antiseptic surgery': 978.
137. 'Rules of antiseptic surgery': 978.
138. 'Rules of antiseptic surgery': 978.

139. 'Antiseptics and anaesthetics in Berlin', *British Medical Journal*, Pt 2 (1890): 469.
140. 'Antiseptics and anaesthetics': 469.
141. 'New antiseptic methods', *British Medical Journal*, Pt 1 (1886): 1119–20.
142. 'Editorial', *Lancet*, Pt 2 (1889): 1013.
143. Joseph Lister, 'On the principles of antiseptic surgery', Virchow-Festschrift Bd. 3 (1891), as reprinted in *The Collected Papers of Joseph Baron Lister*, vol. 2 (Oxford, 1919): 340.
144. Lister, 'On the principles of antiseptic surgery': 340–8.

3   The Sanatorium Treatment for Consumption in Britain, 1890–1914

* I would like to thank John Pickstone, Roger Cooter and Joan Mottram for comments on an earlier draft of this essay and the Wellcome Trust for their financial support.
1. The medical and lay terminology for this disease changed in the period discussed from Phthisis or consumption to Pulmonary Tuberculosis, or just tuberculosis. It is not certain that there was a continuity in the underlying pathology over this transition, although contemporaries assumed such a continuity. In this essay no attempt has been made to standardise the nomenclature, the terms used at the time are given. The term 'consumption' is used in the title as it was still common up to 1913; the National Association for the Prevention of Consumption and other forms of Tuberculosis did not change its name to the National Association for the Prevention of Tuberculosis until 1919.
2. F. B. Smith, *The Retreat of Tuberculosis, 1850-1950* (London, 1988); L. Bryder, *Below the Magic Mountain: A social history of tuberculosis in twentieth century Britain* (Oxford, 1988). Also, the interest in tuberculosis is evident in the new edition of R. and J. Dubos, *The White Plague* (London, 1987).
3. See: N. Rosenberg, *Perspectives on Technology* (London, 1974).
4. The figures are those cited by the anti-tuberculosis movement c. 1900. They are for Great Britain (i.e., England, Wales and Scotland, but not Ireland). At this time it was the convention to multiply the number of deaths by five to estimate the number of sufferers; by 1910 the convention was to multiply by ten.
5. Smith, *The Retreat*: 238–9.

6. *British Medical Journal*, Pt 2 (1908): 505.
7. A. S. Wohl, *Endangered Lives* (London, 1983): 117–32.
8. G. Cronje, 'Tuberculosis and mortality decline in England and Wales, 1851–1910', in R. I. Woods and J. H. Woodward (eds), *Urban Disease and Mortality in Nineteenth Century England* (London, 1984).
9. A powerful medical argument against the infectiousness of consumption was the seeming absence of cross-infection between patients, and between patients and medical staff in specialist hospitals. C. T. Williams, *British Medical Journal*, Pt 2 (1882): 618–21 and J. Thorowgood, *British Medical Journal*, Pt 1 (1885): 889.
10. *British Medical Journal*, Pt 2 (1880): 42.
11. C. Rosenberg, 'The Bitter Fruit: Heredity, Disease and Social Thought in Nineteenth Century America', *Perspectives in American History*, 7 (1974): 189–238. The theory, held by some sanitarians, that consumption was caused by the breathing of foul or re-breathed air was not in any sense an anticipation of later views of communicability by droplet infection; rather it was that such air, and the insanitary places in which it was breathed, weakened and predisposed the body to the disease.
12. C. R. Drysdale, 'Alpine heights and marine climates in the treatment of consumption', *British Medical Journal*, Pt 2 (1892): 628. Also see: J. Pemble, 'The Victorians practised alternative medicine', *The Listener* (22 March 1984): 14–15 and 'When a doctor was identified with God', *The Listener* (29 March 1984): 10.
13. W. Bezley Thorne, *Lancet*, Pt 1 (1895): 119 and A. Ransome, 'On certain bodily conditions resisting Phthisis', *Medical Chronicle* (April 1903).
14. J. M. Grange and P. J. Bishop, '"Über Tuberculose": A tribute to Robert Koch's discovery of the tubercle bacillus, 1882', *Tubercle*, 63 (1982): 3–17.
15. J. Coats, *Glasgow Medical Journal*, 15 (1881): 302.
16. *British Medical Journal*, Pt 1 (1885): 130. On the continuities in the ætiology of consumption around the 'bacteriological revolution' see: N. M. McMurry, '"And I? I am in a consumption": the tuberculosis patient, 1780-1930', unpublished PhD thesis, Duke University (1985): 177–92.
17. *British Medical Journal*, Pt 1 (1885): 897.
18. The variability of the disease allowed doctors wide discretion in their prescriptions; the most popularly used products were alkalis, bromides, morphia, iron, hypophosphites and calcium chloride, with

much attention also given to countering wasting. A. Latham, *The Diagnosis and Modern Treatment of Pulmonary Consumption* (London, 1903); Ch. VII reviews nineteenth century therapies and gives the current 'state of the art' on alternatives to sanatoria. Also see C. Allbutt, 'On the study of tuberculosis: a retrospect', *British Journal of Tuberculosis,* 1 (1907): 5.
19. *British Medical Journal*, Pt 2 (1911): 964–5.
20. J. Sawyer, 'Therapeutic Notes', *British Medical Journal*, Pt 1 (1880): 845. Also see: Smith, *The Retreat*: 45–7.
21. J. Lister, 'Lecture on Koch's treatment of tuberculosis', *British Medical Journal*, Pt 2 (1890): 1372–4.
22. H. Weber, 'Croonian Lectures on the hygienic and climatic treatment of chronic pulmonary phthisis', *British Medical Journal*, Pt 2 (1885): 517–22, 575–6, 641–3, 688-90 and 725–7.
23. R. H. Shryock, *The National Tuberculosis Association, 1904–54: a study of the voluntary health movement in the United States* (New York, 1957): 28.
24. Numerous references could be given but perhaps the most influential writing in this vein was that of A. Ransome; see: *The Treatment of Phthisis* (London, 1896); *The Principles of 'Open-Air' Treatment of Phthisis and of Sanatorium Construction* (London, 1904).
25. R. H. Shryock, *The Development of Modern Medicine* (London, 1974 edn): 305–6.
26. The anti-tuberculosis movement had great difficulty with the notion that the disease was infectious, but not ordinarily so. What was one to make of the following: 'our great bar to progress ... is the inadequate recognition of the essentially infective nature of tuberculosis. Not that tuberculosis is, in the popular sense of the word, infectious, but infective in the sense that the disease cannot exist in the absence of the specific organism, however, strong the predisposition to it may be'. H. A. Calley, *British Medical Journal*, Pt 2 (1898): 1071.
27. A. Ransome, *Researches on Tuberculosis* (London, 1898).
28. This view was reinforced by the remote locations of many institutions and later by the fact that some authorities used the empty wards of isolation hospitals for the treatment.
29. *British Medical Journal*, Pt 2 (1895): 910 and *British Medical Journal*, Pt 1 (1896): 1150 and Pt 2 (1896): 1008.
30. There is little evidence that developments in Britain were influenced by American experiences. The work of Trudeau, Knopf and Krause was acknowledged, but not followed.
31. Older institutions like the Royal National Hospital for Consumption

at Ventnor were sanatoria in the older sense of institutions of recuperation, whereas the new German form was more interventionist and saw itself as a specific cure.

32. This figure is based on information given in: *Sanatorium Annual* (London, 1899-1911); F. R. Walters, *Sanatoria for Consumptives* (London, 1899); C. Reinhardt, *Life in an Open-Air Sanatorium* (London, 1900).
33. For an account of life in a private sanatorium, probably Nordrach-on-Dee at Banchory, though based on experiences just after The First World War, see: W. Somerset Maugham, *Creatures of Circumstance* (London, 1947) 'The Sanatorium'.
34. On events in Scotland, see: D. Lawson, 'The Nordrach treatment for Phthisis in Scotland', *Dublin Journal of Medical Science*, 108 (1899): 324.
35. J. A. Gibson, 'The Nordrach Cure practicable in this country', *Nineteenth Century*, 45 (June 1899): 389–403; R. Thurnam, *The Practitioner* (1899); R. Mander Smyth, *The Practitioner* (1901): 36. There is mention that Dr G. A. Crace Calvert of the Vale of Clwyd Sanatorium was also a former patient at Nordrach.
36. The continuities are greater than these figures suggest as several of the sanatoria which 'closed', did so to allow their proprietor to move premises. For example, Dr D. Johns began, with Arthur Ransome, at Stourfield Park in 1897 and moved to Alderney Manor in 1902.
37. *British Journal of Tuberculosis*, 2, Pt 2 (1908): xxxvii, advertisement for the East Anglian Sanatorium, 'All forms of tuberculosis are admitted, also recovery from other illnesses or surgical operations and patients who require Weir-Mitchell or other rest cures or spinal exercises'; Pt 4 (1908): xxxiii, advertisement of the Kingswood Sanatorium, 'For the treatment of neurasthenia, anemia, debility following operations and incipient lung trouble on Open-Air Principles'.
38. C. Williams, 'The hygienic treatment of consumption independently of sanatoria', *British Medical Journal*, Pt 2 (1898): 947; H. W. Crowe, *Consumption: Treatment at home and rules for living* (London, 1906); G. B. Dixon, 'The domiciliary treatment of tuberculous patients', *British Journal of Tuberculosis*, 9 (1915): 19–22.
39. *Manchester Hospital for Consumption, Report for the Year Ending 31 Dec. 1899*: 11. Some argued that in fact Ventnor had been practising the open-air cure since 1869! See: J. G. Sinclair Coghill, 'The prevention of consumption', *Nineteenth Century*, 45 (June 1899): 304–14.

40. The *OED* dates the change, perhaps conservatively, to 1899, quoting the *Western Gazette* from that year: 'Quite a little shoal of companies for the establishment of sanatoria has recently made its appearance. The latest company is the Harrogate Sanatorium for Consumptives'.
41. J. E. Chapman and N. D. Bardswell, 'Report of eight months' trial of the sanatorium treatment of consumption at the Sheffield Royal Infirmary', *Quarterly Medical Journal*, (February 1900). For the wider picture see: 'The treatment of tuberculosis in Hospitals and Infirmaries', *British Medical Journal*, Pt 2 (1902): 1542, 1599 and 1658.
42. F. S. Toogood, 'Tuberculosis and Metropolitan Pauperism', *British Journal of Tuberculosis*, 2, Pt 2 (1908): 166–74, notes that an enquiry by the Metropolitan Asylums Board showed that twenty London boroughs isolated consumptives and twelve provided open-air treatment. In his own borough of Lewisham, he notes that a full system has been in operation for 10 years! A local survey in South Yorkshire has shown that the Sheffield Poor Law Infirmary at Firth Park had a sanatorium ward c. 1904, P. Speck, *The Institution and Hospital at Fir Vale* (Sheffield, 1978). The Isolation Hospital at Rotherham had shelters for consumptives c. 1910 and in 1912 the Rotherham Union, concerned at the high cost of boarding out patients, established their own sanatorium for twenty-four patients. The *Medical Officer*, 8 (1912): 223. On the treatment in asylums see: D. Blair, 'Isolation and the open-air treatment of the Phthisical insane', *Journal of Mental Science*, 46 (1900): 215–9.
43. Most notably N. D. Bardswell, who went from Sheffield to private sanatoria at Mundesley and Banchory, subsequently to the prestigious King Edward VII Sanatorium at Midhurst and on to head the London County Council anti-tuberculosis service between the wars.
44. Practitioners in Manchester and Liverpool claimed to have pioneered such a campaign as early as the late 1880s, when leaflets stressing the infective nature of tuberculosis had been issued. The key figures were: Ransome, Niven, Tatham and Hope.
45. Bryder, *Below the Magic Mountain*: 2, 15–22.
46. *British Medical Journal*, Pt 2 (1898): 1899.
47. D. Armstrong, *The Political Anatomy of the Body* (London, 1983).
48. D. Watkins, 'The British Revolution in Social Medicine, 1889–1911', unpublished PhD thesis, University of London (1984).
49. J. A. Gibson, 'The open-air treatment of consumption: a personal experience', *Nineteenth Century*, 45 (January 1899): 92; 'The

Nordrach cure practicable in this country', *Nineteenth Century*, 45 (March 1899): 389; 'The cry of the consumptives', *Nineteenth Century*, 45 (October 1899): 641; A. P. Abraham, 'Dr Walther and his cure for consumption at Nordrach', *Windsor Magazine*, 11 (February 1900): 305.
50. J. Burdon Sanderson, 'Our duty to the consumptive bread-earner', Address to the Oxford and District Branch of the BMA (1901). Reprinted in Lady Burdon Sanderson, *Sir John Burdon Sanderson: a memoir* (Oxford, 1911): 299–312.
51. Amongst leading medical figures also prominent in the campaign were Clifford Allbutt, Malcolm Morris and R. W. Philip. See: J. H. Williams, *Requiem for a Great Killer: the story of tuberculosis* (London, 1973). The place of Robert Philip in the introduction of the open-air treatment into Britain is not clear. Certainly he pioneered the dispensary, opening the first British dispensary in Edinburgh in 1887, and he claimed that his Victoria Hospital was run on open-air lines from its opening in 1894. R. W. Philip, 'The Victoria Hospital for Consumption, Edinburgh: its rise and outlook', *Edinburgh Hospital Reports*, 3 (1895): 13–20. *British Medical Journal*, Pt 2 (1898): 217-20. This work attracted little attention south of the border before 1898. See A. T. Wallace, 'Sir Robert Philip: a pioneer in the campaign against tuberculosis', *Medical History*, 5 (1961): 56–64.
52. In many cases the distinction between voluntary, public and private was completely blurred. In 1909 the Winsley Sanatorium of the Bristol and Bath branch of the NAPC had forty-three maintained beds, nine leased to local authorities and sixteen reserved for private patients, twelve at 15*s*. per week and four luxury beds at 35–42*s*. per week. *Report on Winsley Sanatorium, 1909* (Bristol 1910). In 1911 the Crossley Sanatorium had thirty-nine beds for voluntary patients, twenty-five subsidised by local authorities and thirty-six for private patients. Manchester Hospital for Consumption, *Report for the year ending 31 Dec. 1911*.
53. *British Medical Journal*, Pt 1 (1900): 1042.
54. J. R. Bignall, *Frimley: The biography of a sanatorium* (London, 1979).
55. J. V. Pickstone, *Medicine and Industrial Society* (Manchester, 1985): 228–32.
56. Bryder, *Below the Magic Mountain*: 30.
57. Bryder, *Below the Magic Mountain*: 49–53.
58. A. Wood, 'Seventy years of caring', *Intel* (Winter 1980-81): 14–5.
59. On Ransom see: T. H. Jacob, *A History of the General Hospital*

(Nottingham, Bristol, 1951): 234–43. Significantly, the Pathology Laboratory as well as the sanatorium was named after Ransom, showing that it is wrong to think of sanatoria as being in opposition to medical science.
60. G. Stedman Jones, *Outcast London* (London, 1971).
61. J. Cunningham-Browne, *Lancet*, Pt 2 (1910): 1040. On Phthisiophobia in the United States see: McMurry, '"And I?"', ch. 12.
62. Bryder, *Below the Magic Mountain*: 199–226.
63. Cf. F. Mort, *Dangerous Sexualities* (London, 1987).
64. D. J. Chowdry-Muthu, 'Mendip Hills Sanatorium', *British Medical Journal*, Pt 2 (1905): 46. Also see, R. Lowe, 'The early twentieth century open-air movement: origins and implications', in N. Parry and D. McNair (eds), *The Fitness of the Nation: Physical and Health Education in the 19th and 20th Centuries* (Leicester, 1983). The links could probably be extended further to the other outdoor movements (The Holiday Fellowship, Clarion Clubs), heliotherapy and naturism, the view of public parks as the 'lungs of the city', the notion of the Garden City in town planning.
65. H. T. Bulstrode, *On Sanatoria for Consumptives and Certain Other Aspects of the Tuberculosis Question, Supplement to the 35th Annual Report of the Medical Officer for 1905–06*, Cd. 3657 (1908) and *Forty Fifth Annual Report of the LGB for the year 1915–16* (London, 1916): xxi; also cited in A. Newsholme, *International Studies on the Relation between the Private and the Official Practice of Medicine*, vol. III (London, 1931): 226.
66. The main sources were: *The Sanatorium Annual* (1900–12); *The Medical Annual*; *Annual Reports* from individual sanatoria; listings in the back of contemporary books on the treatment.
67. Children's sanatoria before 1914 were not predominantly for the treatment of consumption or pulmonary tuberculosis, but for the treatment of other tuberculous diseases and to treat 'weak' children.
68. The figure of 5500 for England alone was first given in 1916, earlier LGB estimates for both England and Wales gave 5200 beds as being available.
69. It is not surprising that historians using official figures have been misled about the growth of sanatorium beds in the late 1900s as these do show an increase in the period 1907–11 from 1748 (Bulstrode) to 5500 (LGB). My figures show a more modest increase for the same years: 2088 to 2981.
70. P. Boobbyer, 'Municipal Hospitals', *The Medical Officer*, 6 (1911): 219–21.

71. C. C. Easterbrook, 'The sanatorium treatment of active insanity by rest in bed in the open-air', *Journal of Mental Science*, 53 (1907): 723 and 54 (1908): 105–7.
72. *British Medical Journal*, Pt 1 (1905): 83, 214, 260. There was also a debate on what was meant by 'cure', see: 678, 755, 831, 908, 1073. The editor reported extensive correspondence on the topic, especially in January and February 1905.
73. *British Medical Journal*, Pt 1 (1905): 290, 912 and Pt 2: 260.
74. See the remarks of the always outspoken E. D. Marriott of Nottingham, *British Medical Journal*, Pt 1 (1905): 161 and Pt 2: 1160. Earlier, Marriott had been an outspoken critic of isolation hospitals for infectious diseases; see: J. M. Eyler, 'Scarlet fever and confinement: the Edwardian debate over isolation hospitals', *Bulletin of the History of Medicine*, 61 (1987): 7. Much attention was given to the economic costs of the treatment: the question was not, does it work or cure but, is it worthwhile? *British Medical Journal*, Pt 2 (1905): 105 and 1266.
75. In general Medical Officers of Health supported sanatoria; for example, in 1905 their professional body lobbied the Metropolitan Asylums Board to build sanatoria for the poor of London. Of course, these institutions, if controlled by Medical Officers of Health, would have extended their powers. See: *British Medical Journal*, Pt 1 (1905): 905.
76. The near-year-long correspondence was reviewed in an editorial: 'The Sanatorium and After', *British Medical Journal*, Pt 2 (1905): 1057.
77. See: *British Medical Journal*, Pt 1 (1905): 390, 448, 491, 512, 569, 800, 883, 912, 1090, 1358.
78. See discussion at the *Sixth Annual Meeting of the NAPC, 1905*.
79. In the years 1902–6 the proportions of admissions to the Ransom Sanatorium were: early – 41 per cent; secondary – 49 per cent; and advanced – 10 per cent. *Medical Report of the Ransom Sanatorium for the year 1917* (Nottingham, 1918): 12. The first annual report of the Winsley Sanatorium complained that 75 per cent of cases admitted had not been 'early' cases. *Bristol Medico-Chirurgical Journal*, 24 (1906): 278.
80. *The Lancet*, Pt 1 (1906): 1–12.
81. K. Pearson, *A First Study of the Statistics of Pulmonary Tuberculosis* (London, 1907). See subsequent correspondence: *British Medical Journal*, Pt 2 (1907): 414, 555, 701, 1104, 1277, 1623, 1806. It was regularly acknowledged that tuberculosis was a 'class disease'; in

such circumstances, one option would have been to 'eradicate' the vulnerable class by eugenic measures. There was virtually no support for such measures within the medical community. Equally there was little or no support for the alternative of eradicating poverty, and hence the class susceptible, by social reform and economic improvement. Medical efforts focused on the 'technical fixes' of fighting the disease and education.

82. Pearson, *A First Study*. The Bulstrode Report argued that tuberculosis was not a truly infectious disease and did not support notification.
83. *British Medical Journal*, Pt 1 (1908): 275–6 and 403–4. The opinion of successive editorials in the main journals was that the Report had vindicated the treatment, or at least argued that it needed to be given a fairer trial with more suitable early cases and longer stays. In March 1906 the LGB had advised the Metropolitan Asylums Board that there did 'not appear to be sufficient justification for the very heavy outlay' for the scheme of isolation and treatment of consumption proposed by the Society of Medical Officers of Health and the Royal Institute of Public Health. Quoted in G. M. Ayers, *England's First State Hospitals, 1867–1930* (London, 1971): 215.
84. H. Wilson, 'The science and art of sanatorium management', *British Journal of Tuberculosis,* 6 (1912): 37.
85. This shift was evident as early as 1905, see: *British Medical Journal*, Pt 2 (1905): 1057.
86. *Kelling Sanatorium – Historical Synopsis, Kelling Hospital* (mimeo) (nd): 2. Smith, *The Retreat*: 110 says that work therapy was pioneered at Kelling in 1902. This would be difficult unless the sanatorium was built by consumptives, as the sanatorium did not open until 1903! Occupational therapy was introduced in 1904, but actual work for patients was not introduced until 1905, and then only for convalescent patients. At Winsley work was introduced to counter idleness and to ease the transition back to work after discharge.
87. M. Paterson, *Autoinoculation in Pulmonary Tuberculosis* (London, 1911).
88. Bryder, *Below the Magic Mountain*: 46.
89. For example at Kelling in 1911 only 31 per cent of patients were on work therapy. Also Marcus Paterson, the Superintendent at Frimley, based his regime on the highly contentious theory of auto-inoculation and the practice of tuberculin therapy. *British Journal of Tuberculosis*, 5 (1911): 170–1. Indeed, one complaint was that such was the 'lack of harmony' between regimes in different institutions that their credibility had been irreparably damaged.

90. *British Medical Journal*, Pt 2 (1898): 983–6; *British Medical Journal*, Pt 2 (1900): 902.
91. At Benenden, the first rule in the regulations for patients was that 'Patients must obey'. In 1907 'graduated employment' was introduced 'to prevent the slackening and mental rusting which is one of the chief evils of residence in a sanatorium'. *National Association for the Establishment and Maintenance of Sanatoria for Workers, Second Annual Report* (November 1907).
92. 'A danger of the sanatorium', *British Medical Journal*, Pt 1 (1907): 400, notes that sanatoria were in danger of succeeding hydropathic establishments as matrimonial agencies. In 1915 the *British Journal of Tuberculosis*, acknowledging the extent of nocturnal activity, carried an item suggesting that patients be issued with a policeman's whistle and truncheon.
93. At the Crossley Sanatorium the number of patients discharged for 'rule breaking' was 5 per cent in 1905 and 1 per cent in 1906.
94. *British Medical Journal*, Pt 2 (1908): 1211. Of the 77 admissions that year, twenty-three patients discharged themselves and six were expelled for insubordination. Also see: H. Downes, 'The management of sanatoria for working men', *British Journal of Tuberculosis*, 2 (1907): 52–5.
95. V. Drew, *The Mind: its power in the cure of consumption* (London, 1908) and 'The psychology of the consumptive', *British Medical Journal*, Pt 1 (1910): 1253–4 and 1519–20.
96. Wilkinson had begun his advocacy from Australia. He came to Britain in 1909 and established a Tuberculin Dispensary League.
97. *British Medical Journal*, Pt 1 (1911): 295.
98. C. Wilkinson, *Tuberculin in the Diagnosis and Treatment of Tuberculosis* (London, 1912). Tuberculin Dispensaries were established in Paddington, Portsmouth and Sheffield, amongst other places.
99. A. Latham and C. H. Garland. *The Conquest of Consumption: an economic study* (London, 1910); *British Medical Journal*, Pt 2 (1910): 46–7, 349, 406–7, 496, 570, 1544.
100. *British Medical Journal*, Pt 2 (1910): 819–20; A Patient, 'Sanatoria' *British Journal of Tuberculosis*, 5 (1911): 98-106; W. J. Cox, 'Discipline and recreation in the sanatorium', *British Journal of Tuberculosis*, 9 (1915): 23–4.
101. In 1912, the *Lancet* noted that it had been 'the subject of almost ceaseless discussion and controversy', *Lancet*, Pt 2 (1912): 773.
102. The number of sufferers in 1911 was estimated at 300 000 or higher. This increase over the figure of 250 000 cited c. 1900 reflects changing diagnostic categories, as there was almost certainly

a decline in the incidence of the disease during the first ten years of the century.

103. E. P. Hennock, *British Social Reform and German Precedents: the case of social insurance, 1880–1914* (London, 1987); *Lancet*, Pt 1 (1911): 1290. Also see: 'National Insurance and Tuberculosis: A collection of representative opinions', *British Journal of Tuberculosis*, 5 (1911): 161–74 and Bryder, *Below the Magic Mountain*: 36–41.

104. C. H. Garland and T. D. Lister, *Sanatoria for the People: The state campaign against consumption* (London, 1911).

105. In the debates Austen Chamberlain moved a motion to delete Sanatorium Benefit, describing it as an 'excrescence on ... the Bill'. Arthur Balfour, who headed the Committee which oversaw Almroth Wright's Inoculation Department at St Mary's Hospital, which supported the use of tuberculin in certain cases, advocated further research, particularly in bacteriology. *Hansard* (1911), XXV, cols 2064-5; XXVIII, cols 385–445.

106. 'Tuberculosis and the National Insurance Act: Representative Opinion', *British Journal of Tuberculosis*, 6 (1912): comment by T. D. Lister: 134.

107. *Departmental Committee on Tuberculosis, Interim Report* (1912) Cd. 6164; *Final Report* (1913) vol. I, Cd. 6641 and vol. II, Cd. 6654. This scheme was seen by some as implementing for the whole of Britain the system developed in Edinburgh by R. W. Philip since 1887.

108. See the weekly reports in *The Medical Officer* from June 1912.

109. *Forty Third Annual Report of the Local Government Board*, Part III (London, 1914).

110. Bryder, *Below the Magic Mountain*: 76.

4 From Medical Research to Clinical Practice: Serum Therapy for Diphtheria in the 1890s

1. F. Loeffler, 'The History of Diphtheria', in G. H. F. Nuttall and G. S. Graham-Smith (eds), *The Bacteriology of Diphtheria* (Cambridge, 1908).

2. A. Newsholme, *Epidemic Diphtheria, A Research on the Origin and Spread of the Disease from an International Standpoint* (London, 1900): 63–4.

3. C. Salomon-Bayet, *Pasteur et la Révolution Pastorienne* (Paris, 1986) assumes a rapid therapeutic impact.
4. T. McKeown, *The Role of Medicine* (Oxford, 1979): 99. C. Singer, *A Short History of Medicine* (Oxford, 1928): 263.
5. E. Klebs, 'Ueber Diphtherie', *Verhandlungen des Congresses für Innere Medicin* (Wiesbaden, 1883): 139–54, reprinted in *Bulletin of the History of Medicine*, 8 (1940): 509–22.
6. O. Amsterdamska, 'Medical and biological constraints: early research on variation in bacteriology', *Social Studies of Science*, 17 (1987): 657-87.
7. H. H. Eulner, *Die Entwicklung der medizinischen Spezialfächer in Deutschland* (Stuttgart, 1973).
8. B. Latour, *The Pasteurization of France* (Cambridge, Mass. and London, 1987). For an alternative view, see: L. Murard and P. Zylberman, 'De l'hygiène comme introduction à la politique expérimentale (1875–1925)', *Revue de Synthèse*, 3rd series, 115 (1984): 313-41.
9. H. H. Mollaret and J. Brossolet, *Alexandre Yersin ou le vainqueur de la Peste* (Paris, 1985): 58-61.
10. Loeffler, 'The History of Diphtheria': 32.
11. E. Lagrange, *Monsieur Roux* (Brussels, 1954): 118.
12. E. von Behring, 'Serum therapy in therapeutics and medical science', in *Nobel Lectures in Physiology or Medicine 1901–1921* (Amsterdam, 1967): 6.
13. J. Lindemann, 'Immunology in the 1880s: two early theories', in C. M. Steinberg and I. Lefkovits (eds), *The Immune System* (Basel, 1981) vol. 1: 413–22.
14. H. Zeiss and R. Bieling, *Behring, Gestalt und Werk* (Berlin, 1941): 103-6.
15. Trousseau, with Bretonneau, had established the concept of diphtheria as an infectious disease, necessitating isolation of the sick. He had popularised tracheotomy in the 1830s.
16. E. Roux, 'Sur les Sérums antitoxiques. Communication faite au Congrès de Budapest', *Annales de l' Institut Pasteur*, 8 (1894): 723–7.
17. Zeiss and Bieling, *Behring*: 111.
18. O. Heubner, *Die experimentelle Diphtherie. Gekrönte Preisschrift* (Leipzig, 1883); O. Heubner, *Behandlung der Diphtherie mit dem Behringschen Heilserum* (Leipzig, 1895).
19. E. Behring, 'Einleitende Bemerkungen über die ätiologische Therapie von ansteckenden Krankheiten', *Gesammelte Abhand-*

lungen zur ätiologischen Therapie von ansteckenden Krankheiten (Leipzig, 1973).
20. *Le Figaro* (12 October 1894).
21. A. Delaunay, *L'Institut Pasteur* (Paris, 1962): 65–9.
22. Delaunay, *L'Institut Pasteur*: 81–2.
23. Pasteur Institute archives, E. Roux papers, document dated May 1895, no. 9283–92.
24. *Comite zur Beschaffung von Heilserum für Unbemittelte.*
25. P. J. Weindling, *Health, Race and German Politics between National Unification and Nazism* (Cambridge, 1989): 177–81.
26. Bundesarchiv Koblenz (hereafter BAK), R 86/1183.
27. BAK, R 86/1183, Statistik über Diphtherie Heilserum.
28. F. B. Smith, *The People's Health* (London, 1979): 151–2.
29. J. Liebenau, *Medical Science and Medical Industry* (Basingstoke, 1987): 48–56.
30. Virchow interview in *Nationalzeitung*, no. 565 (13 October 1894).
31. Schering archives, ms by P. Korn. I am grateful to Mick Worboys for bringing this source to my attention.
32. BAK, R 86/1646.
33. T. Lenoir, 'A Magic Bullet: Research for Profit and the Growth of Knowledge in Germany around 1900', *Minerva*, 26 (1988): 66–88.
34. H. Buchner, *Acht Vorträge aus der Gesundheitslehre*, 3rd edn (Leipzig, 1909): 127–9; Buchner, 'Ueber Immunität und Immu-nisierung', *Münchener Medizinische Wochenschrift* (1894) nos 37 and 38.
35. A. Gottstein and C. L. Schleich, *Immunität, Infektionstheorie und Diphtherie-Serum* (Berlin, 1894).
36. *Behring zum Gedächtnis. Reden und Wissenschaftliche Vorträge anlässlich der Behring-Erinnerungsfeier. Marburg an der Lahn, 4. bis 6. Dezember 1940* (Berlin, 1942).

5 Vaccine Therapy and Laboratory Medicine in Edwardian Britain

* I would like to thank John Pickstone for comments on earlier drafts of this essay. The work was supported by a grant from the Wellcome Trust.
1. The word 'vaccine' originally referred to the material, derived from cowpox lesions, used in inoculation against smallpox; cowpox was termed vaccinia, from the Latin root for the cow – vacca. The model

provided by smallpox inoculation dominated all attempts to use inoculation methods to protect against diseases, and so the terms of the procedure were carried over. Also, well into this century, the nature of the active constituent of vaccine lymph was unknown, so 'vaccine' became a generic term of all protective materials.
2. A. E. Wright, 'Vaccine Therapy: its administration, value and limitations', *Proceedings of the Royal Society of Medicine*, 3 (1910) Pt 1, General Reports: 2.
3. W. D. Foster, *A History of Medical Bacteriology and Immunology* (London, 1970): 142–5. W. D. Foster, *Pathology as a Profession in Great Britain* (London, 1981): 14.
4. S. C. Dyke, 'Organization of clinical pathology to the present day', in W. D. Foster, *A Short History of Clinical Pathology* (London, 1961): 125.
5. L. Colebrook, *Almroth Wright: Provocative Thinker and Doctor* (London, 1954). These claims were first made by Colebrook in one of his obituary notices on Wright. See: *Obituary Notices of Fellows of the Royal Society*, 6 (1948-9): 300–1 and *British Medical Journal*, Pt 1 (1947): 699 and Pt 2 (1953): 636.
6. H. C. Putnam, 'Vaccine therapy: a report of Sir Almroth Wright's London Clinic', *Boston Medical and Surgical Journal*, Pt 1 (1908): 18. Also see: R. K. Duncan, 'The new microbe inoculation', *Harper's Monthly*, 115 (1907): 204–11.
7. *British Medical Journal*, Pt 1 (1910): 979.
8. A. Landsborough Thomson, *Half a Century of Medical Research*, vol. 1 (London, 1973): 22–5. The MRC came to be chaired by Lord Moulton, who was a friend of Wright and was on the committee that oversaw Wright's department at St Mary's. H. F. Moulton, *The Life of Lord Moulton* (London, 1921). In his diary for 1913, Colebrook noted that Wright had kept himself off the Committee to make himself available to head the National Institute; though he had reservations about whether there would be access to 'sick humanity'. The idea of a small research hospital associated with the Institute seemed ridiculous, what patients would go to a 'research hospital'? In the event, Wright was appointed only head of bacteriology, a post he did not take up because of war duties. After the war he decided to continue at St Mary's with plans for his own research institute. In November 1918, Colebrook noted that Wright had little faith in, 'Fletcher (the new Secretary of the MRC) and a committee out of touch with the actual workers – seems to him to have no promise'. *Diary 1* (1913–19) entry for 28.11.1918.

9. C. J. Lawrence, 'Incommunicable knowledge: Science, technology and the clinical art in Britain, 1850-1914', *Journal of Contemporary History*, 20 (1985): 503–20.
10. S. J. Reiser, *Medicine and the Reign of Technology* (London, 1978) and A. M. Harvey, *Science at the Bedside: Clinical Research in American Medicine, 1905–1945* (Baltimore, 1981).
11. H. J. Parish, *Victory with Vaccines* (London, 1968): 42.
12. Foster, *Pathology, as a Profession:* 14; Foster, *A History of Medical Bacteriology*: 143.
13. H. F. Dowling, *Fighting Infection: Conquests of the Twentieth Century* (London, 1977): 35.
14. P. Keating, 'Vaccine therapy and the problem of opsonins', *Journal of the History of Medicine*, 43 (1988): 290.
15. Most general histories of immunology do begin with practice (i.e., Jenner and vaccination). Interestingly, a recent general history of immunology, which has the unpromising Whiggish title of *Milestones in Immunology*, begins with immuno-therapeutic work in the 1880s and 1890s by Sewall, Behring and Kitasato. See D. J. Bibel, *Milestones in Immunology: a historical exploration* (Madison, Wisc., 1988). For a sample of the secondary literature on the late nineteenth century see: J. Lindemann, 'Immunology in the 1880s: Two early theories', in C. M. Steinberg and I. Lefkovits (eds), *The Immune System*, vol. I (Basel, 1981): 413–22; L. P. Rubin, 'Styles in scientific explanation: Paul Ehrlich and Svante Arrhenius on immuno-chemistry'. *Journal of the History of Medicine*, 35 (1980): 397–425; P. H. M. Mazumdar, 'Immunity in 1890', *Journal of the History of Medicine*, 27 (1972): 312–24; P. H. M. Mazumdar, 'The antigen-antibody reaction and the physics and chemistry of life', *Bulletin of the History of Medicine*, 48 (1974): 1–21; A. M. Silverstein and A. A. Bialasiewicz, 'A history of theories of acquired immunity', *Cellular Immunology*, 51 (1980): 151–67; A. M. Silverstein, 'Cellular versus humoral immunity: determinants and consequences of an epic 19th century battle', *Cellular Immunology*, 48 (1979): 208–21.
16. A. E. Wright, 'The future of medicine', *British Medical Journal*, Pt 1 (1907): 333–4.
17. I. Lowy, 'Immunology and Literature in the early twentieth century: *Arrowsmith* and *The Doctor's Dilemma*', *Medical History*, 32 (1988): 314–32. In contrasting the views of immunology in the two works, Lowy suggests that this was due to the differing positions of medical research in Britain and the USA. A simple explanation

would be that Shaw wrote in 1905 when vaccine therapy was seen to have great potential; Lewis wrote in 1922, when the treatment had been heavily criticised, even discredited.

18. An important implication of this is that we recognise the full range of work done by early clinical laboratories. There has been a tendency to consider them only in terms of their present function – diagnostics. This point will be made in a forthcoming work by Joel Howell on x-rays and electrical departments.

19. This section draws extensively on the two biographies of Wright. Colebrook, *Almroth Wright* and Z. Cope, *Almroth Wright: Founder of modern vaccine-therapy* (London, 1966).

20. He was a civilian rather than a career army medical man, unlike the expected appointee David Bruce. David Bruce was also one of the new generation of pathologists, but was passed over.

21. Active immunity is that produced by the body's own immune system and is usually long term. Passive immunity is that provided by immune materials taken from animals and injected into humans to give short-term protection and in some cases cures. Usually the immune materials are chemicals that counteract the toxins of pathogenic microbes. On Wright's views on different immune states see: A. E. Wright, *Studies in Immunization* (London, 1909).

22. Leonard Colebrook objected to vivisection on religious grounds and his biographer suggests that many at St Mary's were hesistant about vivisection. See W. C. Noble, *Coli: Great Healer of Men* (London, 1974). Witnesses at the Second Royal Commission on Vivisection identified Wright as a critic of serum therapy, which was particularly associated with vivisection, not least because the sera were the products of animals kept in scientific institutes. On the other hand, Wright's vaccines were non-animal products, consisting of the very bacteria causing the infection – indeed, they were associated by some with homoeopathy, because they treated like with like, with very small doses. See: E. Westacott, *A Century of Vivisection and Anti-vivisection* (London, 1949): 309–12, 474–6. I would like to thank Caroline Murphy for this reference.

23. A. E. Wright, *A Short Treatise on Anti-Typhoid Inoculation* (London, 1904); Colebrook, *Almroth Wright*: 30–44; H. J. Parish, *A History of Immunization* (Edinburgh, 1965); D. H. M. Broschel and R. B. Hornick, 'Who introduced typhoid vaccination: Almroth Wright or Richard Pfeiffer?', *Review of Infectious Diseases*, 3 (1981): 1251–4.

24. A. E. Wright, 'Notes on the treatment of furunculosis, cycosis and

acne by the inoculation of a staphylococci vaccine', *Lancet*, Pt. 2 (1902): 874–84.

25. A. E. Wright and S. R. Douglas, 'An experimental investigation of the role of the blood fluids in connection with phagocytosis', *Proceedings of the Royal Society*, 72 (1903): 357–70. All of Wright's major papers from this period were reprinted in Wright, *Studies in Immunization*.

26. Actual counts were made of standard dilutions because of the large number of cells. Blood was usually donated by one of the 'normal' researchers or their assistants!

27. T. D. Brock, *Robert Koch: a life in medicine and bacteriology* (Madison, Wisc., 1988): 195–213; F. B. Smith, *The Retreat of Tuberculosis, 1850–1950* (London, 1988): 57–62.

28. A. Latham, *The Diagnosis and Modern Treatment of Pulmonary Consumption* (London, 1903): 152.

29. Lowered opsonic indices could now be demonstrated in studies of pulmonary patients in sanatoria, where in fact tuberculin therapy was often used. See: *British Medical Journal*, Pt 2 (1905): 1617, 1679 and 1720. Also, *British Medical Journal*, Pt 2 (1905): 1396–7 and Pt 2 (1906): 19–21.

30. *British Medical Journal*, Pt 2 (1904): 560-3 and Pt 2 (1905): 172.

31. W. Bulloch, *British Medical Journal*, Pt 2 (1905): 1604.

32. *British Medical Journal*, Pt 2 (1905): 1396-7 and Pt 2 (1906): 19–23.

33. This met with considerable hostility, not least from surgeons. See: W. Watson Cheyne, *Lancet*, Pt 1 (1906): 78–82 and Leader, *Lancet*, Pt 1 (1906): 237–8.

34. Manchester Royal Infirmary Records, Medical Board Minutes, 5: 279. It was agreed that one of Wright's assistants be employed at the Infirmary for three months. The young pathologist who came, G. Edward Loveday, both practised and taught opsonic methods. He was appointed Hospital Pathologist when the post became vacant at the end of 1906.

35. R. E. Boxill, *Shaw and the Doctors* (London, 1969) and Lowy, 'Immunology and Literature', *passim*.

36. The all-night sessions in the Pathology Department were well known. *St Mary's Hospital Gazette*, 13 (1907): 2–3. Cope suggests that in one year 16 000 opsonic index tests were made, each of which took 30 minutes. Taking a working year as 2000 hours, this equates with four full-time posts to perform the tests. Cope, *Almroth Wright*: 42–50, 102.

37. As early as 1905, there was talk of 'preaching the gospel of opsonins', *St Mary's Hospital Gazette*, 11 (1905): 104. Also see: A. C. Inman, 'Science in Medicine', *Science Progress*, 1 (1906): 238–58.
38. J. G. Emanuel, 'Opsonins and vaccine inoculation', *The Medical Annual*, (1907): 54; *The Practitioner* (1908) carried articles by Wright, Alexander Fleming, R. W. Allen, A. C. Inman and T. J. Horder, amongst others. Also see: J. Courtenay MacWatters, 'Vaccine therapy in General Practice', *The Practitioner*, Pt 2 (1909): 327.
39. R. W. Allen, *Vaccine Therapy* (London, 1907) 2nd edn (1908), 3rd edn (1910), 4th edn (1912); R. W. Allen, *The Opsonic Method of Treatment: a short compendium for general practitioners, students and others* (London, 1907).
40. *British Medical Journal*, Pt 1 (1907): 20.
41. This listing of publications is by no means exhaustive. See: T. W. Stonier, 'A resumé of vaccine therapy', *American Journal of the Medical Sciences*, 141 (1911): 186–213.
42. See Colebrook, *Almroth Wright*: 176-9 and *St Mary's Hospital Gazette*, 13 (1907): 83. The department took over accommodation that the hospital itself was unable to afford to commission. Details are given in: Sir Almroth Wright, 'Brief survey of the history and development of the Inoculation Department, St Mary's Hospital, W. 2.', Pamphlet (nd, c. 1946). I am grateful to Mr Palmer, Librarian at St Mary's Hospital, for bringing this to my attention.
43. *Chemist and Druggist*, 72 (1908): 334.
44. *Chemist and Druggist*, 72 (1908): 334; 74 (1909): 298.
45. See yearly statement, 1908-1915, in the *Lancet*'s 'Student Guide', published before the start of each academic year. For example, Pt 2 (1908): 629; Pt 2 (1911): 576. In July 1909 a letter in the *St Mary's Hospital Gazette* complained that the increased activity in the intellectual sphere at the hospital was making it difficult to put out teams for sports fixtures. *St Mary's Hospital Gazette*, 15 (1909): 96.
46. W. C. Bosanquet and J. W. H. Eyre, *Serums, Vaccines and Toxins in Treatment and Diagnosis* (London, 1909): 354–5.
47. See: *Journal of Clinical Research*, 1 (1907). In 1907 the CRA moved to new laboratories due to increased demand for their services and, it seems, in anticipation of extra demand from vaccine therapy. The early issues of the *Journal of Clinical Research* ran a series of articles on the treatment and there was some debate over its cost. See: *Journal of Clinical Research*, 1(2) (1907): 72.

48. J. W. H. Eyre, 'The new laboratories of the Bacteriological Department at Guy's Hospital', *Guy's Hospital Reports*, 68 (1914): 155–6.
49. *St Thomas's Hospital Reports*, 36 (1907): 324.
50. F. W. Andrewes, 'The beginnings of bacteriology at Bart's', *St Bartholomew's Hospital Journal*, 35 (1928): 117.
51. *Westminster Hospital Reports for 1910*, 192 (1911): 16; B. Russell (ed.), *St John's Hospital for Diseases of the Skin, 1863–1963* (London, 1963): 67.
52. N. H. Schuster, *A Clinical Pathologist Day by Day, 1916–60* (London, 1983): 1–2.
53. D. Moore Alexander, 'Some uses and abuses of vaccine therapy', *Medical Press and Circular*, 147 (1913): 529.
54. In 1931 Parke Davis published the 17th edn of their A. E. Wright, *Vaccine Therapy: The prophylaxis and treatment of bacterial disease by means of vaccines* (London, 1931), which they estimated had gone through 150 000 copies.
55. Allen, *Vaccine Therapy*, abandons the use of the opsonic index in the 3rd edn (1910). On attitudes to opsonic measurement see: *Proceedings of the Royal Society of Medicine*, 3 (1910): 86, 142.
56. For supporters trying to improve the test see: E. E. Glynn and G. Lissant Cox, 'Further observations upon certain sources of error in the opsonic technique', *Journal of Pathology and Bacteriology*, 16 (1912): 283–6. In the United States the objections to the test were fundamental, see: *Journal of Experimental Medicine,* (1906): 651 and (1907): 483; *Journal of Medical Research,* (1907): 521.
57. *Guy's Hospital Gazette*, 25 (1911): 46. The largest user in both 1909 and 1910 was the surgeon, Sir William Arbuthnot Lane.
58. H. Batty Shaw, 'On the present conflict of opinion as to the value of vaccines, including tuberculin', *Medical Chronicle*, 4th series, 26 (1913-4): 188.
59. Foster, *A Short History of Clinical Pathology*: 119–21.
60. The number of specimens fell after the first year to reach a low point in 1902, but grew thereafter. The main area of growth was in an ill-defined category of work – 'Other bacterial cultures' – which may have included vaccine therapy work. The proportion in this category rose from 7.3 per cent in 1904 to a high of 22.6 per cent in 1909, before falling back to 15 per cent in 1913. The latter fall was relative: the actual number of specimens grew, only not as fast as those for other tests. Figures taken from the annual report of the pathology laboratory in *St Thomas's Hospital Reports* (1896–1914).
61. *St Thomas's Hospital Medical School, Prospectus for year commencing October 1912* (London, 1912): 20.

62. *St Thomas's Hospital Reports*, 39 (1910): 323.
63. E. A. Ross, 'General survey and analysis of the cases treated by bacterial vaccines at St Thomas's Hospital', *St Thomas's Hospital Reports*, 37 (1908). The number of out-patient cases listed in the Annual Report peaks in 1907 at 210 and falls to 140 by 1910, only to rise again in the years up to 1914.
64. *42nd Annual Report of the Local Government Board for 1912–13*, Cd. 7181 (1914) Appendix A, No. 9: 223.
65. Andrewes, 'The beginnings of bacteriology': 117.
66. Manchester Royal Infirmary Records, Sub-Committees Minute Book (1895–1908), meeting on 3 July 1906, gave permission to Edward Loveday and Ramsbottom 'to use the laboratory for opsonin cases other than Infirmary cases'. Joel Howell reports similar arrangements at Guy's and the Middlesex in London (personal communication).
67. Both of Wright's biographers mentioned the large sums of money that could easily have been made in private practice.
68. The position had been that young doctors took part-time hospital appointments as clinical pathologists, while they established themselves in medical practice.
69. *Annual Reports of the Lister Institute* (1905–11).
70. *Chemist and Druggist*, 77 (1910): 824, 861, 964. The treatment was discussed at many provincial meetings of pharmacists.
71. The editor, W. d'Este Emery argued that this demise was a result of the successful establishment of the treatment in medicine as a whole.
72. G. Macfarlane, *Alexander Fleming: The man and the myth* (Oxford, 1985): 97, 146; R. Hare, 'The scientific activities of Alexander Fleming, other than the discovery of penicillin', *Medical History*, 27 (1983): 347–72.
73. Wright, *Vaccine Therapy*: i.
74. A. Fleming and G. F. Petrie, *Recent Advances in Serum and Vaccine Therapy* (London, 1934).
75. *British Medical Journal*, Pt 2 (1939): 99–104. Fleming notes (p. 99) that, 'At the present moment vaccine therapy is, in practice, having something of a setback that medical practitioners have become chemically minded because of the sensational results obtained in certain infections by sulphonamide and its allies ... However, it is certain that before long the limits of the new chemotherapy will be better understood, and I hope later to produce evidence indicating that the best results in the treatment of certain bacterial infections will be obtained by a combination of vaccine therapy with the new chemotherapy'.

76. The article is reprinted in Colebrook, *Almroth Wright*: 256–68. In fact, the article was his Introductory Address given at St Mary's in October 1904; see: *St Mary's Hospital Gazette*, 10 (1904): 118–9.
77. Wright, 'The future of medicine': 333.
78. R. W. Allen, *The Opsonic Method of Treatment* (London, 1907): 2.
79. In fact his attacks went further, he even took on public health, see *St Mary's Hospital Gazette*, 13 (1907): 74; and 17 (1911): 59. The *British Medical Journal* referred to Wright's ideas on public health as 'The gospel of Dirt'.
80. *St Mary's Hospital Gazette*, 13 (1907): 91.
81. Lawrence, 'Incommunicable knowledge', *passim*.
82. R. Maulitz, '"Physician *versus* Bacteriologist": The ideology of science in clinical medicine', in C. Rosenberg and M. Vogel (eds), *The Therapeutic Revolution* (Philadelphia, 1979): 91–107.
83. *The Clinical Research Association Limited Handbook* (London, ?1901). Copy held at Royal College of Surgeons, C. AZ, CLI. By 1909, the number of individual subscribers had risen to 5000 with 570 public bodies affiliated. *Journal of Clinical Research*, 3(1).
84. J. Odery Symes, *British Medical Journal*, Pt 1 (1901): 1017. As late as 1923 the authoritative report of the Bacteriological Committee of the Medical Research Council on Diphtheria still maintained that the laboratory was subordinate to the clinic. F. W. Andrewes *et al.* (eds), *Diphtheria: its bacteriology, pathology and immunology* (London, 1923): 235.
85. In the 1900s the best examples would be: Rubert Boyce, Liverpool; Sheridan Delépine, Manchester; J. Martin Beattie, Sheffield. On the latter, see the forthcoming paper in *Medical History* by Steven Sturdy. 'Hospital practice and the organisation of medical science in Sheffield, 1900–1920'.
86. J. D. Howell, 'Patient care at Guy's and the Pennsylvania Hospital, 1900–1920', *Papers of the Joint Conference of the BSHS and HSS* (July 1988): 247–54; S. Jacyna, 'The laboratory and the clinic: The impact of pathology on surgical diagnosis in the Glasgow Royal Infirmary, 1875-1910', *Bulletin of the History of Medicine*, 62 (1988): 384–406.
87. Z. Cope, *The History of the St Mary's Hospital Medical School* (London, 1954): 57.
88. See M. Worboys, 'The sanatorium treatment for consumption in Britain, 1890–1914' (Chapter 3 in this volume).
89. In October 1911, Lord Moulton gave the Opening Address at St Mary's Medical School extolling Wright's work as a model for the future direction of medicine.

90. Wright, 'Vaccine Therapy': 5–6.
91. Wright, 'Vaccine Therapy': 4.
92. *Bulletin of the Committee for the Study of Special Diseases*, 8 (Cambridge) (1907–8): 115.
93. At the RSM, Kingston Fowler's contention was typical; if after several years' use of a drug 'opinion is still unsettled ... it is usually because it has none'. However, there was a report by Freeman of a sophisticated double blind trial, involving 2000 inoculations with whooping cough vaccine. Proceedings of the Royal Society of Medicine, 3 (1910): 97–101.
94. R. Cole, 'The vaccine treatment of infectious disease', *International Clinics*, 17 Pt 2 (1907): 1–13.
95. Proceedings of the Royal Society of Medicine, 3 (1910): 41.
96. The term 'Celtic Siren' was William Osler's. *St Mary's Hospital Gazette*, 13 (1907): 91. In 1910 Conan Doyle made a double-edged comment about vaccine therapy, when he spoke on 'The Romance of Medicine' at St Mary's. He said that it 'transcends romance and seems rather to approach the fairyland of science'. *St Mary's Hospital Gazette*, 16 (1910): 106.
97. H. Stoneman, 'The principles of vaccine therapy: abstract of Sir A. E. Wright's lecture', *British Homoeopathic Review*, 1 (1907): 737–43. Also see Reports on the treatment at London Homoeopathic Hospital by F. A. Watkins in *British Homoeopathic Review*, 1 (1907): 743 and F. A. Watkins, *Vaccine Treatment of Infective Disease* (London, 1906).
98. *Proceedings of the Royal Society of Medicine*, 3 (1910): 54.
99. Report of an address by Wright on 'Auto-inoculation' at the Royal Institution (7 May 1910). *Chemist and Druggist*, 76 (1910): 749.
100. Batty Shaw, 'On the present conflict of opinion': 187.
101. *Proceedings of the Royal Society of Medicine*, 3 (1910): 141.
102. *Proceedings of the Royal Society of Medicine*, 3 (1910): 112.
103. S. West, *Proceedings of the Royal Society of Medicine*, 7 (1914): 115–6.
104. Wright did not attend or speak at this second debate, though some of his colleagues and supporters were present. At the time he was preoccupied with writing the report of his investigations on pneumonia in South African mines, which was expanding to a treatise on the logic of medicine, and to his writing on the suffrage question. Colebrook's *Diary* implies that Wright had lost interest in vaccine therapy and even his grip on research. 'Dougie [S. R. Douglas] and I are something amazed at this turn of affairs – it would seem often that the old Man [Wright] is game to follow any speculative hare if

there be a bit of a footnote for his understanding – and that independent of experimental data. *Diary 1* (1911-19): 99, entry for 27.12.1912.
105. T. J. Horder, *Proceedings of the Royal Society of Medicine*, 7 (1914): 71.
106. Horder, *Proceedings of the Royal Society of Medicine*, 7 (1914): 86.
107. Batty Shaw, 'On the present conflict of opinion': 188.
108. T. J. Horder, *Proceedings of the Royal Society of Medicine*, 3 (1910): 141.
109. There was much interest in bacteriology from the 1890s in the differences between pathogenic bacilli and similar non-pathogenic relatives which were termed pseudo-bacilli.
110. Wright may not have been correct in his prediction that 'The physician of the future will be an immunizator', but his actual practice was prophetic. The physician of the late twentieth century has come to rely on the laboratory at every level, regards clinical research as essential, and makes extensive use of the products of pharmaceutical companies.

6 From the Trenches to the Hospitals at Home: Physiologists, Clinicians and Oxygen Therapy, 1914–30

1. William Osler, *The Principles and Practice of Medicine*, 3rd edn (London, 1898): 137.
2. S. E. D. Shortt, 'Physicians, science and status: issues in the professionalisation of Anglo-American medicine in the 19th century', *Medical History*, 27 (1983): 51–68.
3. Christopher J. Lawrence, 'Incommunicable knowledge: science, technology and the clinical art in Britain, 1850–1914', *Journal of Contemporary History*, 20 (1985): 503-20; Judy Sadler, 'Ideologies of "art" and "science" in medicine: the transition from medical care to the application of technique in the British medical profession', in W. Krohn, W. T. Layton and P. Weingart (eds), *The Dynamics of Science and Technology: Social Values, Technical Norms and Scientific Criteria in the Development of Knowledge* (Dordrecht, 1978): 117–215. Cf. Gerald L. Geison, 'Divided we stand: physiologists and clinicians in the American context', in Morris J. Vogel and Charles E. Rosenberg (eds), *The Therapeutic Revolution* (Philadelphia, 1979): 67–90.
4. Benjamin Moore, 'Administration of oxygen in high percentage',

British Medical Journal, Pt 2 (1909): 839–840; Leonard Hill, 'The administration of oxygen', *British Medical Journal*, Pt 1 (1912): 71–2. Hill also made enthusiastic claims for the beneficial effects of oxygen inhalation on athletes, incidentally prompting a heated debate about the ethics of physiological interference in sporting competitions: Leonard Hill, M. Flack and T. H. Just, 'The influence of oxygen inhalation on athletes', *British Medical Journal*, Pt 2 (1908): 499–500. Haldane was able to find uses for oxygen outside the mainstream of the medical profession, in mine rescue apparatus and as a therapeutic agent in carbon monoxide poisoning: see, for instance, his paper on 'The causes of death in colliery explosions', *Transactions of the Institution of Mining Engineers, London*, 11 (1896): 502–13.

5. A variety of treatments were attempted at the time including morphia, atropine to stimulate the heart, oxygen, bleeding to relieve circulatory embarrassment, the inhalation of alcohol to neutralise chlorine, and the application of cocaine and adrenaline to the trachea and bronchi to reduce inflammation. See, *inter alia*, letters from Rushton Parker, J. D. Mortimer, and W. S. Syme, *British Medical Journal*, Pt 1 (1915): 1027–8; and A. W. Hendry and E. L. Horsburgh, 'Some general notes on suffocation by poisonous gases, with detailed notes on one fatal case', *British Medical Journal*, Pt 1 (1915): 964–5. The most comprehensive and authoritative of the early accounts of gas poisoning was J. Elliot Black, Elliot T. Glenny and J. W. McNee, 'Observations on six hundred and eighty-five cases of poisoning by noxious gases used by the enemy', *Journal of the Royal Army Medical Corps*, 24 (1915): 509–18, who could recommend no treatment more efficacious than the administration of expectorants to help clear oedema fluid from the lungs.

6. Royal Society of London, Minutes of the War Committee (12 May 1915), Royal Society Archives, CMB 36; Minutes of the Chemistry Sub-Committee (30 April 1915), CMB 28.

7. See Walter Morley Fletcher, 'An account of the work done by the Physiology (War) Committee up to December 1917', PRO: MUN5/386/1650/13 – Anti-gas reports (II). Unfortunately the minutes of this committee have not been preserved in the Royal Society.

8. Obituary, 'S. Lyle Cummins', *British Medical Journal*, Pt 1 (1949): 1054–5.

9. D. J. C. Cunningham, 'Claude Gordon Douglas 1882–1963', *Biographical Memoirs of Fellows of the Royal Society of London*, 10 (1964): 51–74.

10. C. J. M[artin], 'Ernest Henry Starling 1866–1927', *Proceedings of the Royal Society, B*, 102 (1928): xvii–xxvii; L. F. Haber, *The Poisonous Cloud: Chemical Warfare in the First World War* (Oxford, 1986): 121–2.
11. Kenneth J. Franklin, *Joseph Barcroft 1872–1947* (Oxford, 1953): 102–3. For an account of some of the kinds of work they performed, see A. W. Crossley, 'The R. E. Experimental Station, Porton' (28 May 1919), PRO: MUN5/386/1650/14. It should be noted that much of the work on poison gases performed by physiologists throughout the country, though designated defensive research, had clear offensive implications. Thus Haldane made studies of the toxicity of cyanide gas, which was being considered by the allies as a possible offensive weapon (see his laboratory notes, National Library of Scotland, NLS MS 20233, ff. 170–7, 188–90), and prepared a 'Memorandum by Dr. Haldane. Capsicum for bombs' (NLS MS 20234, ff. 124–8). Starling was able to recommend the use of mustard gas in shells a year before it was first used by the Germans: Haber, *The Poisonous Cloud*: 116–17.
12. See C. G. Douglas to J. S. Haldane (14 May 1917) NLS MS 20234, ff. 9–12, where he writes: 'I have practically ceased to have anything to do with protective work or with the medical aspect of gas poisoning. However my position [at the Gas Services Central Laboratory] was so unsatisfactory that I could really be of little use. After all if one is working at the medical aspect one must have authority to act in the matter'.
13. Haldane to Douglas (15 May 1916) letter in the possession of D. J. C. Cunningham (hereafter Cunningham Papers).
14. The results of these investigations were submitted to the MRC during the summer of 1915, and eventually published as Joseph Barcroft, 'Report on the condition of the blood in cases of gas poisoning', in *Investigations into the Reaction of the Blood after Gas Poisoning, and the Results of the Administration of Saline and Other Substances*, Report No. 6 of the Chemical Warfare Committee of the MRC (April 1918): 4–6; and Joseph Barcroft and C. G. L. Wolf, 'Report on the change in fixed alkalinity of the blood produced by the administration of saline and other substances', in *Investigations into the Reaction of the Blood*: 6–9.
15. Franklin, *Joseph Barcroft*: 97.
16. Leonard Hill, 'Gas poisoning', *British Medical Journal*, Pt 2 (1915): 80–114.
17. MRC, *Annual Report, 1914–15* (London, 1915): 20.

18. *DNB* (1941–50): 126-8; F. E. Smith, 'John Cadman, Baron Cadman, 1877–1941', *Obituary Notices of Fellows of the Royal Society of London*, 10 (1941): 915–28.
19. See J. S. Haldane, 'The therapeutic administration of oxygen', *British Medical Journal*, Pt 1 (1917): 181–3; J. S. Haldane, 'The symptoms, causes, and prevention of anoxaemia and the value of oxygen in its treatment', *British Medical Journal*, Pt 2 (1919): 65–71; J. S. Haldane, 'Lung irritant gas poisoning and its sequelae', *Journal of the Royal Army Medical Corps*, 33 (1919): 494–507.
20. Memorandum on 'White star poisoning' (11 September 1916) NLS MS 20233, ff. 212–13.
21. Douglas to Haldane (13 November 1916) NLS MS 20233, ff. 223–32, and (1 February 1917) NLS MS 20234, ff. 2–3.
22. Douglas to Haldane (2 September 1916) NLS MS 20233, ff. 206–11, (1 February 1917) NLS MS 20234, ff. 2–3 and (14 May 1917) NLS MS 20234, ff. 9–12; Haldane to Douglas (2 May 1917) Cunningham Papers; C. G. Douglas, 'Examples of gas poisoning treated by oxygen administration by Haldane apparatus', in *The Administration of Oxygen in Irritant Gas Poisoning*, Report No. 10 of the Chemical Warfare Medical Committee of the MRC (October 1918): 11–15.
23. C. G. Douglas and T. R. Elliott, 'The acute lung irritant gases. Symptoms and treatment', in W. G. Macpherson *et al.*, (eds), *Official History of the War. Medical Services: Diseases of the War*, vol. 2 (London, 1923): 383–424, at pp. 414–15.
24. One early critic of this view was Victor Horsley, who adopted a functional stance to berate pathologists for their excessively anatomical and structural perspective. 'What a mass of facts has been accumulated in elucidation of the various changes in the structure of the lungs produced by pneumonia ... but how many workers have been found to investigate the degree and effect of the loss of respiratory function ... Yet this is what kills, this is the whole work of the disease, and this is what must be solved before the treatment of such diseases can be worthy of mention other than as empirical palliation'. Introductory remarks to the Section of Pathology of the British Medical Association, *British Medical Journal*, Pt 2 (1892): 248–9.
25. Osler, *The Principles and Practice of Medicine*, 3rd edn: 137.
26. Osler, *The Principles and Practice of Medicine*, 3rd edn: 137. See also: John Haddon, letter on 'Bloodletting in pneumonia', *British Medical Journal*, Pt 1 (1915): 1069; P. H. Pye Smith, revised A. P. Beddard, 'Lobar pneumonia', in Clifford Allbutt and Humphry Davy

Rolleston (eds), *A System of Medicine by Many Writers*, 2nd edn, vol. 5, *Diseases of the Respiratory System. Disorders of the Blood* (London, 1909): 191–255, at p. 251.

27. A. G. Millar, 'Poisonous gases', *British Medical Journal*, Pt 1 (1915): 1101. Subsequently, though physiologists accepted the theory behind bloodletting, and were sympathetic to the clinicians' claims, they experienced considerable difficulty before they were able to duplicate the beneficial effects in experimental animals: see Douglas and Elliott, 'The acute lung irritant gases': 411–14.

28. A. P. Beddard, 'Acute lobar penumonia and bronchopneumonia', in Allbutt and Rolleston (eds), *A System of Medicine,* vol. 5: 174–91, at p. 190.

29. Osler, *The Principles and Practice of Medicine*, 4th edn (1901): 137; 8th edn (1918): 101.

30. Douglas and Elliott, 'The acute lung irritant gases': 414–15.

31. Douglas and Elliott, 'The acute lung irritant gases': 414–15. Leonard Hill had observed similar problems with what he called the 'ordinary clinical method of administering oxygen', which had prompted him to develop his new breathing mask in 1915: 'Gas poisoning': 804.

32. The classic paper in this vein is J. S. Haldane and J. G. Priestley, 'The regulation of the lung-ventilation', *Journal of Physiology*, 32 (1905): 225–66.

33. Haldane, 'The causes of death in colliery explosions'.

34. But compare, for instance, Barcroft's early attempts to define gas poisoning in terms of a disturbance of the acid-base balance in the blood, as described above. Clearly, it was not immediately obvious, even to a respiratory physiologist, that the principal problem in gas poisoning was oxygen deficiency.

35. Douglas to Haldane (23 May 1917) NLS MS 20234, ff. 13–16. Physiologists experienced similar difficulties convincing doctors of the dangers of oxygen deficiency to aviators, and the needs for breathing apparatus in aeroplanes.

36. Haldane, 'The symptoms, causes, and prevention of anoxaemia': 65.

37. Douglas and Elliott, 'The acute lung irritant gases': 393–4.

38. *The Official History of the War* even reproduced coloured illustrations of the two distinct stages of asphyxia: Macpherson *et al.* (eds), *Official History. Diseases of the War*, vol. 2, opposite p. 393.

39. Douglas to Haldane (30 March 1918) NLS MS 20234, ff. 70–2.

40. Haldane to Douglas (1 April 1918) Cunningham Papers.

41. According to Douglas, 'The rapid consumption of oxygen by the

body was not at first realized, this being so great as to require continuous enrichment of the inspired air with at least two or three litres of oxygen each minute if the respiratory demands of the body, even when resting, were to be met': Douglas and Elliott, 'The acute lung irritant gases': 414. Cf. William Ewart, who reinforced recommendations about intermittent oxygen therapy by reminding his readers 'how relatively small is the bulk of oxygen which corresponds to the ordinary intake of air': Ewart, 'Bronchitis', in Allbutt and Rolleston (eds), *A System of Medicine*, vol. 5: 71–127, at p. 113.

42. Later in the war, for instance, Haldane encouraged surgeons to try oxygen in cases of shock, which he suspected might also be due to respiratory deficiency. 'Baker tried O_2 in shock', he told Douglas, 'but didn't see any good, *except that the lips turned red*. To my mind this evidence is quite enough'. Haldane to Douglas (14 April 1918) Cunningham Papers, emphasis in the original.

43. W. G. Macpherson, *Official History of the War. Medical Services: General History*, vol. 1 (London, 1921): 181.

44. MRC, *Annual Report, 1916–17* (London, 1917): 7.

45. Quoted in A. Landsbrough Thomson, *Half a Century of Medical Research*, vol. 1: *Origins and Policy of the Medical Research Council (UK)* (London, 1973): 14.

46. Landsbrough Thomson, *Half a Century*: 14–21.

47. *Notes on the Pathology and Treatment of the Effects of Pulmonary Irritant Gases*, Report No. 1 of the Chemical Warfare Medical Committee of the MRC (January 1918).

48. Joel Howell, ' "Soldier's heart": the redefinition of heart disease and speciality formation in early twentieth-century Great Britain', *Medical History*, Supplement No. 5 (1985): 34–52.

49. Martin Stone, 'Shell shock and the psychologists', in W. F. Bynum, Roy Porter and Michael Shepherd (eds), *The Anatomy of Madness: Essays in the History of Psychiatry*, vol. 2: *Institutions and Society* (London, 1985): 242–71.

50. In physiological terms, one of the major effects of heart disease was understood to be a reduced supply of oxygenated blood to the tissues, which produced a condition similar to partial asphyxia. Damage to the heart might in turn result from the effects of gas poisoning, which caused severe congestion in the blood vessels of the lungs and threw a strain on the heart muscle. Such an aetiology was suspected, for instance, by Douglas: see Douglas to Haldane (2 September 1916) NLS MS 20233, ff. 206–11; (15 September 1916) NLS MS 20233 ff. 214–16; (23 May 1917) NLS MS 20234, ff. 13–

16. Haldane, on the other hand, was inclined to attribute chronic gas cases to persistent impairment of lung function, and he later extended this model to include conditions like DAH: Haldane to Douglas (15 May 1916, 27 September 1916, 8 February 1917) Cunningham Papers.
51. Howell, '"Soldier's heart"': 41.
52. See obituary, 'Jonathan Campbell Meakins', *British Medical Journal*, Pt 2 (1959): 959–60, 1027–8.
53. MRC, *Annual Report, 1916-17*: 74, 79. Haldane describes their work in his letters to Douglas (18 July 1917, 1 August 1917, 8 August 1917) Cunningham Papers.
54. See, *inter alia*: MRC, *Annual Report, 1917-18* (London, 1918): 60; J. S. Haldane, J. C. Meakins and J. G. Priestley, *The Reflex Restriction of Respiration After Gas Poisoning*, Report No. 5 of the Chemical Warfare Medical Committee of the MRC (April 1918); J. S. Haldane, J. C. Meakins and J. G. Priestley, *Investigations of Chronic Cases of Gas Poisoning*, Report No. 11 of the Chemical Warfare Medical Committee of the MRC (October 1918); J. S. Haldane, J. C. Meakins and J. G. Priestley, 'The effects of shallow breathing', *Journal of Physiology*, 52 (1919): 433–53; H. W. Davies, J. S. Haldane and J. G. Priestley, 'The response to respiratory resistance', *Journal of Physiology*, 53 (1919–20): 60–9; H. W. Davies and J. G. Priestley, 'Relation between the general nervous system and symptoms of "D. A. H." in neurasthenia patients', *British Medical Journal*, Pt 2 (1919): 438–9.
55. Joseph Barcroft, G. H. Hunt and Dorothy Dufton, 'The treatment of chronic cases of gas poisoning by continuous oxygen administration in chambers', in *Polycythaemia After Gas Poisoning and the Effect of Oxygen Administration in Chambers in the Treatment of Chronic Cases*, Report No. 4 of the Chemical Warfare Medical Committee of the MRC (April 1918): 13–64; Joseph Barcroft, G. H. Hunt and Dorothy Dufton, *Treatment of Patients Suffering from "Effort Syndrome" by Continuous Inhalation of Oxygen*, Report No. 12 of the Chemical Warfare Medical Committee of the MRC (October 1918).
56. MRC, *Annual Report, 1918–19* (London, 1919): 60.
57. In particular, the MRC stressed the importance of the respiratory and other research at Taplow, where 'physiological laboratory methods ... used in close conjunction with clinical studies' had identified 'a common prime factor in several morbid or invalid states previously unconnected in clinical studies': MRC *Annual Report, 1917–18*: 59–60. On the role of war-related investigations in the re-defini-

tion of heart disease in terms of physiology rather than anatomy, and the place of this shift in the creation of cardiology as a modern discipline, see Christopher J. Lawrence, 'Moderns and ancients: the "new cardiology" in Britain 1880–1930', *Medical History,* Supplement No. 5 (1985): 1–33.

58. *The Symptoms and Treatment of the Late Effects of Gas Poisoning,* Report No. 3 of the Chemical Warfare Medical Committee of the MRC (April 1918): 6. See also Douglas and Elliott, 'The acute lung irritant gases': 389. Compare this view with the following advice on shell-shock: 'To the soldier's mind it was as much an entity as scarlet fever, with the further addition that, being incurable, shell-shock was more to be dreaded ... To explain to a man that his symptoms were the result of disordered emotional conditions due to his rough experiences in the line, and not, as he imagined, to some serious disruption of his nervous system produced by bursting shells, became the most frequent and successful form of psychotherapy': William Johnson, 'Neurasthenia and the war neuroses', in Macpherson *et al.* (eds), *Official History, Diseases of the War,* vol. 2: 1–67, at p. 9. The psychological aspect of DAH was dealt with in the same way: 'The patient is naturally alarmed about the state of his heart ... It must be explained to the patient that he has no "heart disease" in the ordinary acceptance of that term ... The patient can usually be satisfied if it can be explained to him that his nerves are too sensitive and that this causes his heart to beat too fast, and then that the sensitiveness of the nerves is due to the infection, or the gassing, or the strain through which he has recently passed': John Hay, 'Disorders of the cardio-vascular system', in Macpherson *et al.* (eds), *Official History of the War. Medical Services: Diseases of the War,* vol. 1 (London, 1922): 504-38, at p. 532.
59. Howell, ' "Soldier's heart" ': 46.
60. See the statements made by the Ministry of Reconstruction, particularly to the various representative medical bodies who felt that the MRC should be established as an organ of the Ministry of Health. PRO: MH78/68 and MH78/80.
61. See, for instance, J. S. Haldane, A. M. Kellas and E. L. Kennaway, 'Experiments on acclimatisation to reduced atmospheric pressure', *Journal of Physiology,* 53 (1919–20): 181–206. This work, carried out in steel chambers constructed by the Committee at the Lister Institute, was a physiological study which looked only very indirectly at the medical uses of oxygen – a fact which indicates the latitude that the research councils enjoyed under the Privy Council.

62. MRC, *Annual Report, 1918–19*: 60.
63. MRC, *Annual Report, 1919–20* (London, 1920): 70.
64. MRC, *Annual Report, 1918–19*: 60.
65. A report of this work, by Frank Shufflebotham and G. H. Sowry, was given to the Royal Society of Medicine early in 1920 as part of a discussion on 'Oxygen therapy', *Proceedings of the Royal Society of Medicine*, 13 (1920), Section of Therapeutics and Pharmacology: 59–95.
66. Royal Commission on University Education in London, *Final Report*, Cd. 6717, P.P. 1913, *xl*: 297–544.
67. See papers on 'Grants to medical schools, 1917–1920', PRO: ED24/1961.
68. See George Newman, *Some notes on medical education in England. A memorandum addressed to the President of the Board of Education*, Cd. 9124 (London, 1918), and George Newman, *An outline of the practice of preventive medicine. A memorandum addressed to the Minister of Health*, Cmd. 363 (London, 1919).
69. George Graham, 'The formation of the medical and surgical professorial units in the London teaching hospitals', *Annals of Science*, 26 (1970): 1–22; Arthur W. Chapman, *The Story of a Modern University. A History of the University of Sheffield* (London, 1955): 321–25; A. Logan Turner, *Story of a Great Hospital. The Royal Infirmary of Edinburgh, 1729–1929* (Edinburgh, 1937): 310–11.
70. The initial grants are listed in MRC, *Annual Report, 1919–20*: 70–1; and *Annual Report, 1920-21* (London, 1921): 76–7.
71. See J. C. Meakins, 'The therapeutic value of oxygen in pulmonary lesions', *British Medical Journal*, Pt 1 (1920): 324–6; J. C. Meakins and H. W. Davies, 'Observations on the gases in human arterial and venous blood', *Journal of Pathology and Bacteriology*, 23 (1920): 451–61; J. C. Meakins, 'Observations on the gases in human arterial blood in certain pathological conditions, and their treatment with oxygen', *Journal of Pathology and Bacteriology*, 24 (1921): 79–90; J. C. Meakins and H. W. Davies, *Respiratory Function in Disease* (Edinburgh, 1925).
72. H. C. Cameron, *Mr. Guy's Hospital 1826–1948* (London, 1954): 295-6.
73. Obituary, 'Edward Palmer Poulton', *British Medical Journal*, Pt 2 (1939): 886.
74. Obituary, 'George Herbert Hunt', *British Medical Journal*, Pt 1 (1926): 170–1. At the time, Hunt was in charge of the hospital's physical exercise department, where he 'did much work and investi-

gation in the treatment of heart and lung conditions by graduated physical exercise', thus building on another of the physiologists' therapeutic responses to war-time disability.

75. They were assisted by J. M. H. Campbell, at that time a medical registrar and later a Beit Memorial Fellow at the hospital, who would himself be appointed to the Clinical Committee. See *Munk's Roll*, vol. 6, ed. Gordon Wolstenholme (London, 1982): 86–8. Their publications include J. M. H. Campbell, G. H. Hunt and E. P. Poulton, 'An examination of the blood gases and respiration in disease, with reference to the cause of breathlessness and cyanosis', *Journal of Pathology and Bacteriology*, 26 (1923): 234–96. In collaboration with Argyll Campbell, a member of staff at the MRC's National Institute for Medical Research, Poulton later wrote a book that forcefully presented his own findings, and those of other clinical physiologists, on the treatment of a wide range of respiratory conditions: A. Campbell and E. P. Poulton, *Oxygen Therapy and Carbon Dioxide Therapy* (London, 1934).

76. Adams conducted his investigations with Noah Morris, professor of physiology at Anderson College of Medicine, and later professor of materia medica and therapeutics in Glasgow University. Morris appears to have had more enthusiasm for the work than Adams, who resigned from the Oxygen Committee during 1921–2. See obituary, 'Noah Morris', *British Medical Journal*, Pt 1 (1947): 866–7. Their published papers included: D. K. Adams and N. Morris, 'Anoxaemia and the administration of oxygen', Proceedings of the Physiological Society, *Journal of Physiology*, 54 (1920–21): cii–civ, cvi–cviii; N. Morris, 'Anoxaemia and the administration of oxygen', *Journal of Physiology*, 56 (1922): 283–93.

77. MRC, *Annual Report, 1920–21*: 76.
78. MRC, *Annual Report, 1920–21*: 77.
79. MRC, *Annual Report, 1921–22*: (London, 1922): 69.
80. The most complete obituary notice of Fraser is by Arthur S. MacNalty, in *Journal of Pathology and Bacteriology*, 90 (1965): 701–11.
81. MRC, *Annual Report, 1921–22*: 117; *Annual Report, 1925–26* (London, 1926): 57.
82. Percival Hartley, 'Henry Stanley Raper 1882–1951', *Obituary Notices of Fellows of the Royal Society of London*, 8 (1952–3): 567–82. During the war, Raper had served as Director of the War Office Anti-Gas Department at the Royal Army Medical College, Millbank.
83. MRC, *Annual Report, 1923–24* (London, 1925): 71. By the follow-

ing year, the Committee could report that 'Progress has been made and it may be provisionally concluded that the existence of an anoxaemic type of infantile atrophy has been established': MRC, *Annual Report, 1924–25* (London, 1925): 76.
84. MRC, *Annual Report, 1927–28* (London, 1929): 121.
85. E. P. Poulton, *Taylor's Practice of Medicine*, 12th edn (London, 1922): 61.
86. J. C. Meakins, *The Practice of Medicine* (London, 1937): 195.
87. William Osler, *The Principles and Practice of Medicine*, 9th edn, ed. Thomas McCrae (London, 1920): 103.
88. William Osler, *The Principles and Practice of Medicine*, 13th edn, ed. Henry A. Christian (London, 1930): 32.
89. See, for example, T. K. Monro, *Manual of Medicine* (London, 1925), which lists oxygen alongside stropanthus, digitalis and pituitrin. From an extreme example of this view of the action of oxygen, see Henry A. Ellis, letter, 'The therapeutic uses of oxygen', *Lancet*, Pt 1 (1920): 569.
90. J. J. Conybeare (ed.), *A Textbook of Medicine by Various Authors* (Edinburgh, 1929): 419.
91. See, for example, Geoffrey Bourne, 'A simple portable apparatus for continuous oxygen administration', *British Medical Journal*, Pt 2 (1922): 40–1; John Ryle, letter, 'The therapeutic administration of oxygen', *Lancet*, Pt 1 (1922): 1269–70. It is worth noting that both Bourne and Ryle were in close contact with the new clinical physiology – Bourne was chief assistant in charge of the cardiology department at Bart's, and Ryle was an assistant physician at Guy's – which presumably encouraged them to favour oxygen therapy, if not the Haldane mask. See also the 'Discussion on the therapeutic administration of oxygen and carbon dioxide', *Proceedings of the Royal Society of Medicine*, 25 (1931–32): 621–6, where the physiologists rehearsed their arguments for using the Haldane mask, and Reginald Hilton replied in favour of the nasal catheter.
92. Poulton, *Taylor's Practice of Medicine*: 61.
93. For long-term treatment in oxygen chambers, the cost of the gas was as high as £1 3s. 6d. per person per day, and the MRC was 'convinced that, while the methods for administration of oxygen are being improved and its utility in certain conditions of disease is becoming more apparent ... the high cost of oxygen for medical purposes will nevertheless be a great obstacle to its extended employment': MRC, *Annual Report, 1921–22*: 69.
94. Poulton used figures collected by the MRC to compare the effi-

ciency of the Haldane mask with the nasal catheter in Campbell and Poulton, *Oxygen Therapy and Carbon Dioxide Therapy*: 124–6.
95. Poulton also 'arranged for a service by which an apparatus could be sent with a skilled technician to a hospital or private house anywhere in the country': obituary, 'Edward Palmer Poulton': 886.
96. This has been shown, in particular, by Daniel Fox in his comparative study of the regionalisation of British and American health services, *Health Policies, Health Politics: The British and American Experience* (Princeton, N. J., 1986).

7 X-ray Technology in Obstetrics: Measuring Pelves at the Yale School of Medicine

* I have profited from the comments of Margaret Jacob, Eda Kranakis, Stuart Blume, Malcolm Nicolson, John Pickstone and several of my departmental colleagues in writing this essay. I also want to thank Drs Steer, Cheney and Molumphy, who kindly answered my questions during my stay in the USA in 1987 and the librarians of the Yale Medical Historical Library for finding material on the subject. During this period the Department of History and Sociology of the University of Pennsylvania offered me its hospitality. My stay in the USA was partly made possible by a grant from the Catharina van Tussenbroek Fund, to which I am also indebted.
1. For a thorough discussion of the literature on technological development, see: S. S. Blume, *Insight and Industry. On the Dynamics of Technological Change in Medicine* (Cambridge, Mass., forthcoming, 1991).
2. Hendrik van Deventer, *Manuele Operatien. Eerste deel zijnde een Nieuw Ligt voor Vroedvrouwen* (Lugduno Batavorum, 1701).
3. William Smellie, *Collection of Cases*, I (London, 1754) Collect. XXI nr. 1. Several authors, especially feminist writers, have pointed out that these kinds of criteria for involvement of 'men-midwives' in this traditionally female practice have been instrumental in the general takeover of the field of midwifery by doctors and the subsequent medicalisation of pregnancy and delivery. Among the numerous works on this subject I want to mention only two: D. C. Wertz, 'What birth has done for doctors: a historical view', *Women and Health*, 8(1) (1983): 7–25; Dorothy Wertz argues that pelvimetry was actually one of the first techniques that provided medicine with a demonstrable claim to be a science and therefore with a claim to

expert superiority over midwife practices. Jean Donnison, *Midwives and Medical Men* (London, 1978) sets out the rivalries between doctors and midwives in Britain. See also the references in n. 15 below.

4. G. A. Michaelis, *Das Enge Becken* (Leipzig, 1851).
5. H. Varnier, J. Chappuis, C. Chauvel and M. Funck-Brentano, *Annales de Gynécologie*, 45(3) (1896): 185.
6. A. Pinard and H. Varnier, XIIth International Medical Congress in Moscow, Session of 18 August 1897, *Zentralblatt für Gynekologie*, 21 (1897): 1145.
7. M. Levy and L. Thumin, *Deutsche Medische Wochenschrift*, 32 (1897): 507.
8. A. Albert, Gynakologische Gesellschaft zur Dresden, Sitzung vom 15 Dez. 1898 und 16 Nov. 1899, *Zentralblatt für Gynekologie*, 23 (1899): 418; 24 (1900): 264; and *Verhandlungen der Deutschen. Gesellschaft für Gynekologie* (1899): 356.
9. J. Fabré, 'Radiographie Métrique', in *L'obstétrique*, 5 (1900): 27.
10. J. Whitridge Williams, 'Medical education and the midwife problem in the United States', *Journal of the American Medical Association*, 58 (1912): 1–7.
11. A. Flexner, *Medical Education in the United States and Canada. Carnegie Foundations for the Advancement of Teaching*, Bulletin no. 4 (New York, 1910).
12. Grace L. Meigs, *Maternal Mortality – from all conditions connected with childbirth in the U.S. and certain other countries*, US Children's Bureau Pat. 19 (Washington, D.C., 1917). The first cause of maternal death was tuberculosis.
13. R. Stevens, *American Medicine and the Public Interest* (New Haven and London, 1973): 200.
14. J. B. DeLee, 'The profylactic forceps operation', *American Journal of Obstetrics and Gynecology*, 1 (1920): 34–44.
15. Apart from the promise to women of a safe delivery in the 'comfortable', germ-free environment of the hospital (cf. Wertz and Wertz, 1979, cited below), there were of course other factors involved in this 'move' to the hospital, not least the growth in the number of hospitals and the declining number of midwives. Other factors that might be mentioned are the lowering of hospital costs and state subsidies of hospitalisation. See Richard W. Wertz and Dorothy C. Wertz, *Lying-In: A History of Childbirth in America* (New York, 1979) ch. 5; Morris J. Vogel, *The Invention of the Modern Hospital: Boston 1870-1930* (Chicago, 1980); Frances E. Kobrin, 'The Ameri-

can midwife controversy: A crisis of professionalisation', *Bulletin of the History of Medicine*, 40 (1966): 350–63; J. Waltzer Leavitt, *Brought to Bed. Childbearing in America 1750–1950* (New York, Oxford, 1986) ch 7. From the work of Judith Leavitt we get some picture of how women themselves looked upon the promises of hospital deliveries. See also: J. W. Leavitt and W. Walton, 'Down to Death's Door: Women's Perception of Childbirth in America', in J. Waltzer Leavitt (ed.), *Women and Health in America*, (Madison and London, 1984): 156–66.

16. One of the hospital-based technologies that was especially discussed was the use of anaesthesia during labour and delivery ('twilight sleep'). In fact women themselves, mostly of middle- and upper-class backgrounds, campaigned for the introduction of this technology in American hospitals in the first decades of the twentieth century. Their campaign was successful. In the 1920s 'twilight sleep' became widely used and (for example) in Boston all hospitals were using it by 1938; cf. Wertz and Wertz (1979): 148 ff.

17. See, e.g., B. P. Watson: 'Can our methods of obstetric practice be improved?', *Bulletin of the New York Academy of Medicine*, 6 (1930): 647-33; T. Cianfrani, *A Short History of Obstetrics and Gynecology* (Springfield, Ill., 1960).

18. Howard C. Taylor, Jr (ed.), *The Recruitment of Talent for a Medical Specialty: A Report to the American Gynecological Society* (St. Louis, 1961).

19. Sociologists of medicine have suggested that the introduction of technology can play an important role in the image building of a profession. See, e.g., E. Friedson, *The Profession of Medicine* (New York, 1970); S. J. Reiser, *Medicine and the Reign of Technology* (Cambridge, 1979). It should be mentioned here that this readiness will have played a role, too, in the struggles between specialists and GPs and between specialists and midwives for authority in supervising pregnancy and delivery. See, e.g., J. Antler and D. M. Fox, 'The movement toward a safe maternity: physicians' accountability in New York City, 1915–1940', *Bulletin of the History of Medicine*, 50 (1976): 569–95; Kobrin (1966); on the struggle between generalists and specialists see Stevens (1973), esp. pp. 146–7.

20. Other important factors were: Inadequate and improper pre-natal care; the attendant's inadequacy in judgement or skill; inadequate hospital standards; incompetence of midwives. *White House Conference on Child Health and Protection. Fetal Newborn and Maternal Morbidity and Mortality; Report of the Subcommittee on Factors*

and *Causes of Fetal, Newborn and Maternal Morbidity* (New York, 1933); New York Academy of Medicine, Committee on Public Health Relations, *Maternal Mortality in New York City: A Study of All Puerperal Deaths, 1930-1932* (New York, 1933): 213–16.

21. Surprisingly enough, Joseph DeLee, who set the scene for an interventionist practice of obstetrics, raged against the 'abuse' of caesarean section, claiming that it kept the mortality rate high and would bring 'discredit to the art and influence lay men in high places to think that domicilian obstetrics is desirable', J. DeLee and J. P. Greenhill (eds), *Obstetrics and Gynecology Yearbook* (Practical Medicine Series) (Chicago, 1936): 223. See also n. 27 in Antler and Fox (1976), for a list of articles and reports bringing up this point.

22. According to Antler and Fox, the Sheppard-Towner Act 'aroused great public interest in maternity, enabling many women to learn what to ask and expect of their physicians and alerting the medical profession to their demands'. Antler and Fox (1976): 572. Obstetricians themselves would address women directly in attempts to attract them to the hospital for a safe and scientifically conducted delivery. See, e.g., Paul de Kruif's series on childbirth in *Ladies' Home Journal*, 49 (March 1932): 6–7 and 124–5 and 53 (December 1936): 12–13 and 64–8.

23. C.f. Wertz and Wertz (1979) and Leavitt (1984) and (1986) and the remarks made in nn. 15 and 16 above.

24. H. Thoms, 'A statistical study of the frequency of funnel pelves and the description of a new outlet pelvimeter', *The American Journal Of Obstetrics and Diseases of Women and Children*, 72(1) (1915): 40.

25. H. Thoms, 'Outlining the superior strait of the pelvis by means of the X-ray', *American Journal of Obstetrics and Gynecology*, 3 (1922): 257-63, at p. 258.

26. Thoms, 'Outlining the superior strait': 261.

27. Discussion on the paper 'The clinical significance of x-ray pelvimetry', read by Dr. Herbert Thoms, reported in *American Journal of Obstetrics and Gynecology*, 3 (1922): 599–600, at p. 600.

28. Thoms raises the question about the safety of his method in his 1922 landmark paper. Since he had worked primarily with dried pelves to develop his method, he refers to Edelberg's 1914 research to address the issue:

> Edelberg ... showed that the danger of injury by this means was negligible. He observed the condition of a child which had

been conceived while the mother was under the effect of full x-ray treatment ... The child was born at term, fully developed, with all signs of maturity. Nothing pathologic was noticed and the child was well nourished. (Thoms, 'Outlining the superior straits': 263.)

29. Figures for the clinic admissions of the Department of Obstetrics and Gynaecology in 1926-7 show that the increase of patients is due partly to the Private Pavilion but mostly to referrals of local physicians (*Annual Report* of the department, n. 43). These patients can be expected to be relatively less affected by rickets. This was confirmed by Dr Molumphy, former assistant of Thoms, in an interview I had with him in New Haven (May 1987). From 1925–6 the pediatrics department at Yale reserved one afternoon every week to examine babies in order to prevent rickets. *Annual Report* of the Dean of Yale School of Medicine 1925–26, Department of Pediatrics, Archival Collection Yale Medical Historical Library, unlabelled folder.
30. H. Thoms, 'The diagnosis of rachitic pelves by x-ray', *American Journal of Obstetrics and Gynecology*, 14 (1927): 45-8, at p. 45.
31. This whole operation is a 'paper' one. Not one clinical case is reported in the 1930 article establishing this method of foetal cephalometry. Thoms himself was able to carry out clinical research on this subject only much later, when the method of x-ray measurement was generally applied at Yale. Again, apparently, the meaning and value of x-ray measurement were secured first, even if clinical material to support the claim was lacking. See H. Thoms, 'Fetal cephalometry in utero', *Journal of the American Medical Association*, 95 (1930): 22–4.
32. H. Thoms, 'Shortening of the transverse diameter of the superior strait', *American Journal of Obstetrics and Gynecology*, 19 (1930): 539–44, at p. 539.
33. H. Thoms, 'Occipito posterior position and the transversely contracted pelvis', *American Journal of Obstetrics and Gynecology*, 24 (1932): 50–6, at p. 56.
34. H. Thoms, 'What is a normal pelvis?', *Journal of the American Medical Association*, 102 (1934): 2075–6.
35. Report of Arthur Morse, 'The Organisation and Administration of the Yale Women's Clinic and Its Service to the Community' (July 1923) Yale Coll. Man. and Archives, Coll. Nr YRG 27-A-(5–9), Box 153, folder 3290.

36. Letter from Dr Creadick to Arthur Morse (12 January 1923) Yale Coll. Man. and Archives, Coll. Nr YRG 27-A-(5–9), Box 31, folder 491: 1.
37. Arthur Morse, 'A Report Concerning the Available Teaching Material in the Women's Clinic and Its Relation to the Introduction of Students and Nurses and the Clinical Training of Assistants' (1923) Yale Coll. Man. and Archives, Coll. Nr YRG 27-A-(5–9), Box 152, folder 3290.
38. Arthur Morse to James R. Angell in a letter (22 September 1924) Yale Coll. Man. and Archives, Coll. Nr. YRG 27-A-(5–9), Box 31, folder 492.
39. Morse, 'A Report Concerning the Available Teaching Material': 6. Also: Letter from Arthur Morse To Dean Angell (16 March 1925): 4; Yale Coll. Man. and Archives. Coll. Nr YRG 27-A-(5–9), Box 31, folder 492.
40. Dean M. C. Winternitz to Dean James R. Angell in a letter (7 February 1925) Yale Coll. Man. and Archives, Coll. Nr YRG 27-A-(5–9), Box 31, folder 492.
41. See Brooks M. Kelley, *Yale: A History* (New Haven, 1974).
42. Letter from Arthur Morse to Dean Winternitz (15 July 1926) Yale Coll. Man. and Archives, Coll. Nr. YRG 27-A-(5–9), Box 153, folder 3290.
43. *Annual Report* of the department for the year 1926–27, submitted by E. M. Stone (in the absence of Arthur Morse who was on leave in Europe), no title, no date: 1–2; Yale Coll. Man. and Archives, Coll. Nr YRG 27-A-(5–9), Box 153, folder 3290.
44. *Annual Report*, 'Obstetrics and Gynecology 1927-1928': 2-3; Yale Coll. Man. and Archives, Coll. Nr YRG 27-A-(5–9), Box 153, folder 3290.
45. *Annual Report* of the Dean of Yale School of Medicine, Department of Obstetrics and Gynecology 1934–35: 35, Archival Collection Yale Medical Historical Library, unlabelled folder.
46. *Annual Report* of the Dean of Yale School of Medicine, Department of Obstetrics and Gynecology 1935–36, Archival Collection Yale Medical Historical Library, unlabelled folder.
47. The article continues as follows:

> Much of the progress of medicine in modern times is dependent on the development of mechanical devices which aid the senses of man and provide knowledge from which the brain may reason logically. Of this nature is the device which will enable the ob-

stetrician to compute almost mathematically hazards of childbirth in almost any special instance.

Clipping from an unidentified newspaper with no date (but presumably dating from 1932), present in a collection of miscellaneous papers from Thoms (unarchived material), in the possession of the Medical Historical Library at Yale.

48. The development of the two rival research programmes is compared in A. Hiddinga, 'De ontwikkeling van pelvimetrisch onderzoek in de verloskunde', *Jaarboek voor de Geschiedenis van Bedrijf en Techniek*, 6 (1989): 70–96. The resulting classificatory schemes from the two programmes are analysed in: A. Hiddinga, 'Geordende bekkens. Klassifikatie-modellen in de verloskunde', in *Tijdschrift voor Vrouwenstudies*, 42 (1990): 158–75.

49. To illustrate this point, and to show the expectations addressed to x-ray pelvimetry, we can take a quotation of DeLee from a review of a paper on x-ray pelvimetry:

> It [the reviewed paper] will bring the practice of obstetrics on a higher plane (thank God above the reach of midwives, male and female) it will show the profession and the public that obstetrics is one of the leading medical sciences and a really fine art and that the obstretician has (or should have) the dignity accorded to the great surgeon. This will do more to reduce maternal and fetal mortality than all the committees of maternal welfare could do without it. (J. DeLee and J. P Greenhill (eds), *Obstetrics and Gynecology Yearbook* (Practical Medicine Services) (Chicago, 1938): 39).

50. J. L. Wagner and M. Zubkoff, 'Medical Technology and Hospital Costs', in J. B. McKinlay (ed.), *Issues in Hospital Administration* (Cambridge, Mass., 1982): 161–87, at p. 172.

51. This remark referred especially to Thoms's stubbornness and tenacity.

52. The questionnaire sent to members of the South Atlantic Association of Obstetrics and Gynecology in 1947 indicated that 73 per cent of the obstetricians employed x-ray studies in 1-10 per cent of the cases; while 50 per cent of the obstetricians used the Thoms method, the other half used various other methods, but 11 per cent did not know which method was used. Hunter Jones, 'The value of x-ray

studies of the pelvis in obstetrics', *American Journal of Obstetrics and Gynecology*, 54(5) (1947): 777.

53. See e.g., Louise B. Russell, *Technology in Hospitals: Medical Advances and their Diffusion* (Washington, 1979).
54. At the same time, there are some indications that in the leading research hospitals the frequency of use was much higher than among smaller hospitals or among private obstetricians. This is in accordance with conclusions from the literature on adoption and diffusion of technology, where it is stated that big, leading hospitals are generally inclined to purchase and use new technologies. In 1947 at the New York Lying-In Hospital, pelvic x-ray studies were made on 7.6 per cent of the clinic cases and 20.3 per cent of the private patients. In the Sloane Hospital for Women, also in New York, some 20 per cent of the clinic cases and 30 per cent of the private patients were x-rayed. Hunter Jones (1947): 780.

   A clear indication of how important costs were in the decision to use x-ray pelvimetry on a patient, is given by the following quotation from a radiologist working in this field:

   > With the use of modern screens and films this much raying [he speaks of six exposures] involves no risk of injury to mother or fetus, but it does involve considerable expense either to the patient or the institution. Whether an obstetrician should do without roentgen-ray examination ... is a question that depends upon the obstetric problem and the economic status of the particular patient in the case under consideration. (Paul C. Hodges, in Fred L. Adair (ed.), *Obstetrics and Gynecology, by the departmental staff of the University of Chicago and other contributors* (Philadelphia, 1940) vol. 2, ch. 21: 704).

   See for a treatment of institutional variations and the role of costs Russell (1979).
55. Hunter Jones (1947): 777-8.
56. However, there were some doubts voiced by obstetricians. Surprisingly enough, e.g., by Joseph B. DeLee who had been very positive about Thoms's work and totally supported his policy to x-ray every primipara. But in the 1937 issue of the *Yearbook of Obstetrics and Gynecology* he commented in his review of a paper on roentgen pelvimetry:

   > As regards the dangers of radiation for photography alone, we have always thought it safe, but Hüssy of Basel has disturbed our

complacency. He found that mild x-raying of beans did not affect the first generation, but that in subsequent ones hypoplastic and deformed beans occurred. Several fears of injury to the child or its child have been expressed in Germany regarding photographic doses, and Martins, whose name is well known to roentgenologists urges a restriction of x-raying to the minimum necessary. (J. DeLee and J. P. Greenhill (eds). *Obstetrics and Gynecology Yearbook* (Practical Medicine Series) (Chicago, 1937): 33–4).

57. Paul C. Hodges and Jane E. Hamilton, 'Pelvic roentgenography in pregnancy', *Radiology*, 30 (1938): 157. See also Hodges's remark quoted in n. 54 above.
58. J. P. Medelman, 'History of the Section on Radiology', *Journal of the American Medical Association*, 178(8) (1961): 785–92, at p. 791.
59. A. Stewart, J. Webb, D. Giles and D. Hewitt, 'Malignant disease in childhood and diagnostic irradiation in utero', *Lancet*, Pt 2 (1956): 447. Ann Oakley discusses the impact of Stewart's paper on the British practice of x-ray pelvimetry, where it had an equally dramatic effect. A. Oakley, *The Captured Womb* (Oxford, 1986): 104–5.
60. This phrase was used in interviews I had with both Dr Charles Steer, a contemporary of Thoms working at Columbia University and Dr Paul Molumphy, a colleague of Thoms at Yale in the early 1950s. The rationale of x-ray pelvimetry research was carried further, though, in the work started in the late 1950s with ultrasound technology. See A. Hiddinga and S. S. Blume, 'Technology, science and medical practice: the origins and transformation of cephalopelvimetry', forthcoming in *Science, Technology and Human Values* (1991).
61. Thoms, for example, remarked in 1937:

> From our somewhat extended experience, we have for some time been convinced that, ... a knowledge of the true capacity of the bony pelvis as revealed by x-ray may increase the incidence of caesarean section as an operative choice over the difficult forceps or breech extraction.

> H. Thoms, 'A ten-year survey of caesarean section at the New Haven hospital', *American Journal of Obstetrics and Gynecology*, 51

(1946): 880–4, at p. 884.
Also: Hunter Jones (1947): 778–80.
62. In almost all his papers Thoms mentions the low costs of his method as one of its advantages over other ones. See also n. 54 above. In a lecture at a meeting of radiologists in 1934, Thomas A. Groover remarked that

> The average hospital has comparatively little paying practice that it can hand over to the radiologist. Less than 25 percent of the revenue we derive from hospital departments comes from the bona fide hospital patients. The average hospital cannot possibly run a first class radiological department on revenue derived exclusively from that source even, if the radiologist does not get a cent of it. (R. E. Scammon and L. A. Calkins (eds), *Yearbook of Radiology* (Chicago, 1934)).

63. We find complaints about the difficulty expressed many times in discussions of papers on x-ray pelvimetry, as for example in one of the sessions of the First American Congress on Obstetrics and Gynecology (Ohio, 1939), ten years before Jones' survey. One of the speakers there is referred to in the following terms:

> He is on the side of the majority of the clinicians when he states 'many have experienced technical and interpretative difficulties with the roentgenographic methods of pelvimetry, which for the present I believe limit their value'. (G. S. Schauffler, 'General summary of the sessions on obstetrics of the First American Congress on Obstetrics and Gynecology', *American Journal of Obstetrics and Gynecology*, 39 (1940): 535–9).

64. DeLee in his review of a German paper on x-ray pelvimetry in J. DeLee and J. P. Greenhill (eds), *Obstetrics and Gynecology Yearbook* (Practical Medicine Series) (Chicago, 1933): 25–6.
65. DeLee in his review of a paper by Thoms and co-worker in DeLee and Greenhill (eds), *Obstetrics and Gynecology Yearbook* (1938): 33.
66. By 1950 we find the fears about competition with radiologists to have grown and the issue addressed more directly and detailed:

> To have the roentgenologist dictate obstetrical procedure is ... unwise ... This is particularly true, if instead of stating the avail-

able measurements on the report, the roentgenologist takes the liberty of predicting whether or not he thinks a delivery through the pelvis would be hazardous or difficult ... It is then wise to have the x-ray plates filed in the office of the obstetrician, so that he may compare them with the patient clinically and use them for constant reference. (I. Dyer, 'Clinical evaluation of x-ray pelvimetry', *American Journal of Obstetrics and Gynecology*, 60 (1950): 302–13, at pp. 310–11).

67. According to another colleague of Thoms, Dr Charles Cheney (interview), routine roentgen pelvigrams of primiparous women were taken at Yale twice during pregnancy: in the very beginning and towards the end. This practice was changed by Cheney in 1946 so that only one pelvigram in the last stage of pregnancy remained and finally in 1957 all routine x-ray pelvimetry at Yale stopped.
68. Letter of Thoms to 'Maurice' (22 January 1958) Archival Collection Yale Medical Historical Library, non-archivated material.
69. In the 1938 edition of DeLee's textbook we find the following remarks 'All these methods are very technical and specialistic and the practicing [sic] obstetrician has no need to learn them' (p. 281) and 'It [the x-raying of contracted pelves] is of more scientific interest than practical use. The information it gives was more easily obtainable by palpation' (p. 757). J. B. DeLee, *The Principles and Practice of Obstetrics*, 7th edn (Philadelphia and London, 1938). In Williams' textbook, H. J. Stander, *Williams Obstetrics* (7th edn, New York and London, 1936), only the traditional hand methods of x-ray pelvimetry and the simple calliper-like instruments are discussed. X-ray pelvimetry is mentioned, but the methods are not explained or discussed.

8 The Politics of a Spatial Innovation: Fracture Clinics in Inter-War Britain

* An earlier version of this essay appeared in *Medical History* in 1987. It is a pleasure to again thank the Wellcome Trust for their generous support, and John Pickstone, both for his incisive comments and for arranging the larger project on the history of orthopaedics to which this essay refers.
1. 'British Orthopaedic Association: first founders' lecture', *Journal of*

Bone and Joint Surgery, 41B (1959): 231–6, reprinted in Sir H. Platt, *Selected Papers* (Edinburgh, 1963): 116–25, at p. 124.
2. Delevingne, *Interim Report of the Inter-Departmental Committee on the Rehabilitation of Persons Injured by Accidents* (London, 1937): 11.
3. See 'Report of the Committee on Treatment of Simple Fractures', *British Medical Journal* (30 November 1912): 1505–41.
4. See: Frederick Watson, *The Life of Sir Robert Jones* (London, 1934).
5. 'Report of the Orthopaedic Committee of the Royal College of Surgeons' (4 July 1918), in MS *Committee for Temporary Purposes*, vol. 6 (1907–22): 313, Royal College of Surgeons.
6. See R. Jones and G. R. Girdlestone, 'The cure of crippled children, proposed national scheme', *British Medical Journal* (11 October 1919): 457–60. On the orthopaedic hospitals, see Central Council for the Care of Cripples, *Directory for 1935* (London, 1935).
7. See the report on 'Orthopaedic after-care, 1932-47', Greater London Record Office (hereafter GLRO): PH/HOSP/1/66; see also, 'BMA Report of Committee on Fractures', *British Medical Journal* Supplement (16 February 1935): 53–62, at p. 57. On Sinclair's 92-bed segregated fracture service at No. 8 Stationary Hospital, Wimereux, France, see Robert Jones, 'Introduction' to M. Sinclair, *Fractures* (London, 1931): xxxiii–iv; and Harvey Cushing, *From a Surgeon's Journal, 1915–1918* (London, 1936) entry for 4.6.1917: 113.
8. H. Platt, 'Orthopaedic surgery in Boston', *Medical Chronicle*, 58 (March 1914): 473–9. On medicine in Boston at this time see also Morris Vogel, *The Invention of the Modern Hospital: Boston, 1870-1930* (Chicago, 1985): 63–5; and Clement A. Smith, *The Children's Hospital of Boston* (Boston, 1983), ch. 12.
9. The average was thirty new fracture cases a week: H. Platt, 'On the organisation of a fracture service', *Lancet* (17 September 1921): 620–1, reprinted in H. Platt, *Selected Papers*: 1–5. On the Ancoats Hospital, see John Pickstone, *Medicine and Industrial Society: a history of hospital development in Manchester and its region, 1752-1946* (Manchester, 1985): 145–6 *et passim*.
10. J. L. Thornton, 'Orthopaedic surgeons at St. Bartholomew's Hospital, London', *St Bartholomew's Hospital Journal*, 59 (1955): 195–204, at p. 200; and n. 22 below.
11. See: P. Gray, 'Grangethorpe Hospital Rusholme, 1917–1929', *Transactions of the Lancashire and Cheshire Antiquarian Society*, 78 (1975): 51–64. On the orthopaedic lessons of the war, see Robert Jones, 'Orthopaedic surgery in its relation to the war', *Recalled to*

Life, vol. 1 (1917): 50–9; S. M. Smith, 'Fractures of the lower extremity', in W. G. Macpherson, *et al.* (eds), *Medical Services in the History of the Great War: Surgery of the War*, vol. 2 (London, 1922): 339–80, at p. 353, and G. M. Levick, *et al.*, 'Organisation for orthopaedic treatment of war injuries', in Macpherson *et al.* (eds), *Medical Services*: 381–408.

12. On the profound significance of record-keeping in the reform of American hospitals, see David Rosner, *A Once Charitable Enterprise: hospitals and health care in Brooklyn and New York, 1885–1915* (New York, 1982): 55 *et passim*. On the history of record-keeping (largely as pioneered at the Massachusetts General Hospital), see S. J. Reiser, 'Creating form out of mass: the development of the medical record', in E. Mendelsohn (ed.), *Transformation and Tradition in the Sciences: essays in honor of I. Bernard Cohen* (New York, 1984): 301–16.

13. Platt, 'On the organisation of a fracture service', *Lancet* (17 September 1921): 620–1.

14. 'BMA Report of Committee on Fractures'. Of the seventeen members of the Fracture Committee, ten were members of the BOA; of the others, only H. S. Souttar, the Chairman, Henry Brackenbury, and Bishop Harman did not have a vested interest in the campaign for fracture clinics.

15. Delevingne, *Interim Report of the Inter-Departmental Committee on the Rehabilitation of Persons Injured by Accidents* (London, 1937), *Final Report* (London, 1939). On the setting up of the Delevingne committee, see Public Record Office (hereafter PRO): ED/50/173; for discussion of the reports, see below.

16. 'Lady Jones lecture on crippling due to fractures: its prevention and remedy', *British Medical Journal* (16 May 1925): 909–13.

17. H. Platt, 'British Orthopaedic Association: first founders' lecture': 124; and see Bristow's attack on the London teaching hospitals in W. R. Bristow, 'The influence of war surgery on treatment of fractures in Great Britain', *Journal of the American Medical Association* (3 December 1927): 1920–4.

18. See, for example, on the 'treatment of fractures by unqualified persons', *British Medical Journal* (13 June 1874): 777–8; and on the urgent need for 'continuity in fracture treatment', *Annual Report of the City Orthopaedic Hospital* (London, 1905): 12. See also: *Third Report from the Select Committee of the House of Lords on Metropolitan Hospitals, together with ... Minutes of Evidence* (London, 1891) Cmd. 457: 34.

19. 'Discussion on the treatment of fractures: with special reference to its organization and teaching', *British Medical Journal* (22 August 1925): 317–31.
20. Platt in the 'Discussion on the treatment of fractures' at Bath: 325.
21. See, for example, Melvin Henderson, 'Leadership in orthopaedic surgery', *Journal of Bone and Joint Surgery*, 16 (1934): 495–8.
22. Gask, 'Discussion on the treatment of fractures' at Bath: 318. Gask's point of reference was Sir T. Clifford Allbutt's *The Historical Relations of Medicine and Surgery* (London, 1905).
23. W. McAdam Eccles, in the 'Discussion on the treatment of fractures' at Bath: 329.
24. See editorials in *The Practitioner*, 137 (1936): 402–3, and *The Medical Officer* (26 June 1937): 255. The *Lancet* in its editorial on 'The fracture problem' rightly criticised the BMA 'Report of committee on fractures' for completely avoiding the question of 'the function of the GP in his duty to his middle-class patients [sustaining fractures] (16 February 1935): 383–4. The criticism was never dealt with.
25. See Anne Digby and Nick Bosanquet, 'Doctors and patients in an era of national health insurance and private practice, 1913–1938', *Economic History Review*, 41 (1988): 74–94, at p. 91.
26. See Frank Honigsbaum, *The Division in British Medicine* (London, 1979): 146–8, *et passim*.
27. Hey Groves, 'On the treatment of fractures: a problem of organization', *British Medical Journal* (1 December 1928): 993–5.
28. See A. Marwick, 'The Labour Party and the welfare state in Britain, 1900-1948', *American Historical Review*, 73 (1967–8): 380-403, esp. at pp. 386–90.
29. See Lyndall Urwick, *The Meaning of Rationalisation* (London, 1929); in specific relation to hospitals, see Rosner, *A Once Charitable Enterprise*.
30. TUC and Labour Party, *The Labour Movement and the Hospital Crisis* (London, 1922): 7. See also, 'The Labour Party and the hospital problem, conference at Caxton Hall', *British Medical Journal, Supplement* (3 May 1924): 213–22; and TUC and Labour Party, *The Labour Movement and Preventive & Curative Medical Services: a statement of policy with regard to health* (London, nd): 6. On the Dawson Report, see the minutes of the Consultative Council, PRO: MH/73/38-49, and Honigsbaum, *Division in British Medicine*, ch. 6.
31. 'Accidents and hospitals', *Lancet*, (23 October 1926): 864. See also *Lancet* (26 February 1927): 463; 'Medical practitioners and road

accidents', *British Medical Journal*, Supplement (25 July 1931): 62–5; and 'Emergency treatment for road accidents', *British Medical Journal* (4 August 1934): 213–14. For the implications of motor accidents for cottage hospitals, see *Lancet* (27 June 1931): 1410.

32. *Board of Trade: Statistical Abstracts* (London, 1934): 293 and W. Plowden, *The Motor Car and Politics in Britain 1896–1970* (Harmondsworth, 1973): 271 and Appendix D, 'Road casualties, 1928-69': 483.
33. Quoted in Plowden, *The Motor Car*: 252.
34. Of 19 286 fractures cases in 1937, 14.9 per cent were the result of road traffic accidents: Delevingne, *Interim Report*: 8. An LCC inquiry into 1068 fracture cases treated at four of their hospitals during summer and winter sample periods in 1936 and 1937 revealed that industry was responsible for 13.5 per cent (145), road traffic for 22.9 per cent (245), while 63.4 per cent (678) were the result of other causes. GLRO:PH/HOSP/1/72. In Manchester, where there were about 5000 fracture cases annually in the early 1930s, 70 per cent were domestic, 18 per cent street accidents, and 12 per cent industrial: see Pickstone, *Medicine and Industrial Society*: 287. For similar American statistics, see Roy N. Anderson, *The Disabled Man and his Vocational Adjustment* (New York, Institute for the Crippled and Disabled, 1932): 12.
35. Plowden, *The Motor Car*: 276. For statistics on the victims treated and the costs recovered in Manchester and Salford in 1932, see Pickstone, *Medicine and Industrial Society*: 273.
36. See C. V. Mackay, 'Dr. Böhler's fracture clinic in Vienna', *British Medical Journal* (19 April 1935): 522; H. Platt, 'Orthopaedics in Continental Europe, 1900–1950', in H. Platt, *Selected Papers*: 65–88, at pp. 84–5; H. Platt, 'The evolution of the treatment of fractures', *Manchester University Medical School Gazette*, 17 (1938): 56–62, at p. 59; E. M.. Bick, *Source Books of Orthopaedic Surgery*, 2nd edn (Baltimore, 1948): 293–4; W. H. Ogilvie, 'Physiology and the surgeon', *Edinburgh Medical Journal*, 43 (1936), reprinted in W. H. Ogilvie, *Surgery: orthodox and heterodox* (Oxford, 1948): 147–67, at p. 152; Friedrich Lorenz, *Lorenz Böhler; der Vater der Unfallchirurgie* (Vienna, 1955); and the 'BMA Report of Committee on Fractures': 57–8.
37. *Lancet*, (16 February 1935): 383.
38. This was often expressed (as by Platt and Bristow after their visit to Böhler's Clinic on the behalf of the BOA in 1929) by the adoption of plaster-of-Paris fixation. T. Porter McMurray, one of Jones's suc-

cessors in Liverpool, continued to use splints rather than plaster in the 1930s, 'never having visited Böhler's clinic': H. Platt, interview with the author (7 November 1984). However, there were also important commercial interests behind the promotion of gypson; see: Richard Bennett and J. A. Leavey, 'Gypsona and the fracture clinics', in R. Bennett and J. A. Leavey, *A History of Smith & Nephew, 1856-1981* (London, 1981): pp. 19–22.

39. Böhler, *The Treatment of Fractures*, 4th English edn, trans. by Ernest Hey Groves from the 4th enlarged and revised German edn (1933) (Bristol, 1935): 14ff.
40. *Lancet*, (23 October 1926): 864.
41. Hey Groves, trans., *The Treatment of Fractures*: i.
42. Delevingne, *Final Report*: 14ff. See also, for 1935, statistics on the incidence of fractures at 33 LCC hospitals, in GLRO:PH/HOSP/1/72.
43. T. P. McMurray in the discussion on the paper by H. E. Moore, 'Avoidable wastage in connexion with industrial injuries', *Liverpool Medico-Chirurgical Journal*, 41 (1933): 19–50, at p. 38.
44. See BOA, *Memorandum on Fracture and Accident Services Committee* (London, 1943); see also, G. R. Girdlestone, 'A regional orthopaedic and accident service', *British Medical Journal* (23 April 1949): 720–2; and J. Trueta, *Gathorne Robert Girdlestone* (Oxford, 1971): 79. For a survey of the subsequent development of orthopaedic and accident services, see Nuffield Provincial Hospitals Trust, *Casualty Services and Their Setting: a study in medical care* (Oxford, 1960); BOA, *Casualty Departments: the Accident Committee* (July 1973), which led to the BMA's, *Medical Staffing of Accident and Emergency Services: a report prepared* [by W. Lewin] *on behalf of the Joint Consultants Committee and Presented in April 1978* (London, 1978).
45. Platt, 'Orthopaedics in Continental Europe': 85.
46. Meeting of 22.12.1930, at Dr T. Carnwath's office at the Ministry of Health (a copy of the minutes of which were circulated in County Hall): GLRO:PH/HOSP/1/66. The meeting was apparently forgotten by 1936 when Sir Frederic Menzies, head of the medical services for the LCC, held a meeting on fracture clinics with Sir Malcolm Delevingne and, referring to the practice of the Metropolitan Life Assurance Company of New York, 'suggested that some of the bigger London insurance companies might be approached. He thought that grants from industry or insurance companies would greatly facilitate the establishment of special fracture units by the LCC': GLRO:PH/HOSP/1/73.

47. See T. L. Hazlett and W. W. Hummel, *Industrial Medicine in Western Pennsylvania 1850-1950* (Pittsburg, 1957): 73–5, 244–7.
48. Moore, 'Avoidable wastage'; H. E. Moore, 'Observations on the after-care of industrial casualties', together with extracts from Moore's *Annual Reports* of the Crewe Hospice (1927-30), submitted to T. Carnwath, Ministry of Health (22 December 1930) in GLRO:PH/HOSP/1/66; and see the obituary on Moore by R. Watson-Jones, *Journal of Bone and Joint Surgery*, 34B (1952): 708. On the quiet but important role of Gwynne Maitland in the fracture movement, see his obituary, also by Watson-Jones, *Journal of Bone and Joint Surgery*, 31B (1949): 130–1.
49. 'BMA report of Committee on Fractures': 60.
50. Hey Groves, 'A Surgical Adventure: an autobiographical sketch', reprinted from *Bristol Medico-Chirurgical Journal*, 50 (1933): 22.
51. See J. C. Nicholson, 'Fracture of the Neck of the femur, a personal experience', *British Medical Journal* (27 August 1938): 464–6, and Hey Groves's comments on this article, *British Medical Journal* (17 September 1938): 633–4.
52. Hey Groves, 'Should medicine be a mendicant? A review of our hospital service', *Lancet* (10 November 1930): 1107. See also the argument for state-funded orthopaedic institutions in Ireland managed by private societies and under the control of orthopaedists: W. C. Somerville-Large, 'Study of a national orthopaedic system', *Irish Journal of Medicine and Science*, 6th series (1937): 161–72, and W. Somerville-Large, 'The orthopaedic problem in Ireland', *Irish Journal of Medicine and Science*, 6th series (1935): 82–8.
53. Hey Groves, 'Should medicine be a mendicant?': 1106, 1051. Among other consultants at this time to call for state aid without state control was H. S. Souttar, surgeon to the London Hospital, who was subsequently the Chairman of the BMA Fracture Committee and an important figure in NHS discussions. See Arthur Newsholme, *Medicine and the State* (London, 1932): 48-9.
54. Letter from James Russell to Professor R. C. Alexander of Dundee (26 December 1944) Dundee University Library Archives, MS 16/17 (4). I am grateful to David Cantor for drawing my attention to this correspondence.
55. Voluntary Hospitals Committee for London, *Organised Fracture Services for London: Report by Fracture Sub-Committee* (June 1939): 6.
56. See letters from E. H. Young to Oliver Stanley, in PRO: MH 55/61. I am grateful to Helen Jones for this reference.

57. See, for example, General Federation of Trade Unions, *Report of Conference on Institutional Treatment of Fractures, 7 October 1936 at Onward Hall, Manchester* (London, 1936).
58. For the Joint TUC-BMA Memorandum, see *British Medical Journal* Supplement (18 December 1937): 367–71; or 'Appendix D' to the Memorandum of Evidence by the TUC to the *Royal Commission on Workmen's Compensation* (London, 1939, 1940): 445–8; TUC General Council's *Report to the Blackpool Congress* (1938) paras 88–103; and 'Joint Committee of BMA and TUC (1936–9)', 3 vols, BMA archives. On the Committee itself, as viewed largely from the perspective of the TUC (and without reference to the fracture issue), see R. Earwicker, 'A study of the BMA-TUC joint committee on medical questions, 1935-1939', *Journal of Social Policy*, 8 (1979): 335–56.
59. Watson-Jones was the author of the popular textbook *Fractures and Other Bone and Joint Injuries*, first published in 1940, and was subsequently the editor of the British volumes of the *Journal of Bone and Joint Surgery* and president of the BOA (see obituary in *British Medical Journal*, 26 August 1972: 533). His early career owed much to Robert Jones and he studied fracture treatment under Platt at Ancoats before establishing his fracture clinic at the Royal Liverpool Infirmary. In 1943, he was appointed director of the orthopaedic and accident department of the London Hospital. He visited Russia in 1943 at the same time as the TUC delegation: see his 'Russian surgeons and Russian surgery', *British Medical Journal* (28 August 1943): 276, and Walter Citrine, *Two Careers: a second volume of autobiography* (London, 1967): 167–8, 171. However, when it came to implementing the NHS, by which time his private practice in London was large and world-famous, he declared: 'We want freedom from medical control, and that freedom demands private practice. I saw the abolition of such freedom in Russia, and it has meant the end of medical progress in that country': 'The consultant's vote', *British Medical Journal* (7 February 1948): 264–7, at p. 266.
60. See, in particular (G. Tomlinson) *Report of the Inter-Departmental Committee on the Rehabilitation and Resettlement of Disabled Persons* (1943) Cmd. 6415; and see Honigsbaum, *Division in British Medicine*, ch. 24: 'The BMA–TUC alliance and the Beveridge Report'.
61. Honigsbaum, *Division in British Medicine*: 240.

62. Delevingne, *Interim Report*: 7; and *Final Report*: 26. St Bartholomew's Hospital had a fracture service from 1927, but the in-patient treatment of fractures was shared by general surgeons, with only special cases being referred to the Orthopaedic Department. At St Thomas's and the Westminster hospitals a similar situation existed in the 1930s. A segregated fracture service adhering to the principles of continuity of treatment, unity of control under an orthopaedist, and after-care was established at the Manchester Royal Infirmary (under Platt) in 1936. See BMA 'Report of Committee on Fractures': 56ff.
63. Delevingne, *Final Report*: 76.
64. See Arnold Wilson and H. Levy, *Workmen's Compensation*, vol. 2 (Oxford, 1941): 203.
65. Delevingne, careful always to avoid both controversy and dissent in his Committee, was well aware of 'the difficulty which existed as to whether an orthopaedic surgeon or a general surgeon should undertake the treatment of fractures'. At a meeting with Sir Frederick Menzies (6 May 1938) he let it be known 'that it was the intention of his committee to use the term "fracture surgeon" only'. GLRO: PH/HOSP/4/28.
66. For references and for discussion on both planning bodies, see Pickstone, *Medicine and Industrial Society*: 287ff, 301ff, and Neville M. Goodman, *Wilson Jameson: architect of national health*, (London, 1970): 133 *et passim*.
67. Examples would include Geoffrey Jefferson (neurosurgery), John Morley (general and abdominal surgery and paediatrics), E. D. Telford (general surgery and orthopaedics), James Spence (paediatrics) and (although without the provincial background) E. Rock Carling (a London consultant with a special interest in radiology).
68. Representative were the twenty-one members of the Delevingne Committee, who were drawn from the Ministries of Health, Labour and Pensions, the Department of Health for Scotland, the BMA, the British Hospitals Association, the Accident Officers' Association, LMS Railways, Midland Colliery Owners' Mutual Indemnity, Durham Miners, National Union of Railwaymen, and the Scottish TUC General Council. BOA interests were represented by W. A. Cochrane and Ernest Hey Groves.
69. In 1978 174 out of 228 major Accident/Emergency departments in British hospitals were under the control of orthopaedic surgeons. See W. Lewin, *Medical Staffing*.

9 Cortisone and the Politics of Drama, 1949–55

* It is a pleasure to thank Roger Cooter, Mary Fissell, Daniel Fox, Bill Luckin and John Pickstone for reading and commenting on this essay. All errors remain my own.
1. 'Rheumatics flock to Golders Green', *Hampstead News* (26 May 1949). M. G. Good, 'The rheumatic diseases', *Scotsman* (7 March 1949). Good offered a 'cure' of rheumatism. One patient refused to pay on the grounds that she had not been cured, and was supported by a Sheriff's court, 'Couple sued by specialist', *Manchester Daily Telegraph* (18 January 1952). 'No-cure specialist loses £398 fee', *Daily Express* (3 September 1952). 'Specialist loses £398 claim', *Scotsman* (3 September 1952).
2. 'She has 2,000,000 bees and goes around stinging people', *Sunday Dispatch* (30 October 1949). 'Truth with a sting' *Surrey Comet* (1 October 1949). 'Bee in his bonnet', *Daily Herald* (3 September 1949). 'Bee stings and rheumatism', *Smallholder* (7 October 1949).
3. 'Remedy for rheumatism', *Everybody's* (5 November 1949).
4. 'Homeopath cures mother and son', *South London Press*, (14 October 1949).
5. 'Rheumatism pains? Lumbago? Try this ...', *Sunderland Echo* (4 June 1949).
6. 'Rheumatism "rubbed out" in minutes', *Scottish Daily Express* (14 April 1949).
7. 'He has discarded his crutches', *Crewe Chronicle* (30 April 1949).
8. 'Pros and cons on curative value of molasses', *Natal Witness (Pietermaritzburg)* (19 July 1949).
9. '"Hot-house" from rheumatism', *Manchester Guardian* (2 June 1949).
10. 'Many worker ills traced to tension', *New York Times* (31 May 1949).
11. 'Nettles cure for rheumatism', *Irish News* (12 November 1949).
12. 'They thought it was the "Elixir"', *Sunday Express* (23 October 1949).
13. 'Belgians to offer new line of drugs', *New York Times* (20 February 1949).
14. 'Nylon for arthritis', *Textile Review* (21 October 1949), 'Nylon on the knee', *Time* (24 October 1949).
15. 'How about wolf fat for rheumatic pain?', *St John's Evening Telegraph (Newfoundland)* (2 February 1949).
16. 'Manufacture of cortisone', *British Medical Journal*, Pt 2 (1949):

974. See also 'Plant sources of cortical hormones', *British Medical Journal*, Pt 2 (1949): 481. This article estimated that 40 cattle were needed to obtain the 100 mg needed daily for the treatment of a single patient over one year.
17. 'Treatment of rheumatoid arthritis with steroids', *British Medical Journal*, Pt 2 (1950): 874–5, at p. 874.
18. Edward C. Kendall, *Cortisone* (New York, 1971).
19. P. S. Hench, E. C. Kendall, C. H. Slocumb and H. F. Polley, 'The effect of a hormone of the adrenal cortex (17-hydroxy-11-dehydrocorticosterone: compound E) and of pituitary adrenocorticotropic hormone on rheumatoid arthritis: Preliminary report', *Proceedings of the Staff Meetings of the Mayo Clinic*, 24 (1949): 181–97. For American responses to cortisone see Daniel M. Fox and Marcia Meldrum, 'The miracle of cortisone: Historicism and chronic disease in Britain and the United States' and Harry M. Marks, 'The political biography of a drug: Cortisone the middle years', papers given to the annual meeting of the American Association for the History of Medicine, Baltimore, Maryland (May 1990).
20. Radio-script '"Science Survey". The story of cortisone by a Fellow of the Royal College of Physicians', transmission: 20 October 1949; 10.30–10.45 pm, Home Service in Studio 3B. Arthritis and Rheumatism Council (ARC) archives, File 'Cortisone committee'.
21. 'The problem of rheumatism. Two day conference in London', *British Medical Journal* Pt 2 (1949): 810–11, at p. 810.
22. 'Cortisone in the treatment of rheumatism', *British Medical Journal* Pt 2 (1949): 24–5, at p. 24.
23. 'Suprarenal hormone in rheumatoid arthritis', *Lancet*, Pt 1 (1949): 1058.
24. 'New treatment for rheumatism', *Times* (23 April 1949).
25. Quoted in Rudolf Klein, *The Politics of the National Health Service* (London and New York, 1983): 36.
26. Mary E. Fissell, *Patients, Power and the Poor in Eighteenth-Century Bristol* (Cambridge, 1991). Simon Schaffer, 'Natural philosophy and public spectacle in the eighteenth century', *History of Science*, 21 (1983): 1–43.
27. Thomas Richards, *The Commodity Culture of Victorian England. Advertising and Spectacle, 1851–1914* (Stanford, 1990). I am grateful to Professor Richards for sending me a copy of part of his book prior to publication. For other histories of advertising see T. R. Nevett, *Advertising In Britain: A History* (London, 1982). Frank S. Presbrey, *The History and Development of Advertising* (Garden City,

NY, 1929). E. S. Turner, *The Shocking History of Advertising!* (London, 1952). Blanche B. Elliot, *A History of English Advertising* (London, 1962).
28. Frederic John Poynton and Bernard Schlesinger, *Recent Advances in the Study of Rheumatism*, 2nd edn (London, 1937): 288.
29. Poynton and Schlesinger, *Recent Advances*: 287.
30. Poynton and Schlesinger, *Recent Advances*: 288.
31. D. L. LeMahieu, *A Culture for Democracy. Mass Communication and the Cultivated Mind in Britain Between the Wars* (Oxford, 1988).
32. Charles W. Buckley, 'British Spas and Their Waters. A Foreword', *The Prescriber*, 17 (1923): 95–9, at p. 95.
33. LeMahieu, *A Culture for Democracy*. On how Buckley's comments fitted into inter-war professional disputes within the spa towns see David Cantor, 'The contradictions of specialization: Rheumatism and the decline of the spa in inter-war Britain', *Medical History Supplement*, No. 10 (1990): 127–144.
34. W. S. C. Copeman, *The Treatment of Rheumatism in General Practice,* 2nd edn (London, 1935): 77.
35. David Vincent, *Literacy and Popular Culture. England 1750–1914* (Cambridge, 1989): 159–71.
36. W. S. C. Copeman, 'The fight against rheumatism', *The Spectator* (14 August 1937). On the magical properties of radioactivity see Spencer R. Weart, *Nuclear Fear. A History of Images* (Cambridge, Mass., and London, 1988).
37. 'Ray treatment for rheumatism', *Yorkshire Post* (30 July 1937). 'Research into rheumatism. Early start under new scheme. Fraudulent "cures" to be tested', *Staffordshire Advertiser* (28 August 1937).
38. The major exception is Warren Crowe's textbook which advocated vaccine therapy as the mainstay of treatment. H. Warren Crowe, *Handbook of the Vaccine Treatment of Chronic Rheumatic Diseases* (Oxford and London, 1930); 2nd edn (1932); 3rd edn (1939).
39. G. D. Kersley, *The Rheumatic Diseases. A Concise Manual for the Practitioner* (London, 1934): 69.
40. Copeman, *The Treatment of Rheumatism*: 119.
41. Copeman, *The Treatment of Rheumatism*: 209.
42. Kersley, *The Rheumatic Diseases*: 69.
43. H. Rolleston, 'Introduction', *The British Journal of Physical Medicine and International Review*, 1 (1938): 331.
44. Copeman, *The Treatment of Rheumatism*: 118.
45. L. S. P. Davidson, 'Rheumatoid arthritis', in W. S. C. Copeman (ed.), *Textbook of the Rheumatic Diseases* (Edinburgh, 1948): 120–

42, at p. 133. See also C. W. Buckley, 'The rheumatic diseases ' (nd, ?1936/7/8): 2, typescript in William Willcox Papers, File 'Rheumatism Campaign II', Royal College of Physicians' archives.
46. Copeman, *The Treatment of Rheumatism*: 204.
47. D. Armstrong, 'The doctor-patient relationship: 1930–1980', in P. Wright and A. Treacher, *The Problem of Knowledge. Examining the Social Construction of Medicine* (Edinburgh, 1982): 109–22.
48. Copeman, *The Treatment of Rheumatism*: 207. See also W. S. C. Copeman, 'Occupational therapy in the treatment of arthritis', *The British Journal of Physical Medicine and International Review*, 1 (1938): 350–2, at p. 351.
49. James Fenton, 'Presidential address on propaganda in relation to rheumatism', *Journal of the Royal Sanitary Institute*, 59 (1938): 349–55, at p. 351. For other reports of Fenton's speech see David Cantor, *These Rheumatic Isles: A Social History of British Rheumatism Charities, 1920-1986* (forthcoming), where it is set in the context of medical attempts to raise the profile of rheumatic diseases and to appeal for funds for the newly created Empire Rheumatism Council. For another view of disease as a dramaturgical event see Charles E. Rosenberg, 'What is an epidemic? AIDS in historical perspective', *Daedalus*, 118 (1989): 1–17.
50. Fenton, 'Presidential address': 350.
51. Fenton, 'Presidential address': 351.
52. Fenton, 'Presidential address': 351.
53. Turner, *The Shocking History*: 168.
54. 'Modern retailing', *The Ironmonger* (23 June 1934): 46; 'Trade descriptions', *The Ironmonger* (30 June 1934): 51. For an account of the institutional developments to control advertising in 1930s see Nevett, *Advertising*, ch 8. On Selfridge see A. H. Williams, *No Name on the Door: A Memoir of Gordon Selfridge* (London, 1956). Reginald Pound, *Selfridge: A Biography* (London, 1960).
55. 'Selfridge's £10 reward to discoverers of mis-statements', *Mens Wear* (9 September 1933). 'Selfridge's "£10 offer" has cost £1,320' *The Draper's Record* (14 March 1936): 68.
56. 'Rules for retail trading', *Times* (28 March 1935). Note also that another vice-president of the ERC, J. Spedan Lewis, attended the inaugural meeting of the Retail Standards Association, 'The shopper's Magna Carta', *Times* (28 March 1935). According to Nevett (*Advertising*, p. 74), John Lewis and Bourne and Hollingsworth remained opposed to the idea of advertising. The Bourne family were prominent donors to the ERC, and W. S. C.

Copeman, the ERC medical secretary, married the daughter of the store-owner. (Information from W. S. C. Copeman's son, Peter Copeman.)
57. 'New organization will form standards for retail trading', *World Press News* (4 March 1935).
58. 'The robot doctor', *World's Faire* (25 November 1933).
59. 'Selfridge's health and housing exhibition', *Westminster Record* (19 March 1935).
60. Bill Luckin, *Questions of Power. Electricity and Environment in Inter-War Britain* (Manchester, 1990). Paul Boyer, *By the Bomb's Early Light. American Thought and Culture at the Dawn of the Atomic Age* (New York, 1985). Weart, *Nuclear Fear*. On the use of radiation in rheumatic diseases see 'A-power now used in Bath research', *Bath and Wilts Chronicle* (13 June 1952) and *Bath Weekly Chronicle* (14 June 1952).
61. See for example, 'New wonders for the doctor's bag', *Co-operative News* (9 December 1950).
62. Hench, *et al.* 'The effect of a hormone of the adrenal cortex'.
63. For example see 'Insulin ... M and B ... penicillin ... and now another new medical miracle COMPOUND E', *Daily Mail* (14 June 1949). '"Wonder cure" means 2-year wait', *Evening News* (12 July 1949). 'Wonder "Drugs"', *Evening News* (26 November 1949). 'Manchester tries out super-drug', *Evening Chronicle (Manchester)* (23 October 1950). 'Cortisone, the latest wonder drug', *Middlesborough Evening Gazette* (13 November 1950).
64. Public Record Office (hereafter PRO): MH71/105.
65. J. Charles to J. Pater (11 August 1950) MRC archives, File 1063/31d/2. See also, W. S. C. Copeman, *Cortisone and A.C.T.H. in Clinical Practice* (London, 1953).
66. 'The Nuffield endowments', *British Medical Journal*, Pt 2 (1951): 414–5, at p. 414.
67. H. Himsworth to W. A. Sanderson (14 October 1949) MRC archives, File 1063/38.
68. Extract from MRC minutes (21 October 1949), MRC archives, file DC/11/archives. For a discussion of the MRC's previous neglect of rheumatoid arthritis see Cantor, *These Rheumatic Isles*.
69. The Nuffield Foundation appointed a rheumatism clinical trials committee in January 1948. This committee was set up to plan and initiate properly-controlled trials of different methods of therapy and relief for chronic rheumatic complaints. It was financed with a grant of £1000 p.a. for two years from the Foundation's Oliver Bird Fund, a special fund set up to research into rheumatism. The committee

survived until Merck offered 1 kg of cortisone jointly to the Nuffield Foundation and the MRC. Subsequently in 1950 the Nuffield Foundation and the MRC merged their committees. For details see, Nuffield Foundation archives, Minute book of the Nuffield Foundation Oliver Bird Fund. Rheumatism Clinical-Trials Committee (January 1949 to June 1950) and the Medical Research Council/Nuffield Foundation Joint Committee on Clinical trials of Cortisone and A.C.T.H. (January 1951 to March 1963).
70. 'The Nuffield endowments', *British Medical Journal*, Pt 2 (1951): 414–5, quotation at p. 414.
71. J. H. Glyn, *Cortisone Therapy Mainly Applied to the Rheumatic Diseases* (London, 1957): 125.
72. Letter to Secretary (26 October 1951), MRC archives, file 1063/31/9.
73. Cantor, *These Rheumatic Isles*.
74. 'Now there is just a little hope', *Daily Mail* (28 February 1950).
75. Charles Webster, *The Health Services Since the War. Volume I. Problems of Health Care. The National Health Service Before 1957* (London, 1988); Klein, *The Politics of the National Health Service*.
76. H. Jephcott, 'Cortisone' (1 July 1952), Glaxo archives, Box D1/1/40.
77. This account derives from, D. H. Aldcroft, *The British Economy, Volume I. The Years of Turmoil 1920-1951* (Brighton, 1986).
78. H. Jephcott, 'Cortisone' (1 July 1952), Glaxo archives, Box D1/1/40.
79. H. Jephcott, 'Cortisone' (1 July 1952), Glaxo archives, Box D1/1/40. For further discussion of the impact of cortisone on Britain's dollar problem see David Cantor, 'Cortisone and the pursuit of empire' (forthcoming).
80. W. S. C. Copeman and R. M. Mason, *Rheumatism, Fibrositis, Arthritis, Lumbago, Sciatica, "Slipped Disc", Gout, Spondylitis* (London, 1954): 117. But see Glyn, *Cortisone Therapy*: 122 for an alternative account, in which the author suggests that doctors were as much to blame as the lay press for exaggerating the benefits of cortisone.
81. W. S. C. Copeman, 'Rheumatic and collagen diseases', in Copeman, *Cortisone and A.C.T.H. in Clinical Practice*: 35.
82. Copeman, 'Rheumatic and collagen diseases': 38.
83. H. P. Himsworth, 'Complications of cortisone and A.C.T.H. therapy', *British Medical Journal*, Pt 2 (1951): 116. H. P. Himsworth, 'Complications of cortisone and A.C.T.H. therapy, *Lancet*, Pt 2 (1951): 78.

84. For responses to Himsworth's request see MRC archives, file 1063/31/12b.
85. H. A. Layzell to IS (4 February 1953), PRO: MH55/1057. For reasons of confidentiality patients, their relatives and their physicians are identified only by their initials and the date of the document.
86. H. M. Walker to H. Jephcott and others (19 May 1952), Glaxo archives, Box D1/1/40.
87. H. M. Walker to A. N. Richards (30 October 1952), A. N. Richards Papers, University of Pennsylvania archives Box 30.
88. D. W. Richards to A. N. Richards (24 November 1952), A. N. Richards papers, Box 30.
89. D. W. Richards to A. N. Richards (24 November 1952), A. N. Richards papers.
90. D. W. Richards to A. N. Richards (24 November 1952), A. N. Richards papers.
91. D. W. Richards to A. N. Richards (24 November 1952), A. N. Richards papers.
92. H. M. Walker to A. N. Richards (22 September 1952), A. N. Richards Papers, Box 30. Johnston in fact contemplated organising a trial of cortisone involving general practitioners in Scotland. J. M. Johnston to A. N. Richards (15 November 1952), A. N. Richards Papers.
93. H. M. Walker to A. N. Richards (11 August 1952), A. N. Richards papers.
94. D. W. Richards to R. T. Major (25 August 1952), A. N. Richards Papers.
95. O. Savage, 'The adrenal hormones', in W. S. C. Copeman (ed.), *Textbook of the Rheumatic Diseases,* 2nd edn (Edinburgh and London, 1955): 91–120, at p. 118. J. J. R. Duthie and L. S. P. Davidson, 'Rheumatoid arthritis', in Copeman, *Textbook,* 2nd edn: 165–215, at p. 190.
96. 'Don't pester your doctor', *Middlesborough Evening Gazette* (2 February 1956). 'Cortisone warning as drug is freed', *Observer* (11-12 October 1955). 'Cortisone can be dangerous', *Oxford Mail* (10 December 1955). 'You can't fool around with cortisone', *Evening News* (1 February 1956). 'Problem drug has its uses', *Gloucester Citizen* (2 February 1956). 'From today doctors may prescribe this wonder drug – Cortisone comes off ration', *Birmingham Gazette* (5 December 1955).
97. Note that the science journalist Paul de Kruif discusses the transforming power of cortisone in 'The mysterious power of cortisone', *Readers Digest* (December 1952).

98. BM (27 June 1954), PRO: MH55/1057.
99. BM (27 June 1954), PRO: MH55/1057.
100. IS (17 January 1953), PRO: MH55/1057.
101. IS (28 October 1952), PRO: MH55/1057.
102. IS (17 January 1953), PRO: MH55/1057. 'Midland black market starts in new American miracle drug', *Sunday Mercury* (23 December 1951). '"Miracle" drug cost patients £90'. *Sunday Mail* (9 December 1951). Merck were also aware of a 'gray-market' in cortisone in which patients purchased private supplies of their own. R. T. Major to C. M. Anderson and others (12 August 1952), A. N. Richards papers, Box 30.
103. AT (4 November 1956) and NN (25 March 1953), PRO: MH55/1057.
104. ECE (23 November 1953), PRO: MH55/1057. RG (4 February 1954), PRO: MH55/1057.
105. CL (28 April 1953), PRO: MH55/1057.
106. BM (19 May 1954), PRO: MH55/1057.
107. JHR (20 April 1953), PRO: MH55/1057.
108. IS (17 January 1953), PRO: MH55/1057.
109. IS (17 January 1953), PRO: MH55/1057. For another criticism of the priority of war and destruction over cures for ill-health see R. M. Lines, 'Rheumatism', *Everybody's* (16 December 1950).
110. IS (28 October 1952), PRO: MH55/1057.
111. IS (17 January 1953), PRO: MH55/1057.
112. LDM (6 July 1955), PRO: MH55/1057. For an alternative view of the value of physical methods see ECH, 'Rheumatism bogey', *John Bull* (25 March 1950).
113. LDM (6 July 1955), PRO: MH55/1057.
114. BM (27 June 1954), PRO: MH55/1057. For another possible example of public mistrust of clinical trials see FB (17 December 1953), PRO: MH55/1057.
115. Duthie and Davidson, 'Rheumatoid arthritis': 190.
116. F. H. K. Green to H. Himsworth (1 September 1950), MRC archives, File 1063/31/9.
117. IS (28 October 1952 & 17 January 1953), PRO: MH55/1057.
118. TBD to Sir Weldon Dalrymple-Champneys (15 December 1952), PRO: MH55/1057.
119. TBD to Sir Weldon Dalrymple-Champneys (15 December 1952), PRO: MH55/1057.
120. TBD to Sir Weldon Dalrymple-Champneys (15 December 1952), PRO: MH55/1057.
121. H. A. Layzell to IS (8 January 1953), PRO: MH55/1057.

122. Sociologists have pointed out that the doctor-patient consultation is structured by the social class of the patients. In 1974 Cartwright and O'Brian found that the average consultation with a middle-class patient took 6.2 minutes as against 4.7 minutes for a working-class patient. They argued that doctors had less sympathetic and understanding relationships with their working-class patients yet, paradoxically, found them more satisfying. A. Cartwright and M. O'Brian, 'Social class variations in health care and the nature of general practitioner consultations', in M. Stacey (ed.), *The Sociology of the NHS*, Sociological Review Monograph, no. 22, Keele (1976).
123. '"Miracle" drug cost patients £90', *Sunday Mail*, (9 December 1951).
124. Chris J. Ham, 'Power, patients and pluralism', in Keith Barnard and Kenneth Lee (eds), *Conflicts in the National Health Service* (London and New York, 1977): 99-120. Rudolf Klein and Janet Lewis, *The Politics of Consumer Representation*, Centre for Studies in Social Policy (London, 1976).

For a general discussion of the idea of professions as conspiracies against the laity see H. Perkin, *The Rise of Professional Society. England Since 1880* (London and New York, 1989) ch. 10.

10 Psychiatry in District General Hospitals: History, Contingency and Local Innovation in the Early Years of the National Health Service

* My former colleague Michael Jones began to explore the MRHB mental hospital policy when he was working on the history of mental hospitals in the Manchester region. I am also grateful to the late J. S. B. Mackay, Dr Arthur Pool and especially Dr F. N. Marshall for the interviews they gave. My present colleagues, especially Roger Cooter and Mary Fissell, have provided helpful criticism of previous drafts, and Hugh Freeman also made available some of his own research on psychiatry in general hospitals. Responsibility for the interpretations here advanced is mine.

1. The standard, welfarist, history of British asylums is Kathleen Jones, *A History of the Mental Health Services* (London, 1972). The leading sceptical works are by Andrew Scull: *Museums of Madness: the Social Organisation of Insanity in Nineteenth Century England* (London, 1979) and *Decarceration. Community Treatment and Deviance, a Radical View*, 2nd edn (London, 1984). Joan Busfield has pro-

vided a very useful overview: *Managing Madness. Changing Ideas and Practice* (London, 1986).
For mental health services under the early NHS also see Charles Webster, *The Health Services Since the War*, Vol. I, Problems of Health Care: The National Health Service Before 1957 (London, 1988).
Developments in the 1950s and 1960s are outlined by Jones, *A History* and by Busfield *Managing Madness*. The latter provides a useful critique of Scull's accounts. But all these works concentrate on the run-down of asylums, and on the prospect and reality of 'community care'; they say relatively little about psychiatry in general hospitals.

2. Dr Hugh Freeman is writing a PhD thesis for Manchester University on the development of general hospital psychiatry in Britain. His interview with George Godber, formerly of the Ministry of Health, is published in *Psychiatric Bulletin*, 12 (1988): 513–20.
3. Later published as C. P. Blacker, *Neurosis and the Mental Health Services* (London, 1946).
4. Busfield, *Managing Madness*: 332-6. Dr Arthur Pool (interview with author, 18 January 1983), worked on malarial therapy for GPI while he was a junior psychiatrist at Rainhill Asylum near Liverpool (1927–31). He also used the many 'medical cases' in the asylum for his coaching sessions with Liverpool medical students. Among English asylums, Whittingham (near Preston, Lancashire) was among the first to use malarial therapy. See A. R. Grant, The treatment of general paralysis by malaria, *British Medical Journal*, Pt 2 (1923): 698.
5. For the structure of non-psychiatric hospitals especially around Manchester see J. V. Pickstone, *Medicine and Industrial Society. A History of Hospital Development in Manchester and its Region 1752-1946 (Manchester, 1985)*.
6. Hugh Freeman, 'The Portsmouth Mental Health Service, 1926–1952. An historical note', *The Medical Officer*, 107 (1962): 149–51.
7. Manchester city used 'Springfield' in this way: the old workhouse building adjacent to 'Crumpsall', the former workhouse hospital.
8. Pickstone, *Medicine and Industrial Society* describes the war-time planning as it affected (and would have affected) hospitals in the Manchester region.
9. Webster, *The Health Services Since the War*.
10. On the hospital service – local authority split see Pickstone, *Medicine and Industrial Society* and Webster, *The Health Services Since*

the War. On consequences for the occasional integrated mental health scheme see Freeman 'The Portsmouth Mental Health Service'.
11. Salford was notable for its community mental health services, developed by the Medical Officer of Health in spite of the administrative divides. See Hugh Freeman, 'Mental health services in an English county borough before 1974', *Medical History*, 28 (1984): 111–28.
12. For example, the Manchester RHB's spending on capital projects 1948–74, was 25 per cent below the national average on a *per capita* basis.
13. The early years of the Manchester RHB will be described in a forthcoming volume. Also see Christopher Ham, *Policy Making in the National Health Service. A Case Study of the Leeds Regional Hospital Board* (London, 1981), which includes a discussion of mental health services. The Leeds RHB inherited the asylums of the West Riding Mental Hospitals Board, which in some ways resembled its Lancashire counterpart. The Leeds RHB proved, however, to be very sceptical of general hospital psychiatry, fearing adverse effects on the asylums. The contrast is worth further study.
14. Manchester RHB *Minutes* (25 July 1950)(1950–1): 140. The Board, through its Medical Advisory Panel, overruled the TAP on this issue. There was also a Mental Health Sub-Committee which reported to the Planning Committee, but it was not responsible for decisions about medical staffing. Its attempts to back the TAP on staffing asylums were overruled (p. 139). The possibility of the Mental Health TAP being directly referred to the Mental Health Sub-Committee had been ruled out. In the first instance, the TAP report was to go to the Medical Advisory Panel (Manchester RHB *Minutes* 24 January 1950)(1949-50): 510.
15. Manchester RHB *Minutes* (28 February 1950)(1949-50): 595.
16. Manchester RHB *Minutes* (31 March 1950)(1950–1): 28.
17. Manchester RHB *Minutes* (23 May, 27 June, 25 July 1950)(1950-1): 75, 103, 134, 139, 140, 142–5, 151. The report quoted is on pp. 142–5.
18. J. M. Leyberg, 'A district psychiatry service – the Bolton pattern', *Lancet*, Pt 2 (1959): 282–4; H. L. Freeman, 'Oldham and district psychiatric service', *Lancet*, Pt 1 (1960): 212–21. For further references and a comprehensive account of the Burnley and Blackpool units, see J. Hoenig and M. W. Hamilton, *The Desegregation of the Mentally Ill* (London, 1969).
19. Mike Jones and I conducted an informal interview with Dr Mackay,

which confirmed the opinions of Dr Pool (interview) and Dr Freeman (personal communication), that Mackay was the chief advocate of psychiatric services, and the major protagonist of the District General Hospital scheme. According to Pool (interview) Dr Marshall, the Senior Administrative Medical Officer (SAMO), was 'a very non-committal person, he was primarily interested in everything but mental illness'. Of course, the fact that Marshall was ready to leave psychiatry to Mackay – and, indeed, his apparent lack of interest – is quite compatible with a possible role in the initiative to treat psychiatry primarily as a general hospital service. But we must allow that Marshall may, in retrospect, have overestimated his own part in the formulation of the new policy.

20. Manchester RHB *Minutes* (28 February 1950)(1949–50): 595 and Pickstone, *Medicine and Industrial Society*: 291–2.
21. Interview with Dr F. N. Marshall (26 November 1982) (J. V. Pickstone and Joan Mottram).
22. W. E. Le Gros Clark and W. Mansfield Cooper, 'John Sebastian Bach Stopford, 1888-1961', *Biographical Memoirs of Fellows of the Royal Society*, 7 (1961): 271–9.
23. On Jefferson and Kletts see W. Brockbank, *The Honorary Medical Staff of the Manchester Royal Infirmary, 1830–1948* (Manchester, 1965).
24. Marshall, interview (1982).
25. Marshall, interview (1982); Pool, interview (1983).
26. See Pickstone, *Medicine and Industrial Society*: 204–8, 286–9, 300–4. Platt's attitudes to specialism are explored in Roger Cooter, *Medicine and Society in Peace and War* (London, forthcoming in this series).
27. Theresa P. Macdonald, Christie Hospital and Holt Radium Institute: foundation, development and achievements, 1892–1962', MSc thesis, UMIST (1977).
28. Manchester RHB *Minutes* (27 January 1948)(1947–9): 47–8. Dr G. T. James, a deputy SAMO with experience as a psychiatrist, was initially responsible for mental health. After he left, Dr Mackay took over administrative responsibility, aided by Dr Pyle who remained as superintendent of Mary Dendy hospital for the mentally handicapped. James appears to have favoured the new policy, but left before it was implemented (interviews with Marshall, 1982 and Pool, 1983).
29. On 31 March 1949 applications were considered for four psychiatric posts, all in asylums; the Committee was unable to recommend any

of the applicants. Manchester RHB *Minutes* (26 April 1949) (1949–50): 54.
30. The history of the Lancashire asylums is to be described in a forthcoming volume.
31. The notorious critique of asylums published by M. Lomax, *Experiences of an Asylum Doctor, with Suggestions for Asylum and Lunacy Law Reform* (London, 1921), was based on Prestwich Asylum, near Manchester.
32. Marshall (interview, 1982) recalled Stopford and himself being called to see the Chairman of the Board of Control, who complained about Manchester's failure to build new mental accommodation.
33. Marshall (interview, 1982) recalled this criticism and the interest of Walter Briggs, the Blackburn surgeon.
34. Marshall interview (1982).
35. Marshall interview (1982).
36. The initial proposal was apparently for appointments at Blackburn-Burnley-Whittingham, Bolton-Bury-Prestwich and Rochdale-Oldham-Prestwich. Prestwich HMC did not like these arrangements (Manchester RHB *Minutes* (1950–1): 192, 253. Pool was appointed to Oldham-Rochdale-Prestwich (p. 215) and the other posts readvertised (p. 253).
37. Dr Arthur Pool, interview (1983). The medical superintendent of the Oldham hospital (Boundary Park) was Dr Robert Marr, a friend of Pool from medical school.
38. See n. 18 above.

# INDEX

Abel-Smith, Brian, 200n6
accident services, 16, 150, 154–8, 160–2
  see also fracture clinics; orthopaedics
ACTH, 174, 177, 181, 182
advertising, 166–8, 171–3
Albert, A., 130
Allen, R. W., 91, 95, 96
American Board of Obstetrics and
  Gynecology, 131
Amsterdamska, O., 223n6
anaesthetics, 2, 10, 247n16
Anbar, Michael, 201n9
Ancoats Hospital, Manchester, 148, 149,
  150, 159, 262n59
Anderson, Thomas, 20, 21
ante-natal care (US), 131–2, 134, 137–40,
  142, 145
antibiotics, 15
antisepsis, 2, 3, 13, 15, 17–46
  Germany, 34, 44–5
  in military surgery, 43–4
  opponents, 21, 23–6, 29–35, 37–41
  reception, 22–6
  as science, 25–6, 33, 45
  see also asepsis; carbolic acid; Lister,
    Joseph; wound treatment; innovation; economics of medicine
anti-toxin, for diphtheria, see serum therapy
  for diphtheria
Armstrong, David, 171
Army Medical Service, 88, 107, 112
arthritis, 8
asepsis, 2, 35, 44–6
  see also antisepsis
assessment, of innovation, see innovation,
  assessment of
Association of the British Pharmaceutical
  Industry, 175
asylums, 8, 186–92, 195–8, 276n30
  acute units, 189–90
  NHS, 190
  tuberculosis treatment, 216n42
  voluntary admissions, 190
  see also psychiatry
aureomycin, 173
Austria, 44, 155–7
Austrian National Insurance Company, 155

bacteriology, 4–6, 40, 49–51, 66, 72, 81–2,
  85–6, 88, 95, 97–102, 234n109
  see also Koch; Pasteur
Balfour, Arthur, 98, 222n105
Barcroft, Joseph, 104, 107, 108, 115
Batty Shaw, H. W., 94, 102, 230n58
Beattie, J. M., 85, 232n85
Behring, E. von, 74–82
  Nobel Prize, 75
Berlin, 34, 45, 50, 73, 75–7, 90
Bevan, Aneurin, 166, 175, 190
Beveridge Report, 161
Bibel, D. J., 226n15
Bickersteth, E. R., 29
Bijker, W. E., 201n10
biochemists, 6, 174
Birmingham, 208n54
Black, Campbell, 25
Blackburn, 191, 276n36
Blacker, C. P., 187, 273n3
Blacker Report, 193
Blackpool, 191, 197, 274n18
Board of Control, see psychiatry
Böhler, Lorenz, 155, 156
Bolton, 57, 191, 274n18, 276n36
Bosanquet, N., 200n4
Boston (USA), 149, 150
Bourne, Geoffrey, 244n91
Boyce, Rubert, 232n85
Brackenbury, Sir Henry, 152, 257n14
Bradford, 56, 63
Briggs, Walter, 194, 276n33
Bristol, 159
Bristow, W. Rowley, 151
British Hospitals Association, 263n68
British Medical Association (BMA), 21, 23,
  49, 55, 92, 147, 217n50
  meeting, Bath (1925), 152
  'Report on Fractures', 150, 158, 160, 162
British Medical Journal, 24, 35, 45, 56, 62
British Orthopaedic Association (BOA),
  148
Brockbank, W., 275n23
Brompton Hospital, 50, 56, 91
Bryant, Thomas, 18, 24
Bryder, Linda, 47, 57, 63, 217n56, n57, n62
Buckley, Charles, 168, 173
Bulstrode Report, 59, 62, 65, 220n82
Burnley, 191, 197, 276n36
Bury, 57, 276n36

Busfield, Joan, 272n1

Caldwell, W., 140
Callender, George, 19–20, 31, 32, 36–7, 38, 46
Calvert, Frederick Crace, 21, 24–5, 204n19
Calvert, G. A. Crace, 215n35
Cambridge, 18, 115
Cameron, Hector, 29
Campbell, J. M. H., 243n75
Canada, 203n3
cancer, 11, 25, 171, 173, 194–5
carbolic acid
  as disinfectant, 19, 21, 23, 31
  lotion, 32, 35, 44
  spray, 27–35, 38–40, 42, 44–6
  as surgical dressing, 21–8, 43; diffusion, 21–6
  in treatment of cancer and syphilis, 25
  see also antisepsis
cardiazol, 187
cardiology, 240n57
Carling, E. Rock, 263n67
Carlisle, 21
Carnwath, T., 260n46, n48
Cave Committee, 153
cephalometry, 136
Chamberlain, A., 222n105
Cheshire, 196
child health movement, 81, USA, 131
Christie Cancer Hospital and Holt Radium Institute, 194–5, 275n27
Clarke, Tom, 172
classification, see x-ray pelvimetry
cleanliness, 4
  in wound treatment, 19–20, 30–3, 38–46
clinical pathology, 85, 87, 97–100
clinical research/science, 5, 7, 94–5, 97, 99, 107, 117–23, 153
  and cortisone, 174–5
  in obstetrics, 125, 131, 136–42, 145
  see also laboratories
Clinical Research Association (CRA), 92, 97
clinician–scientist relations, see scientist–clinician relations
Cochrane, W. A., 263n68
Colebrook, Leonard, 85–7, 95, 103, 225n8, 227n22
Coleman, James S., 200n2
community care, see psychiatry
consumption, 4
  contagion, 49–50, 55
  incidence, 48

medical neglect before 1880, 48
mortality and morbidity rates, 48, 65
reception of Koch's bacillary theory, 49–50
'seed' and 'soil', 50–1
tubercle baccilus, 49, 50–1
consumption, treatment of
  climatic, 48
  climatic to open-air, 50, 51–2
  discipline, 63–4
  medical, 50
  open air to sanatorium, 63
  opposition to sanatorium treatment, 61–3
  tuberculin, 50, 64
  work therapy, 63
continuity of treatment, 150, 153
  see also scientific management
Coote, Holmes, 30–1
Cooter, Roger, 275n26
Copeman, W. S. C., 168–71, 176, 177, 267n56
Cormack, Sir John Rose, 29
cortisone, 8, 12, 16, 165–84
  and atomic bomb, 173
  clinical trials of, 177–8, 181, 182, 184
  and dollar crisis, 175–6
  and general practitioners, 178–9
  as a 'miracle' drug, 165, 173, 176, 181, 271n102
  and NHS expenditure, 175
  patients, criticisms, see patients
  and rheumatology, 165–7, 173–9
  shortage of, 166, 177–81, 183
  side effects of, 177–9
  as a 'wonder' drug, 173, 176, 179
  see also patients; drama and spectacle; innovation; discovery; economics of medicine; pharmaceutical (and other medical) companies
Cowan, Ruth Schwartz, 201n10
crippled children, 148, 153
Cronje, G., 213n8
Crossley Sanatorium, 56–7
Culyer, A. J., 200n6
Cunard Steamship Company, 158
cures, 2–8
  diphtheria, 72–4, 79, 81, 82
  rheumatism, 165–6, 169–74, 176, 183
  tuberculosis, 47, 52, 56, 90–1, 219n72
  vaccines, 84, 92
  see also cortisone; tuberculin; consumption, treatment of

Davidson, L. S. P., 170, 182

Davy, Henry, 37
Dawson Report, 154
DeLee, Joseph, B., 131, 144, 248n21, 251n49
Delépine, Sheridan, 232n85
Delevingne (Inter-Departmental) Committee, 150, 160, 161–2
Delevingne, Sir Malcolm, 150
Department of Health for Scotland, 177–8
Department of Science and Industrial Research (DSIR), 117
department stores, 168, 173
Deventer, Hendrick van, 125–6
diabetes, 6, 166
diffusion, of innovation, *see* innovation
Digby, Anne, 200n4
diphtheria, 4, 5, 72–83
    bacillus first recognised, 74
    mortality, 73, 77
    see also serum therapy for; cures
disability, 6
discovery, 1, 9
    anti-rabies serum, 74
    cortisone, first isolated, 166
    diphtheria bacillus, 74
    tubercle bacillus, 49
    x-rays, 124
    see also innovation; invention
disease concepts, 4–6, 12, 17–19, 38, 42, 73–4, 81
    structural v. functional, 109–12, 116
    see also germ theory
disinfectant, 19, 24, 75
    see also carbolic acid
Dixey, Richard, 203n2
Douglas, Claude Gordon, 107, 109, 110, 111
drama and spectacle, 8, 165–84
    control of, 166, 169
    and cortisone, 165–7, 173–84
    and enthusiasm, 166–73, 179, 180, 182, 183
    and exaggeration, 166, 167, 168, 177, 184
    medical attitudes to (pre-1949), 167–73
    and optimism, 166, 168, 169, 171, 177
    and sensationalism, 166, 172, 173, 177
    and spectacle, 167–9, 172, 173, 175, 182
    and surprise, 166
drugs, in psychiatry, 186, 198
    see also cortisone; pharmaceutical (and other medical) companies; oxygen therapy; serum therapy for

diphtheria; vaccine therapy
Dubos, J., 212n2
Dubos, R., 212n2
Dundee
    fracture service, 159
    sanatoria, 58
Durham, sanatoria, 56
Duthie, J. J. R., 182
Dyke, S. C., 85–7, 95, 225n4

economic history, 2, 3, 9, 10, 13
economics of medicine, 1, 3, 8, 10, 131
    antisepsis, 30
    cortisone, 166, 168, 174–8, 180–4
    fractures, 152, 154, 156–63
    private v. public practice, 121–2, 154
    sanatorium treatment, 54–7, 61, 63–5, 219n74
    serum therapy for diphtheria, 76, 78–82
    vaccine therapy, 92–3, 100
    x-ray pelvimetry, 144, 252n54
Edgerton, David, 202n10
Edinburgh Royal Infirmary, 27, 29, 30, 33
Edinburgh University, 118–119
Ehrlich, Paul, 74, 76, 78, 79, 80, 85, 98
Empire Rheumatism Council (ERC), 169, 172, 173, 175, 267n49
Erichsen, John, 41
erysipelas, 18, 38
    see also hospital diseases
Eversfield Hospital, 54
Exeter, 18
Eyler, J. M., 219n74

Fabré, J., 130
Fennell, Mary L., 11, 201n8
Fenton, James, 171–2
Fergusson, Sir William, 40
First World War, *see* war
Fleming, Alexander, 96, 231n72, n75
Fletcher, Walter Morley, 114
Flexner, Abraham, 131
folk-lore, 169
Forster, John Cooper, 30
Foster, W. D., 85–7, 95, 103, 225n3
Fox, Daniel M., 265n19
Fox, Nicholas J., 203n2
fracture clinics, 7, 13, 15
    American influence, 149–50
    BMA, 150
    diffusion, 150–2, 157, 160–2
    innovation, 146–7
    Labour Party supports, 160–1

fracture clinics *cont.*
  opponents, 148–9, 152
  product champions, 151
  rhetoric, 150–3
  role of World War I, 148–50
  state support for, 157–9, 161–2
  TUC supports, 160–1
  see also orthopaedics; innovation; economics of medicine; inventions; state involvement
France, 4, 21, 32, 50, 72–80, 107–9, 128, 148
Freeman, Christopher, 201n7
Freeman, Hugh, 188, 272, 273n2, n6, 274n18–19
Friedson, E., 247n19
functional neuroses, 14

gangrene, hospital, 18, 21, 34, 38
  see also hospital diseases
Gamgee, Sampson, 33
Gask, George E., 149, 152, 153
Geddes, Patrick, 200n1
general practitioners, 148, 168, 170, 184, 271n122
  and cortisone, 178–9
  and fracture treatment, 147, 152
  and sanatoria, 54, 61
  and vaccine therapy, 95
germ theories, 4, 21, 25, 27–31, 33–46
  Koch's theory, 40, 74
  Pasteur's theory, 20–2, 40, 73
  microbes, 4, 5
  'seed' and 'soil', 40, 51
  see also miasma; also disease concepts
Germany, 3, 4, 50, 65, 126, 129–30, 203n3
  antisepsis, 34, 44–5
  diphtheria treatment, 72–82
  sanatoria, 48, 52–3
  see also surgeons, Germany
Glasgow Royal Infirmary, 20, 26, 29, 37
Glasgow University, 20, 119
Glaxo, *see* pharmaceutical companies
Glyn, J. H., 175
Godwin, Surgeon-Major C. H., 43
Goldin, Grace, 204n12
Goodenough Committee, 193
Grangethorpe, 149
Grant, A. R., 273n4
Gray, P., 256n11
Greer, A. L., 11
Groves, Ernest Hey, 153, 156, 158, 160, 263n68

Guy's Hospital, London, 18, 24, 30, 37, 91, 92, 94–5, 101, 119, 210n97, 244n91

Haldane, John Scott, 104, 106, 107, 108, 110, 111, 115
Ham, Christopher, 274n13
Hamilton, David, 203n2
Hamilton, M. W., 274n18
Hamilton, Robert, 29, 31, 205n28, 207n53
Hanbury, Sir James, 43
Harman, Bishop, 257n14
Harvey, A. M., 86, 226n10
healing power of nature, 5, 19
Hench, Philip, 166, 173, 176, 177
Hennock, E. P., 222n103
Heubner, Otto, 76, 77
Hill, Leonard, 104, 108
Himsworth, Harold, 174
Hiroshima, 173
history of technology, 2, 3, 13, 202n11
Hobson, L. J., 37
Hoenig, J., 274n18
Holmes, Timothy, 19
homoeopathy, 100, 227n22, 264n4
Horder, T. J., 102
Horsley, Victor, 237n24
hospital diseases, 18–20, 21, 26, 30, 33–5, 37–8
  see also erysipelas; gangrene, hospital; pyaemia
hospitals, 3, 5, 7, 8, 11–12, 16–19
  and medical profession, 17
  children's, 75–7, 82
  environment, 30, 31
  Germany, 34, 44–5
  London, 18, 24, 30, 35–6, 91–5, 118–20, 149, 159, 160, 187
  maternity (US), 131–2, 138–45
  military, 7, 12, 115
  mortality statistics, 18–19, 26, 32, 37
  municipal, 188–90, 196
  orthopaedic departments, 149
  Poor Law, 188–90, 196–7, 273n7; and fracture treatment, 151; open-air treatment at, 55–6
  private, 3
  psychiatric, *see* asylums; psychiatry
  sanitary reform of, 18–19
  town/country, 19
  voluntary (charity), 11, 30, 159–60, 187–90; open-air treatment at, 54–5
  see also sanatoria
Hounshell, David A., 201n10
Howell, Joel, 97, 227n18, 231n66

Howse, H. S., 37
Hughes, T. P., 201n10
Hunt, George Herbert, 107, 118, 119
Hutchinson, Jonathan, 37
hygiene, *see* public health

Ideal Homes Exhibition, 171
immunisation, 72, 82
  passive, 4
immunology, 75, 81–2, 86
innovation
  assessment of, 3–5, 8–9, 14, 16; antisepsis, 23–6; cortisone, 177–8; fracture clinics, 146, 151, 161, 163–4; sanatorium treatment, 59, 61–3, 66; serum therapy for diphtheria, 73, 79; vaccine therapy, 99–100, 103; x-ray pelvimetry, 143; technical, limitations of, 105–6; depends on theoretical and practical context, 109–13, 120–3
  decline/failed, 64, 84
  diffusion, 1, 9; cortisone, 174–5; fracture clinics, 150–2, 157, 160–2; psychiatry in general hospitals, 197–8; sanatorium treatment, 59–66; serum therapy for diphtheria, 73, 75–9; x-ray pelvimetry, 128–30, 133–4; vaccine therapy, 91–3
  models of, 1, 9, 47–8, 66–7, 82–3, 124–5
  opponents of, 1, 9; antisepsis, 21, 23–6, 29–35, 37–41; fracture clinics, 148–9, 152; sanatoria, 61–3; serum therapy for diphtheria, 80–2; vaccine therapy, 100–2; x-ray pelvimetry, 133, 143–5
  reception, 1, 9; antisepsis, 22–6; oxygen therapy, 108–17, 120–3
  rhetoric: fracture clinics, 150–3; sanatoria, 55–8; vaccine therapy, 96–8, 100
  routinisation: oxygen therapy, 122; vaccine therapy, 93; x-ray pelvimetry, 143, 145; psychiatry in general hospitals, 198
  social construction (of problems and innovations), 2–7, 14, 16–17, 20, 82, 125, 137–9
  see also invention; discovery; political contexts; side-effects; standardisation; local conjunctures; product champion
Institute for Infectious Diseases, 74–6

insulin, 6, 86, 187
insurance companies, 156–8
International Congress of Rheumatic Diseases (1949), 176
international congress on medical science
  Amsterdam (1879), 37, 39
  Berlin (1890), 50
  London (1881), 37
  London (1913), 85
  Moscow (1897), 128
invention, 1, 9
  cortisone treatment, 166–7
  fracture clinics, 146–7, 149–50, 163–4
  oxygen therapy, 105–9, 112–13, 122–3
  sanatorium treatment, 48–59
  serum therapy for diphtheria, 72, 75–6
  tuberculin, 50, 74, 90
  vaccine therapy, 89–90
  see also innovation; discovery
Italy, 50

Jacyna, L. S., 97
James, G. T., 275n28
Jefferson, Sir Geoffrey, 149, 193–4
Jephcott, H., 175–6
John Lewis, 173
Johnston, J. M., 178
Jones, Kathleen, 272n1
Jones, Michael, 272, 274n19
Jones, Robert, 148, 151

Keith, Thomas, 41
Kelling Sanatorium, 56, 58, 220n86
Kendall, E. C., 176
Keogh, Alfred, 107, 114
Kersley, George, 169
King Edward VII Sanatorium, 58, 63
King's College, London, 35
King's College Hospital, 24, 29, 35–6, 38, 40, 91, 210n96
Kitasato, Shibasaburo, 75, 76
Klebs, E., 74, 75
Kletts, Norman, 194, 197
Koch, Robert, 4, 40, 49, 75–7, 79, 80
  Institute for Infectious Diseases, 74–6
  'principles', 74
  tuberculin therapy, 50, 74, 81, 90
Korean War, 176

laboratories, 4–6, 13, 72–4, 84–8, 91, 99, 101–5, 107, 109, 112, 114, 123
  and children's hospitals, 75
  sanatoria as, 51
  laboratory science *v.* social medicine, 77
  and serum therapy, 72–4, 82

Labour Party, 153, 154, 160, 161
Lancashire
  psychiatric services, 8, 190–8
  sanatoria, 57
Lancashire County Council, 196
Lancashire Mental Hospitals Board, 195–6
Lancaster Moor Hospital, 196
Lancet, 23, 24, 33, 34, 35, 45, 62
Lane, Sir William Arbuthnot, 101, 230n57
Latour, B., 223n8
Lawrence, Christopher J., 203n2, 226n9, 234n3
Lawrence, William, 31, 204n13
Leavitt, J. W., 246n15
Leeds, 25, 120
Leeds Regional Hospital Board, 274n13
LeMahieu, D. L., 168
Lemaire, Jules, 205n26
Lenoir, T., 224n33
Lewisham, 216n42
Leyberg, J. M., 274n18
Liebenau, J., 200n4, 224n29
Lindemann, J., 223n13
Lister, Joseph, 2, 12, 13, 17–46, 214n21
  abandons spray, 45
  antiseptic principle, 21–2; reception of, 22–6; modifies, 23, 35, 44, 46
  and asepsis, 35, 45–6
  early career, 20
  influence of Pasteur on, 20–2
  moves to London, 36
  publishes statistics, 26
  see also antisepsis; carbolic acid
Lister, Joseph Jackson, 20
Litzmann, C. C. Th., 126–7
liver extract, 6, 166
Liverpool, 23, 29, 31, 56, 93, 216n44
Liverpool Royal Infirmary, 29, 262n59
Liverpool Sanatorium, 58
Lloyd George, David, 65
local conjunctures, 9, 73, 138–41
  see also innovation
local government, 56, 62, 187, 190, 195–6
Lockwood, C. B., 101
Loeffler, F., 74, 75
Lomax, M., 276n31
London, 35, 36, 216n43
  see also hospitals, London
London County Council, hospitals, 159, 160, 187
London Hospital, 24, 93, 94
London Midland and Scottish Railway (LMS), 158, 263n68
Loveday, Edward, 95

Lowy, I., 86–7, 226n17
Luckin, Bill, 202n10
Lund, Edward, 29

McCloskey, Donald N., 200n3
MacCormac, William, 36, 39
Macdonald, T. P., 275n27
McGaw, Judith A., 201n10
Mackay, J. S. B., 192, 272n1, 274n19
Mackenzie, Donald, 201n10
McKeown, T., 74
McKinlay, John B., 201n8
McMurray, T. Porter, 259n38, 260n43
McMurry, N. M., 213n16
magic, 169
Maitland, Gwynne, 158
malarial therapy, 187
Manchester, 21, 29, 38, 56, 95, 120, 193–5, 205n28, 216n44
  psychiatric services, 186, 189, 193, 273n7
Manchester Hospital for Consumption (Bowdon), 54
Manchester Joint Hospitals Advisory Board, 162
Manchester Regional Hospital Board, 272, 274n13
  organisation, 189, 191, 193–6, 274n14
  economy, 191, 196, 274n12
  Medical Advisory Panels, 191, 193
  Mental Health Sub-Committee, 274n14
  psychiatry, 185, 189–99
  and specialisation, 194–5
  Technical Advisory Panel (TAP), psychiatry, 191–2, 194, 274n14
Manchester Royal Infirmary, 29, 37, 91, 93, 193, 263n62
Manchester University, 193–4
market forces, 147
  see also economics of medicine
Marks, Harry M., 265n19
Marr, Robert, 276n37
Marriott, E. D., 219n74
Marshall, F. N., 193–4, 197, 272, 275n19, 275n21, 276n32, 33
Mary Dendy Hospital, 275n28
Maudsley, Henry, 187
Maulitz, R., 232n82
Mayo Clinic, 166, 176
Meakins, Jonathan Campbell, 115, 117–20
measurement, 13, 76, 89, 91, 126–30, 133–7, 140–1, 145
media, use of, 78, 79, 165–8, 172–3, 179
medical administration, 117

medical education, 6, 7, 11, 35, 72, 104, 118–19, 123, 131, 193, 195
  obstetrics, 130
medical entrepreneurs, 53–5, 91–3
Medical Officers of Health, 55, 66, 162, 188, 197, 219n75
medical profession, 17, 45
  metropolitan/provincial split, 101
  and pharmaceutical industry, 72
  physician/surgeon tensions, 100
  specialisation, 147
  specialists v. GPs, 247n19
  and state, 72, 154
  see also surgeons; obstetrics; paediatrics; orthopaedics; psychiatry; rheumatologists; scientist–clinician relations; specialisation/professionalisation
Medical Research Committee/Council (MRC), 65, 85, 107, 108, 113–21, 225n8
  and cortisone, 174, 175, 177–8, 182, 184, 269n69
medical statistics, 18, 26, 32, 34, 37, 94
  consumption, 48, 59, 62, 65–6
  diphtheria, 73, 76–7
  fractures, 150–1, 156–7, 164
  see also hospitals, mortality statistics
Medical Times and Gazette, 24
Meldrum, Marcia, 265n19
mental health service, see psychiatry
Menzies, Sir Frederick, 263n65
Merck, see pharmaceutical companies
Metropolitan Asylums Board, 219n75
Metropolitan Life Assurance Company of New York, 260n46
miasma, 17–19, 30
  see also disease concepts
Michaelis, G. A., 126–7
microbes, see germ theory
microbiology, see bacteriology
Middlesex Hospital, 93
military hospitals, see hospitals, military
military surgery, see surgery, military
Miners' Welfare Commission, 161
Ministry of Health, 117, 174–5, 177–80, 182–4
  and psychiatry, 186, 189, 196
Ministry of Transport, 155
'modernist' ideology, 153
  see also scientific management
Moloy, H., 140
Moore, H. E., 158
Morley, John, 263n67

Morris, Noah, 243n76
Morse, Arthur, H., 138–9
mortality, 48, 72–3
  diphtheria, 72–3
  hospital, 18–20, 26, 32, 37, 203n6;
    see also hospital, mortality statistics
  infant/child, 72–3
  maternal, 7; US, 131–2, 141
  motor vehicle accidents, 155
Mount Vernon Hospital, 91
Mumford, L., 200n1
Murard, L., 223n8

Nagasaki, 173
National Association for the Prevention of Consumption, 55–6, 58–60, 212n1
national health insurance, 3
National Health Service (NHS), 8, 9
  consultant staffing, 190, 198
  and cortisone, 166, 173–5, 179–81, 183
  hospital/community split, 8, 186, 190, 198
  policy, 185, 189–90, 198–9
  Regional Hospital Boards (RHBs), 189
  see also Manchester Regional Hospital Board
National Institute of Medical Research, 85
National Insurance Act, 67
  GP remuneration, 152
  Lloyd George, David, 65
  (local) NI committees, 152
  sanatorium benefit, 47, 65–6
National Sanatorium, Bournemouth, 54
'neo-technics', 4, 14, 200n1
Newcastle Infirmary, 37
Newsholme, Arthur, 73
Nightingale, Florence, 19
Noble, David, 201n10
Norfolk and Norwich Hospital, 54–5
North Staffordshire Infirmary, 118
Nottingham, 56, 188
Nuffield Foundation, 174, 268n69
Nuffield Provincial Hospitals Trust, 162, 194
Nunneley, Thomas, 25, 204n8
Nussbaum, Johann, 34, 44

Oakley, Ann, 253n59
obstetrics
  specialisation, 131
  specialists v. midwives, 247n19, 251n49
  specialists v. radiologists, 254n66
  standards of practice (US), 130–1, 138
  and technology, 132, 144, 247n19

obstetrics *cont.*
  at Yale, 133, 138–41
  see also x-ray pelvimetry; patients,
    women; specialisation/
    professionalisation
Oldham, 33, 191–2, 197, 276n36, n37
Oliver Bird Fund, 268n69
open-air treatment, *see* consumption,
  treatment of
opponents of innovation, *see* innovation,
  opponents of
opsonic index, 90
opsonins, 89
orthopaedics, 12
  generalists *v.* specialists, 148–9, 152
  in London, 149, 151
  in Manchester, 149
  and state, 157–9, 162
  and voluntary hospitals, 159
  United States, 149
  see also fracture clinics; crippled
    children; specialisation/
    professionalisation
Osler, William, 104, 109
ovariotomy, and antisepsis, 41–2, 204n9
Oxford, 18
oxygen therapy, 12–15
  diffusion, 118–20
  innovation (social *v.* merely technical),
    113, 122–3
  reception, 108–17, 120–3
  role of First World War, 105–17, 122
  routinisation, 121–2
  state involvement, 106–9, 112–20
  see also innovation; invention; war and
    medicine; scientist–clinician
    relations; state involvement;
    scientific medicine, claims for, etc.

paediatrics, 72, 75–7, 82
Paget, Sir James, 24, 37
palaeotechnics, 200n1
Paris, 29, 32, 72–80
Pasteur, Louis, 3, 4, 20–2, 46, 75, 84
  anti-rabies serum, 74
Pasteur Institute, 72–80
Paterson, Rolston, 194–5
patients, 16
  attitudes of, 8, 10, 15
  criticisms of cortisone shortage,
    179–83
  doctor–patient encounter, 171, 181–3,
    272n122
  effect of cortisone on, 179–83

effect of enthusiasm and drama on, 168,
  170–1, 181–3
effect of withdrawal of cortisone on, 177
perceptions of harmful effects of
  cortisone, 167, 181–3
sanatoria, 63–4
shortage of, 138
vaccine therapy, 100
and new 'cures', 173
women, 247n16, 248n22
Pearson, Karl, 62, 64, 219n81, 220n82
pelvimetry, 125–8
  establishment of 'norms', 126–7, 134
  see also x-ray pelvimetry
penicillin, 8, 166, 173
Pernick, Martin, 200n5
pernicious anaemia, 6, 166
pharmaceutical (and other medical)
  companies, 8, 167
  Glaxo, and cortisone, 175–8, 184
  Merck, and cortisone, 174, 177–80, 182,
    184, 269n69
  and serum therum for diphtheria, 72, 76–
    9, 82
  and therapeutic vaccines, 92–6
Philip, R. W., 91, 217n51
physical medicine, 169, 170, 173–5
Pickstone, J. V., 256n9, 273n4, 275n26
Pinch, T. J., 201n10
plaster-of-Paris, 7, 150
Platt, Sir Harry, 146, 148–50, 152, 157,
  162, 194
policy, 2, 11, 185
political contexts, 6, 8–10, 12, 16, 18–20,
  53, 78, 105–6, 122–3, 131–2, 146–8,
  153–4, 160–1, 166–7, 184, 193–4
Pool, Arthur, 192, 197, 272n1, 273n4,
  275n19, 277n36
Poor Law, psychiatry, *see* psychiatry, Poor
  Law
Portsmouth, 188, 190, 221n98
Poulton, Edward Palmer, 119–22
Powell, Enoch, 186, 198
pre-natal care, *see* ante-natal care
Prestwich hospital, 196–7, 276n31, 36
product champions, protagonists, 9–11, 192
  see also Lister, Joseph; Thoms, Herbert;
    Wright, Almroth
psychiatry
  Board of Control, 190, 196–7, 276n32
  community care, 186, 188, 198
  drug therapy, 198
  in general hospitals, 8, 12, 16, 185–7,
    190–9 (*see also* innovation)

and general medicine, 186–9
local government, 187–8
out-patients, 186–90
physical treatments, 187
Poor Law (workhouse), 188–9, 196–8
  (see also hospitals, Poor Law)
profession, 186–7, 189–92, 197
in voluntary hospitals, 188
see also asylums; Manchester Regional
  Hospital Board; Lancashire;
  specialisation/professionalisation
public health, 4, 5, 7, 18, 72–4, 81–2, 97,
  162, 163, 192
laboratories, 82
tuberculosis, 48–9, 55
vaccines, 101
pyaemia, 18, 21, 34, 38
see also hospital diseases

quackery, 8, 168, 169
Queen's Hospital, Birmingham, 33

radiation, 15, 124–5, 143, 268n60
see also x-ray pelvimetry
Ransom, W. B., 56, 61–2
Ransome, A., 214n24
rationalisation, of medical organisation/
  procedures, 2, 7, 9, 14, 132, 153–4,
  162, 164
Reiser, S. J., 86, 201n9, 226n10, 247n19,
  257n12
Retail Standards Association, 172
Retreat, York, 197
rhetoric, see innovation
rheumatic diseases, 16, 165–72, 175, 176,
  181, 184
rheumatoid arthritis, 166, 171, 174–6, 183
rheumatologists, 8, 12, 166–7, 168, 173–9,
  183, 184
see also cortisone; drama and spectacle;
  specialisation/professionalisation
Richards, Thomas, 167
rickets, 6, 134–6, 148
Ricketts, Frederick, 23
Road Traffic Act (1934), 155
Rochdale, 192, 197, 276n36
Rolleston, Humphrey, 169
Rosen, George, 200n4
Rosenberg, Charles E., 202n11, 213n11,
  267n49
Rosenberg, Nathan, 200n3, 212n3
Rosner, David, 257n12
Roth, Julius, 201n9
Roux, Emile, 75–8

Royal College of Surgeons, 148
Royal Medical and Chirurgical Society, 91
Royal National hospital, 54, 91
Royal Society, 106, 107, 114, 117
Royal Society of Medicine (RSM), 92,
  98–102
Royal Victoria Hospital For Consumption,
  54
Ruzek, Sheryl, B., 201n9
Ryle, John, 244n91

Sadler, J., 234n3
safety, see x-ray pelvimetry; side-effects,
  drawbacks, etc., of innovation
St Bartholomew's Hospital, 20, 24, 30, 31,
  37, 38, 91, 93–4, 120, 152, 244n91,
  263n62
St George's Hospital, London, 19, 24
St James Poor Law Hospital, Balham, 148
St Mary's Hospital, London, 24, 91–2, 94
St Thomas's Hospital, London, 19, 24, 55,
  92, 94–5, 263n62
Salford, psychiatric services, 273n11
Salford Royal Hospital, 149
Salvarsan, 80, 98
Samaritan Hospital, London, 18
sanatoria, 3, 4, 13, 14, 187
British, private, 53–4, 59
British, public, 54–9
criticism of, 61–3
diffusion, 59–66
discipline, 63–4
German, 52–3
individual, details of number of beds
  1895–1910, 68–71
innovation, 53–9, 67
invention, 48–53
isolation, 52
number of beds, 53, 56, 59–60, 66
and philanthropy, 56–8
'work therapy', 63
and the working class, 55, 58–9, 64–5
see also consumption, treatment of;
  innovation; economics of medicine;
  invention; state involvement
sanitary reform, 18–19
Schaffer, Simon, 265n26
science, state support of, scientists prefer to
  private medical practice, 121
see also state involvement; Army
  Medical Service; Department of
  Scientific and Industrial Research;
  Medical Research Committee/
  Council; Ministry of Health;

science *cont.*
　University Grants Committee; War Office
scientific management, 14
　in medicine, 150–1, 154, 156, 164
scientific medicine, claims for, etc., 2–7, 11–14, 21–2, 25, 45–6, 72, 74–5, 78, 82–3, 85–7, 96–103, 131–2, 139
　cortisone, 169, 174–6
　and oxygen therapy, 107, 113–15, 117–20
　and proprietary medicines, 167
scientist–clinician relations, 5, 6, 11–12
　diphtheria anti-toxin, 74–7
　oxygen therapy, 104–6, 111–12, 118, 120–3
　vaccine therapy, 96–103
Scotland, 215n34
Scotsman, 30
Scull, Andrew, 272n1
Selfridge, Gordon, 172, 267n54
Selfridges, 172–3
serum therapy for diphtheria, 4–5, 72–83
　assessment of, 73
　Britain, 79
　claims for, 73–4, 76–7
　clinical trials, 76–7
　critics of, 73, 80–2
　diffusion, 73, 75–9
　French and German cooperation, 75
　and pharmaceutical companies, 72, 76, 79–80
　public response, 78
　side-effects, 80
　standardisation, 80
　state support for, 72–80
　see also innovation; discovery; diphtheria; economics of medicine; scientist–clinician relations; state involvement
Sheffield, 55, 85, 118, 216n43, 221n98
shell-shock, 14, 115, 241n58
Shortt, S. E. D., 234n2
Shryock, R. H., 214n23, n25
side-effects, drawbacks, etc., of innovation, 4
　cortisone, 177–9, 184
　serum therapy for diphtheria, 80
　x-ray pelvimetry, 125, 143, 145
　see also innovation
Simpson, Sir James Young, 18, 23
Sinclair, Meurice, 148
Smellie, William, 126
Smith, Clement, A., 256n8
Smith, F. B., 15, 220n86, 224n28

Smith, John, K., 201n10
soldiers' heart, 14
Souttar, H. S., 257n14, 261n53
spas, 266n33
specialisation/professionalisation, 5–8, 12, 13, 16
　laboratory research, 77, 82
　physiologists, 105, 111, 121–3
　obstetrics, 124, 130–2, 137–40, 144
　orthopaedics, 148–9, 151–3, 157–60, 163–4
　rheumatologists, 166–7, 173–6, 179, 183
　sanatoria, 54
　vaccinists, 94–5, 102
　psychiatry, 189, 191–8
Stacey, M., 272n122
standardisation, 13, 76, 80, 93, 127, 131–2
　hospital services, 153–4
　'norms' in pelvimetry, 126–7, 134–7
　see also innovation
state involvement, 3, 8, 11
　and fracture clinics, 157–9, 161–2
　and oxygen therapy, 106–9, 112–20
　and sanatoria, 47, 53, 65–6
　and serum therapy for diphtheria, 72, 74, 76
　see also National Health Service; science state support of
statistics, *see* medical statistics
Staudenmaier, J. M., 202n11
Stevens, Rosemary, 131
Stopford, Sir J. S. B., 193–7, 275n22, 276n32
streptomycin, 166, 173, 178
surgeons, 18, 19, 21, 32, 33, 228n33
　anti-specialisation, 148–9, 152
　and germ theory, 32–3
　France, 21, 32
　Germany, 34, 39, 44–5, 46
　London, 20, 23, 24, 204n9
　London/provincial split, 23, 32–3
　military, 43
surgery, 2, 11, 12
　military, 29
　see also antisepsis; asepsis
Syme, James, 20, 27
syphilis, 25, 80

Tait, Lawson, 41, 42
Taplow, 115
'team work', 150, 153
　see also scientific management
technology, 14, 124, 131–2, 141–4, 150–1
　see also history of technology
Telford, E. D., 263n67

Temin, Peter, 200n3, n4
theatre, 168
therapeutics, 2, 10, 13
  and physiological theory, 110–12, 116
therapeutic vaccines, 84, 91–3
  autogenous, 91
  phylacogens, 93
  stock, 93
therapies, 4–5, 11, 12, 19
  see also consumption, treatment of;
    cortisone; oxygen therapy; serum
    therapy; tuberculin; vaccine therapy
Thiersch, Carl, 34
Thoms, Herbert, 7, 133–45
Thomson, George, 33
Thornton, J. L., 256n10
toxins, 4, 75
Trades Union Congress (TUC), 147, 153, 154, 160, 161
tuberculin, 50, 64, 74, 81, 90, 220n89
tuberculosis, 3, 4, 74, 79, 81, 187
  see also consumption
Turner, E.S., 172

'unification of control', 153
  see also scientific management
United States of America, 7, 8, 10–12, 130–45, 149–50, 165, 166, 176, 203n3, 214n30, 265n19
  orthopaedics, 149–50
University College Hospital, London, 24
University College London, 20
University Grants Committee, 118

vaccination, 4, 72
vaccine therapy, 5, 11–15, 84–103
  and clinical pathology, 93–5
  and general practitioner, 95
  at Guy's Hospital, London, 94, 101
  decline, 95–6, 101
  growth and diffusion, 91–2
  metropolitan/provincial rivalry, 101
  out-patient clinics, 92–3
  at pseudo-bacteriology, 102
  at St Mary's Hospital, London, 91–2, 94
  at St Thomas's Hospital, London, 94–5
  physician–surgeon relations, 100
  see also therapeutic vaccines; innovation;
    economics of medicine; invention
Victoria Hospital for Consumption,
  Edinburgh, 217n51
Vienna Accident Hospital, 155–6
Vincent, David, 169, 266n35
Virchow, Rudolf, 77, 79, 81

vitamins, 6
Vogel, Morris, 246n15, 256n8
Volkmann, R., 34
Voluntary Hospitals Committee for
  London, 150

Wajcman, J., 201n10
Walker, H. M., 177–8
Walker, Jane, 53, 63
war, and medicine, 2, 6–7, 12, 14, 89, 271n109
  antisepsis, 29, 43
  fractures, 148
  impact on clinical laboratories, 86
  oxygen therapy, 105–17, 122
  see also Army Medical Service
Warnecke, R. B., 11, 201n8
Warner, J. H., 200n5
War Office, 106, 114–6
Watkins, Dorothy, 216n48
Watson, Edith, 194, 197
Watson-Jones, Reginald, 160, 262n59
Webster, Charles, 269n75, 273n1
welfare state, 8
  see also National Health Service
Wells, Thomas Spencer, 18, 41, 204n9
Wertz, D. C., 245n3
Westminster Hospital, London, 93, 94, 263n62
Westmoreland Sanatorium, 56
Whittingham Hospital, 196, 273n4, 276n36
Wilkinson, W. Camac, 64, 65
Wohl, A. S., 213n7
Wood, John (King's College), 29, 30, 38, 40
workmen's compensation, 158, 160
wound treatment
  acupressure, 23
  dressings, 19, 21–34, 38–9, 41, 43–5
  putrefaction, 21–2, 25, 28, 33
  suppuration, 21, 23, 28, 31, 33
  tube drainage, 28, 32
  see also antisepsis; carbolic acid;
    cleanliness; asepsis
Wright, Almroth, 5
  'British Pasteur', 83, 100
  career, 87–8
  future of medicine, 86, 96
  Medical Research Council, 98

x-ray pelvimetry, 7, 12, 15, 124–5, 128–45
  as classification tool, 133, 137, 140–2
  costs (financial), 252n54
  criticisms of, 133, 143–5

x-ray pelvimetry *cont.*
  develops in Europe, 128–30
  diffusion, 128–30, 133–4
  measurement, 133–7, 140–1, 249n31
  'normal' pelvis, 134, 137
  routine use, 143, 145
  safety, 129, 134, 143–4
  Yale, 132–45
  see also innovation; obstetrics; safety; measurement; ante-natal care
x-rays, 124–5, 143
  see also x-ray pelvimetry

Yersin, Alexandre, 75

Zylberman, P., 223n8